Bioelectrocatalysis

THE ELECTROCHEMICAL SOCIETY SERIES

The Electrochemical Society
Advancing solid state & electrochemical science & technology

ECS – The Electrochemical Society
65 South Main Street
Pennington, NJ 08534-2839
http://www.electrochem.org

A complete list of the titles in this series appears at the end of this volume.

Bioelectrocatalysis

From Electron Transfer Processes to Emerging Technological Applications

Edited by

Shelley D. Minteer
Missouri University of Science and Technology
Kummer Institute Center for Resource Sustainability
Rolla, USA

Matteo Grattieri
Universit degli Studi di Bari Aldo Moro
Dipartimento di Chimica
Bari, Italy

WILEY

Published by John Wiley & Sons, Inc., Hoboken, New Jersey.

Library of Congress Cataloging-in-Publication Data applied for

Hardback ISBN: 9781394207763

Cover Design: Wiley
Cover Image: © WLADIMIR BULGAR/SCIENCE PHOTO LIBRARY/Getty Images

Set in 9.5/12.5pt STIXTwoText by Straive, Chennai, India

Contents

Preface

Bioelectrocatalysis is a fascinating interdisciplinary research field that couples biological materials to electrodes, obtaining biohybrid systems, which have attracted significant interest and found a variety of applications. The present book is aimed at providing both the fundamental principles at the basis of bioelectrocatalysis and its most recent practical applications. Accordingly, the book is composed of two main sections, with the first one, comprising Chapters 1–3, dedicated to the fundamental aspects of bioelectrocatalysis. Here, the electron transfer processes between biological entities and electrodes, and the approaches to study these processes are described. This section presents a comprehensive approach to the fundamentals of bioelectrocatalysis, starting from the comparison of enzymatic and microbial electrocatalysis to later focusing on the discussion of direct vs. mediated bioelectrocatalysis and the challenges and strategies for promoting extracellular electron transfer. The approaches and techniques utilized to study the electron transfer processes, reaction kinetics and to confirm the type of electron transfer taking place (i.e. direct or mediated) are also presented in detail.

The second section, comprising Chapters 4–9, focuses on the established and emerging applications of bioelectrochemical systems relying on bioelectrocatalysis. Specifically, the chapters are dedicated to (i) glucometers and continuous glucose monitors, (ii) *ex-situ* biosensors, (iii) biofuel cells and biosolar cells, (iv) used water treatment, (v) electrosynthesis, and (vi) wearable and implantable devices. It is interesting to note that bioelectrochemical systems have played an important role in all these fields, and the recent research progresses have opened up exciting new opportunities. Electrochemical glucose sensors have already revolutionized diabetes management by enabling accurate, real-time blood glucose monitoring. Self-powered biosensors allow for the cost-effective monitoring of various analytes, including biological oxygen demand, pesticides, explosives, and heavy metals, without requiring external power sources. Biofuel cells and biosolar cells have experienced a significant increase in power densities, achieving multiple W m^{-2}, and have been demonstrated for powering various practical devices. Microbial fuel cells for water treatment are being deployed at a large scale, providing more sustainable approaches for contaminant removal. Bioelectrosynthesis is emerging as a sustainable approach for the production of various high-value products. Skin and ocular systems for wearable and implantable devices are paving the way to revolutionizing approaches for analyte monitoring. All these topics are presented in the book, highlighting the latest results and critically discussing the future challenges and research needs.

The book has been prepared focusing on an introductory level, and many chapters contain "call-out box" where critical background for the topics discussed in the text is provided. The various topics are then discussed, reaching an advanced level, and the combination of fundamentals and applicative aspects in a single book makes this volume a reference for both undergraduate and graduate students approaching the field of bioelectrocatalysis.

November 15th, 2025

Shelley D. Minteer
Missouri University of Science and Technology
Rolla, USA

Matteo Grattieri
University of Bari Aldo Moro, Bari, Italy

List of Contributors

Jose María Abad
Instituto de Catalisis y Petroleoquimica
CSIC, Department of Biocatalysis
Madrid
Spain

Nunzio Giorgio G. Carducci
Department of Chemical Engineering and
Material Science
Michigan State University
East Lansing
MI
USA

Gabriel García-Molina
Instituto de Catalisis y Petroleoquimica
CSIC, Department of Biocatalysis
Madrid
Spain

Matteo Grattieri
Department of Chemistry
Università degli Studi di Bari Aldo Moro
Bari
Italy

and

Istituto per i Processi Chimico Fisici
Consiglio Nazionale delle Ricerche
Bari
Italy

David P. Hickey
Department of Chemical Engineering and
Material Science
Michigan State University
East Lansing
MI
USA

Kavita Jayakumar
Instituto de Catalisis y Petroleoquimica
CSIC, Department of Biocatalysis
Madrid
Spain

Antonio L. De Lacey
Instituto de Catalisis y Petroleoquimica
CSIC, Department of Biocatalysis
Madrid
Spain

Christine Lewis
Biodesign Swette Center for Environmental
Biotechnology
Arizona State University
Tempe
AZ
USA

Fred Lisdat
Department of Biosystems Technology
Technical University of Applied Sciences
Wildau
Brandenburg
Germany

Edmond Magner
Department of Chemical Sciences and Bernal Institute
University of Limerick
Limerick
Ireland

Jacquelyn E. McBride
Lebanon Valley College
Department of Chemistry
Annville
PA
USA

Shelley D. Minteer
Kummer Institute Center for Resource Sustainability
Missouri University of Science and Technology
Rolla
MO
USA

Juan F. Ortiz Medina
Biodesign Swette Center for Environmental Biotechnology
Arizona State University
Tempe
AZ
USA

Jesús A. Perez García
Biodesign Swette Center for Environmental Biotechnology
Arizona State University
Tempe
AZ
USA

Marcos Pita
Instituto de Catalisis y Petroleoquimica
CSIC, Department of Biocatalysis
Madrid
Spain

Michelle Rasmussen
Lebanon Valley College
Department of Chemistry
Annville
PA
USA

Ruggero Rossi
Department of Environmental Health and Engineering
Johns Hopkins University
Baltimore
MD
USA

Daniel Schäfer
Department of Biosystems Technology
Technical University of Applied Sciences
Wildau
Brandenburg
Germany

César I. Torres
Biodesign Swette Center for Environmental Biotechnology
Arizona State University
Tempe
AZ
USA

and

School for Engineering of Matter
Transport, and Energy
Arizona State University
Tempe
AZ
USA

1

Fundamentals of Bioelectrocatalysis

Matteo Grattieri[1,2] *and Shelley D. Minteer*[3]

[1] *Department of Chemistry, Università degli Studi di Bari Aldo Moro, Bari, Italy*
[2] *Istituto per i Processi Chimico Fisici, Consiglio Nazionale delle Ricerche, Bari, Italy*
[3] *Kummer Institute Center for Resource Sustainability, Missouri University of Science and Technology, Rolla, MO, USA*

1.1 Introduction

Bioelectrocatalysis is a fascinating research field where a biological material, spanning from isolated apparatuses to intact organisms (Box 1.1), is utilized as a catalyst for redox reactions taking place at an electrode. The interest in this research field has its foundations on the combination of the advantages provided by the features of biological systems, such as diverse catalytic functions, mild reaction conditions, high activity and selectivity, and the possibility to control the oxidizing and reducing ability of an electrode by modifying its Fermi level [1]. Nowadays, bioelectrocatalysis has several applications, which are presented and discussed in detail in various chapters of the books, including the development of (i) biofuel cells for power production [2, 3], (ii) biosensors [4–9], (iii) systems for drug release [10], (iv) wearable devices [11, 12], (v) systems performing the treatment of contaminated water [13–15], and (vi) systems performing the electrosynthesis of valuable products [16–19].

At the basis of bioelectrocatalysis is the process of electron transfer (ET) between the biological components and the electrodes. Achieving a rapid and efficient ET defines the success or the failure of a bioelectrocatalytic system. Accordingly, the ET process constitutes the major challenge in bioelectrocatalysis, and this chapter is dedicated to introduce the general aspects governing the interaction between biological catalysts and electrodes, providing the basic concepts to approach the book. The detailed ET processes for enzymes and microorganisms are discussed in Chapters 2 and 3, respectively.

Box 1.1 Enzymes as Biological Catalysts

Enzymes play a pivotal role in life, with more than 99% of all the reactions that are crucial to biological systems being catalyzed by enzymes. Like other catalysts, enzymes allow for an increase in the rate of a chemical reaction without being consumed or altered by such reactions, thus overcoming the kinetic barriers that hinder a specific process. Interestingly, when the term "catalysis" was proposed in 1836 by the Swedish chemist Berzelius (1779–1848) from the Greek *kata* (wholly) and *lyein* (to loosen), no distinction was made between a chemical

(Continued)

Bioelectrocatalysis: From Electron Transfer Processes to Emerging Technological Applications, First Edition.
Edited by Shelley D. Minteer and Matteo Grattieri.

Box 1.1 (Continued)

and biological catalysis, and only the term "contact substance" was used for referring to a catalyst [20]. One of the unique properties of enzymes, their high specificity for a specific substrate, was highlighted as early as 1894 by the German chemist Emil Fischer (1852–1919, Nobel Prize in Chemistry in 1902), who hypothesized the popular lock-and-key mechanism of enzyme interaction with a substrate. Shortly later, the kinetics of enzyme catalysis was studied by Victor Henri (in 1903) and followed by Leonor Michaelis and Maud Menten (in 1913), who described the rate equation of enzyme kinetics with the Henri–Michaelis–Menten equation, commonly referred to as the Michaelis–Menten equation:

$$v = \frac{V_{\max}[S]}{K_M + [S]} \tag{1.1}$$

where v is the rate of enzyme-catalyzed reaction (velocity), V_{max} is the limiting maximum value for the velocity, S is the substrate for the reaction, and K_M is a constant often referred to as "Michaelis constant" arising from the three rate constants (k_1, k_{-1}, and k_2) of the following reaction:

$$E + S \underset{k_{-1}}{\overset{k_1}{\rightleftarrows}} ES \underset{k_{-2}}{\overset{k_2}{\rightleftarrows}} E + P \tag{1.2}$$

where E is the enzyme, S is the substrate, ES is the enzyme–substrate complex that is formed when enzyme and substrate come together, and P is the product. k_1 and k_2 are the forward rate constants for the first and second steps, respectively, while k_{-1} and k_{-2} are the backward rate constants for the first and second steps, respectively. Although not all enzymes follow Michaelis–Menten kinetics, it is usually a good approximation for enzymatic electrochemistry systems.

1.2 Bioelectrocatalysts and Electron Transfer

Enzymes can be considered the most commonly utilized biological catalyst for bioelectrocatalysis. They provide high specificity due to the molecular recognition event that takes place in the binding pocket for their substrates, where their redox active center(s), the cofactors, allow performing the catalytic process, converting the substrate into the products. This process is made possible by the 3D structure of the enzyme. However, different from the commonly known metallic electrocatalysts (i.e. platinum group and non-platinum group catalysts) [21, 22] and nonmetallic electrocatalysts (i.e. 2,2,6,6-tetramethylpiperidinyl-N-oxyl) [23], the 3D structure of biological catalysts constitutes a nonconductive layer surrounding the cofactors [24]. As a result, when aiming to "electrically wire" an enzyme to an electrode, the primary goal is to successfully transfer the electrons across this insulating layer. This ET process can be described by utilizing the semi-classical Marcus theory for ET in nonadiabatic systems. According to Marcus theory, the rate constant of ET (k_{ET}) can be expressed as:

$$k_{ET} = \frac{2\pi}{\hbar} \frac{H_{DA}^2}{\sqrt{4\pi\lambda RT}} e^{-(\Delta_r G^0 + \lambda)^2/4\lambda RT} \tag{1.3}$$

where $\hbar$ is the reduced Planck's constant ($1.05457182 \times 10^{-34}$ Js), H_{DA} is the matrix coupling element (also known as electronic coupling parameter), λ is the nuclear reorganization energy (the energy required to force the reactant into the same nuclear configuration as the products without the ET taking place), and ΔG^0 is the thermodynamic driving force of the reaction.

Importantly, the electronic coupling parameter exponentially decays with an increasing distance between the electron donor and acceptor sites (r_{DA}) as described by:

$$H_{\mathrm{DA}}^2 = \left(H_{\mathrm{DA}}^0\right)^2 e^{-\beta(r_{DA})} \tag{1.4}$$

where H_{DA}^0 indicates the electronic coupling parameter for the donor (D) and acceptor (A) when they are at the minimum distance, and β is the electronic coupling attenuation coefficient that varies depending on the medium separating donor and acceptor.

According to Eq. (1.4), and considering that the electron tunneling time for biological redox systems should be in the millisecond to microsecond range for their proper functioning, the maximum center-to-center distance for a single-step electron tunneling should be approximately 24 Å [25, 26], with lower distances resulting in higher rates of ET (Figure 1.1) [28]. This distance should not be taken as a hard limit, and ET at distances higher than 24 Å has been reported [29]. Taking this into consideration, it is interesting to underline the center-to-center distance for two widely employed enzymes in bioelectrocatalysis: laccase and glucose oxidase. Laccases (EC 1.10.3.2) [30] are multicopper oxidoreductase where a Type 1 copper center (located about 6.5 Å below the enzyme surface) accepts one electron from diphenols and related substances and transfers it to a trinuclear copper cluster (where the oxygen reduction process takes place) that is located at a distance of 12–13 Å [31, 32]. Glucose oxidase from *Aspergillus niger* (E.C. 1.1.3.4) is an enzyme that catalyzes the oxidation of D-glucose due to two flavin active sites buried deeply within the enzyme structure, at a distance of approximately 20–27 Å [33–35].

Another important aspect is that the ET rate also depends on the electrochemical overpotential applied. Specifically, knowing that:

$$\Delta G^o = -nFE^o \tag{1.5}$$

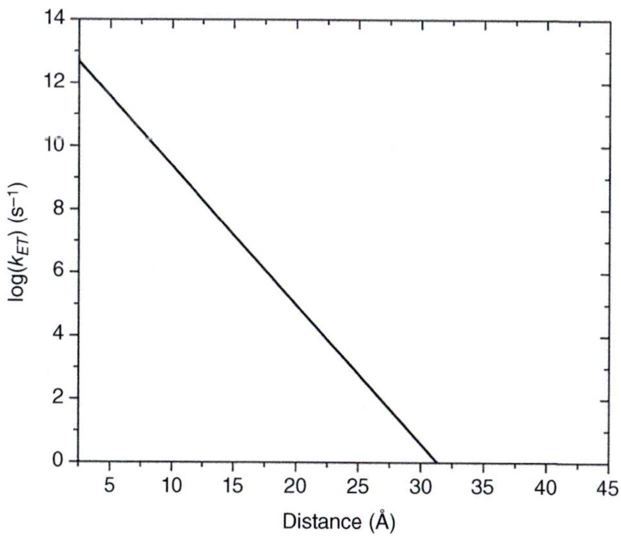

Figure 1.1 Dependence of the rate of single-step electron transfer on the distance between donor and acceptor centers based on semi-classical Marcus theory. *Source:* Hickey et al. [27] / with permission of The Royal Society of Chemistry.

and that we can apply a specific overpotential (η) by modulating the potential applied to an electrode with respect to the redox potential of the redox species involved in the electrode transfer process:

$$\eta = E_{\text{App}} - E^{o} \tag{1.6}$$

we can relate the free energy and the overpotential with Eq. (1.7):

$$\Delta G^{o} \approx -nF\eta \tag{1.7}$$

By combining Eqs. (1.3) and (1.7), and noticing the quadratic nature of the free energy term in Eq. (1.3), a finite range where an increase in overpotential results in an increase in the ET rate can be defined. Passing this overpotential results in a point of inversion in the ET rate, which starts to decrease for increasing overpotentials, indicating what is known as the "Marcus inverted region" [25, 36, 37]. However, such an "inverted region" is not typically observed in bioelectrochemical systems, likely because at high overpotentials, the reactions become limited by mass transfer or intramolecular ET [38, 39].

1.2.1 Enzymatic Bioelectrocatalysis

Based on Eq. (1.4), it becomes evident that the location of the cofactors in the protein structure plays a critical role in the possibility of achieving the transfer of electrons to (and/or from) an electrode. Furthermore, considering the broad variety of dimensions for different redox enzymes, and the resulting distances between the redox cofactors and the enzyme exterior, the orientation of a physisorbed (i.e. non-covalently bound) enzyme on the electrode surface can result in a hindered ET process [40]. At this purpose, a preferential orientation of the enzyme on the electrode surface, with the "electron-entry site" facing the electrode, has been reported to facilitate the ET process, as shown in Figure 1.2, with various examples for the photosynthetic reaction center and photosystems [41, 44], multicopper oxidases [42], and dehydrogenases [43]. Furthermore, it should also be considered that the charged regions present on the protein structure might lead to electrostatic attraction or repulsion with an electrode surface. Accordingly, the control of the surface functionalities present on the electrode surface at a specific pH has allowed tailoring the electrical wiring of biocatalysts, knowing their isoelectric point [45, 46]. Interestingly, it should be underlined that orientation and low donor–acceptor distance do not necessarily ensure successful direct ET, since it has been proposed that the enzyme might be adsorbed in a conformation that eventually blocks substrate access to the cofactors [47].

The engineering of the biotic/abiotic interface with nanostructured materials has also been reported as an effective approach to enhance ET between enzymes and electrode surfaces, resulting in improved stability of the bioelectrodes [48].

Intuitively, the donor–acceptor distance becomes critical in the case of large enzymes, where the cofactors are buried inside the protein structure and thus less accessible to the electrode surface. An example of an enzyme widely used in bioelectrochemical systems that belongs to this category is glucose oxidase from *A. niger* (E.C. 1.1.3.4) [35, 49], which is presented in detail in Chapter 2. Under such conditions, various approaches can be employed to facilitate ET between enzymes and electrodes, with the use of redox mediators for mediated ET (Figure 1.3a) being widely utilized [44, 52–57]. An emerging approach, enabled by advances in synthetic biology, involves protein engineering to modify protein amino acid sequences by inserting immobilization components (Figure 1.3b), such as a single amino acid or a binding protein [51, 58, 59].

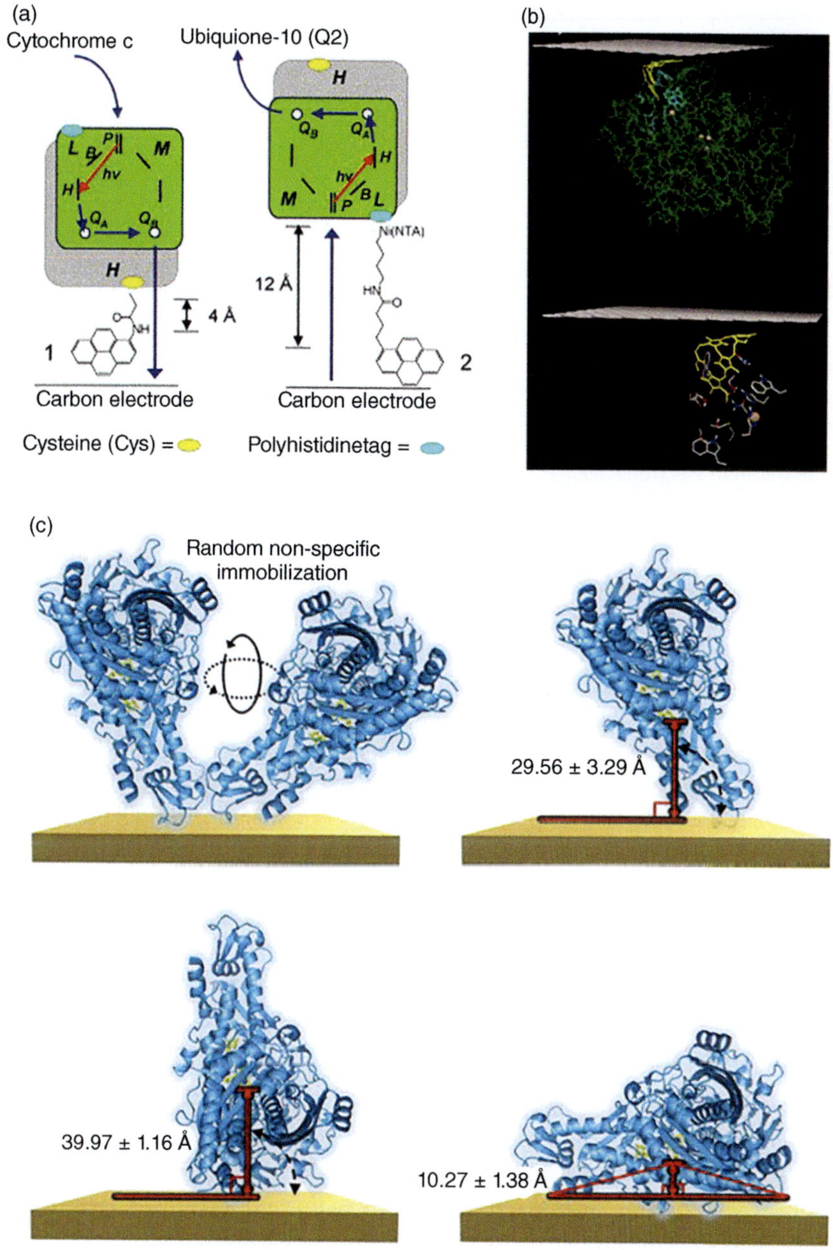

Figure 1.2 Schemes of different approaches utilized for preferential enzyme orientation. (a) Orientation of the photosynthetic reaction center from *Rhodobacter sphaeroides* with the H-subunit side or the special pair (P) side facing the electrode surface. *Source:* Trammell et al. [41] / with permission of Elsevier. (b) Binding of bilirubin to bilirubin oxidase positions and orients the enzyme on a graphene electrode. *Source:* Reprinted with permission from Matanovic et al. [42] Copyright 2016 American Chemical Society. (c) Carbon monoxide dehydrogenase with random or controlled orientation achieved by fusing gold-binding peptides at the N- or C-terminus of the enzyme, with indication of the distance to reach the cofactor from the electrode surface. *Source:* Reproduced from Reginald et al. [43] / Springer Nature / CC BY 4.0.

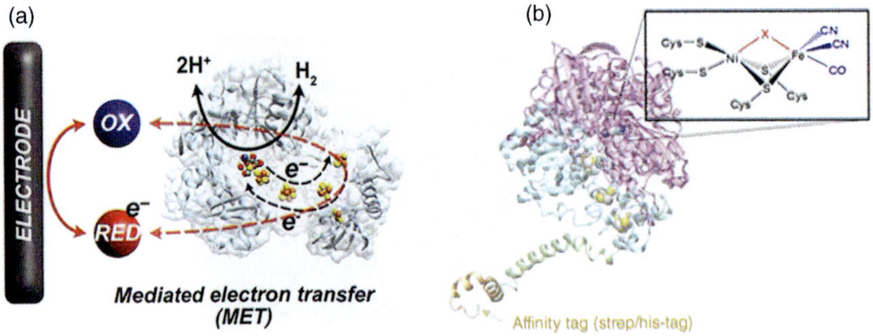

Figure 1.3 (a) Scheme of mediated electron transfer. For example, a Fe–Fe hydrogenase (PDB 6N59) is utilized. Fe = rust, S = yellow, C = gray, O = red, N = blue. *Source:* Cadoux and Milton [50] / with permission of John Wiley & Sons. (b) Modeled protein structure of a membrane-bound (NiFe) hydrogenase containing the membrane subunit helix and the affinity tag added to the C-terminus of the enzyme for adsorption on a tin-rich indium tin oxide thin-film electrode. *Source:* Davis et al. [51] / with permission of American Chemical Society.

Once an efficient ET can be established, the use of enzymes as bioelectrocatalysts offers unique features, such as high specificity toward a specific substrate, enabling their application for the development of technologies such as enzymatic biosensors and wearable devices, presented in Chapters 4 and 9, respectively. Conversely, when aiming at developing power-generating systems, the high selectivity of enzymes might lead to partial fuel oxidation, resulting in a lower energy density. These concepts can be easily explained considering that in biological organisms, the oxidation of a substrate is not accomplished by a single enzyme, but rather by means of complex metabolic pathways that are performed by multiple enzymes (i.e. the tricarboxylic acid cycle or Krebs cycle involving eight enzymes, or the glyoxylate cycle involving five enzymes) [60]. To address this limitation, the use of "enzyme cascades" mimicking natural pathways has been reported over the last decade [61]. This approach will be discussed in detail in Chapter 6, where other limitations of enzyme-based electrochemical systems are also extensively presented, with the current approaches to tackle them, such as the need for protein purification and decreased stability of the enzymes once extracted from a living cell.

1.2.2 Microbial Bioelectrocatalysis

With the aim of obtaining biocatalysts capable of deep substrate oxidation or reducing broadly available substrates (i.e. CO_2 or H^+) to valuable products (i.e. methane, acetate, and hydrogen), the use of intact microorganisms becomes particularly interesting, due to their complex catalytic machinery for multistep reactions [62]. However, when moving to intact organisms, such as bacteria, yeast, and algae, the redox enzymes involved in their metabolisms are confined inside the cellular membrane(s), adding a barrier for the "electric wiring" of these biological catalysts to an electrode surface that is necessary for both electron harvesting and uptake [63, 64]. Furthermore, it should be noted that different from isolated enzymes, microorganisms are living entities and thus a certain amount of energy is required for their sustainment, posing some limitations on the amount of electrons that can be donated to an electrode while retaining their capability to grow and replicate [65].

Chapter 3 will describe in detail the various extracellular ET mechanisms that have been studied in detail in the last two decades for some bacteria that are defined as "model electroactive organisms" [66]. Specifically, two Gram-negative bacteria of the species *Geobacter sulfurreducens* and *Shewanella oneidensis* have developed ET pathways that allow them to exchange the electrons coming from their metabolism with external electron acceptors, and several research efforts have been directed toward clarifying their specific mechanisms of ET [67–77].

Remarkably, even for these model electroactive bacteria, and for all the other microorganisms that have been studied for bioelectrocatalysis [78], the use of redox mediators or artificial redox mediation approaches, together with engineered electrode materials, has allowed greatly enhancing the extracellular ET process, until achieving impressive current densities up to various mA cm^{-2} [3, 79–81]. As discussed in detail in the following chapters, the exogenous redox mediators can be both added as diffusible molecules in solution [82–85] and confined on the polymer backbone of what are defined as "redox polymers" [86–92]. Hybrid conductive-redox polymers have also been reported to facilitate the ET and enhance current generation with different biological catalysts [93–95]. Specifically, based on the location of the redox pendant, and thus the polymer structure, the redox polymers can be classified into branched or unbranched, with the first ones having linker arms that connect the redox pendant to the (conductive or nonconductive) polymer backbone (Figure 1.4a–e), and the second ones having the redox pendants embedded into their conducting polymer backbone (Figure 1.4f–k) [96].

In addition to the use of redox mediators, bacterial cells have been modified with conductive nanomaterials that can act as electron conduits for the extraction (or insertion) of electrons from their redox centers [97, 98]. Among the various nanomaterials, both carbon-based nanomaterials [99–103] and metal nanoparticles [3, 104–107] have been reported to successfully facilitate the extracellular ET, while minimizing damage to the bacterial membrane. Regarding the role of these nanomaterials in facilitating the extracellular electron transfer (EET) process, it was proposed that carbon nanotubes (CNTs) can establish a conductive bridge directly connecting the redox centers in bacterial cells (i.e. membrane-bound cytochromes in the case of *S. oneidensis* MR-1) to the electrode [108]. The interaction of CNTs with bacterial cells could also result in more conductive biofilms, which allows higher current and power densities [109]. Similarly, metal nanoparticles were also reported to create a network of conductive particles that facilitated ET by allowing e^- hopping from the bacterial redox centers to the electrode (Figure 1.5). A recent work by Xia et al. underlined the importance of nanoparticle dimension on lowering the onset potential for electron harvesting and/or increasing current densities, showing that nanoparticles with smaller dimensions decreased the potential at which electrons could be transferred to the electrode [106]. The role of nanoparticle surface functionalization on the extracellular ET was also recently investigated by Lasala et al., showing that nanoparticle surface functionalization with zwitterionic molecules allows obtaining a uniform distribution of the nanomaterials, with a resulting decreased onset potential and enhanced extracellular ET [107].

As previously introduced in Section 1.2.1 for the engineering of enzymes, synthetic biology has also been successfully used to engineer microorganisms for promoting the extracellular ET process [110]. Among the various engineering approaches, the last decades have seen particular efforts of the scientific community focused toward the introduction of heterologous ET pathways from model electroactive bacteria into desired organisms [111]. Starting from the pioneering work of Gescher et al. (2008), with the expression of cymA alone from *S. oneidensis* in *Escherichia coli* conferring iron reduction capabilities [112], several works from the Ajo-Franklin group elucidated the role of different components of the *Shewanella* ET pathway for efficient extracellular ET and of the induction systems in retaining cell growth and metabolism after the expression [113–116].

The complete *S. oneidensis* Mtr-pathway was recently expressed in *E. coli* cells, resulting in a threefold increase in current generation compared to wild-type cells [117]. Furthermore, the engineering of heterologous ET pathways from *Geobacter* species to express the outer membrane protein cytochrome OmcS was also accomplished in photosynthetic organisms, as recently reported by Dong et al., greatly enhancing the electron uptake and faradic efficiency for the reduction of N_2 into ammonia in cyanobacteria-based biocathodes [118].

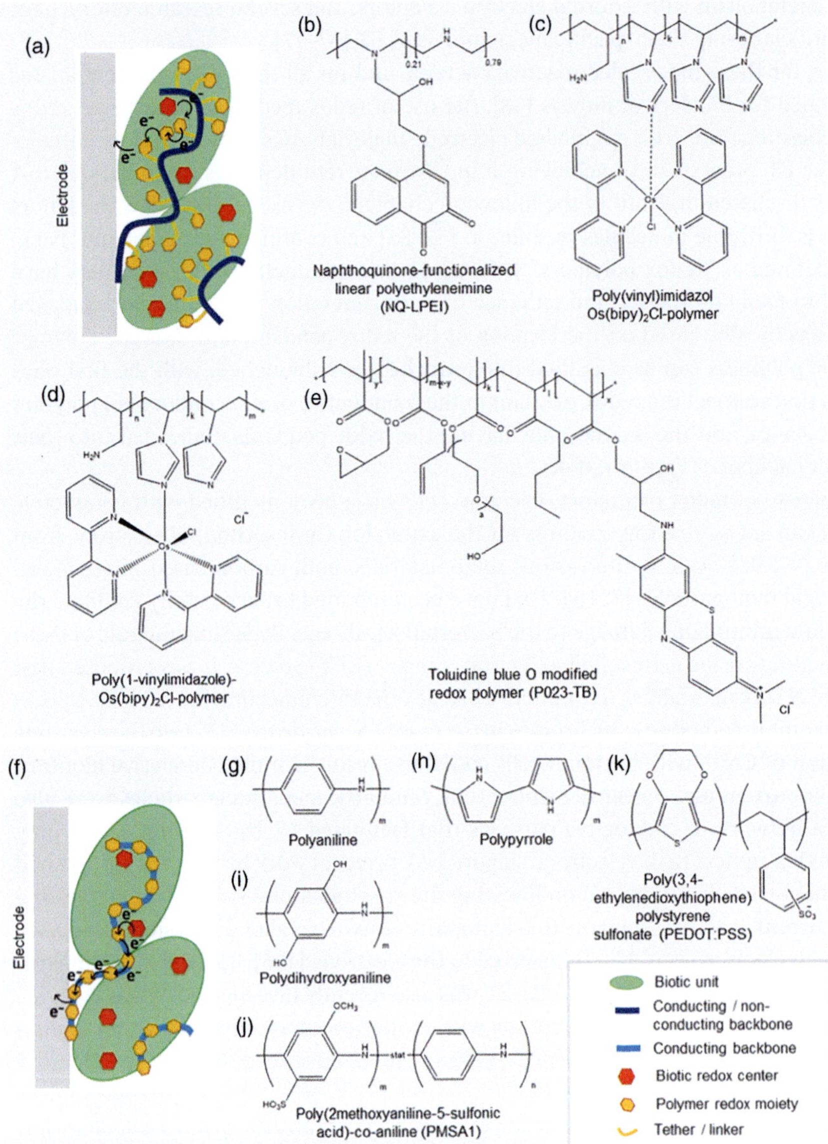

Figure 1.4 Structural classification of branched and unbranched redox polymers. (a) Scheme of the extracellular electron transfer between microorganisms and branched redox polymers. (b–e) Examples of branched redox polymers. (f) Scheme of the extracellular electron transfer between microorganisms and unbranched redox polymers. (g–k) Examples of unbranched redox polymers. *Source:* Reproduced from Weliwatte et al. [96] / Springer Nature / CC BY 4.0.

1.3 Conclusions and Outlook

The present chapter has introduced the field of bioelectrocatalysis, starting from the basis of the ET process, with a focus on both enzymatic and microbial bioelectrocatalysis. The fundamental aspects of the process of (extracellular) ET have been presented, and the artificial approaches that have sparked the interest of the scientific community in addressing the intrinsic limitations

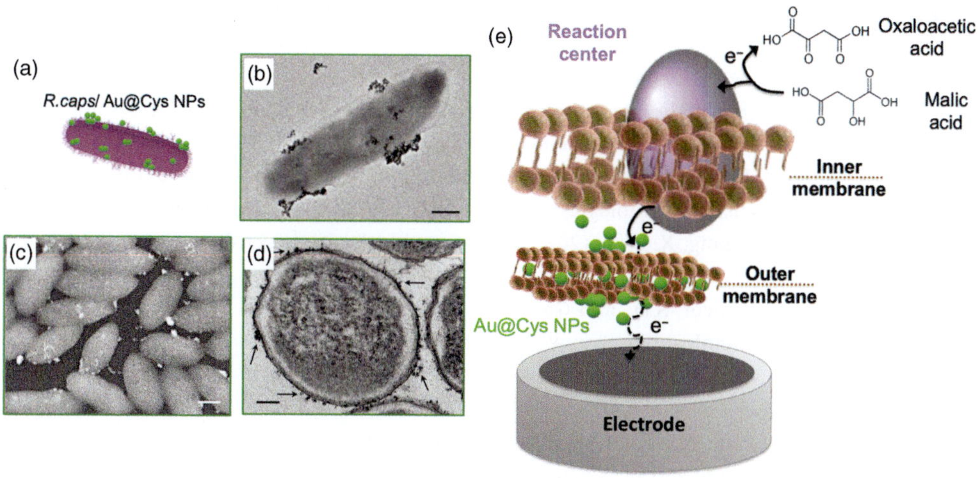

Figure 1.5 Role of metal nanoparticles in facilitating EET, (a) representation of a bacterial cell functionalized with distributed nanoparticles, (b) transmission electron microscopy (TEM) micrograph of one bacterial cell with Au-nanoparticles (scale bar 200 nm), (c) scanning electron microscopy micrograph of multiple bacterial cells exposed to Au-nanoparticles (scale bar 500 nm), (d) TEM image of a thin section of a bacterial cell that has been exposed to Au-nanoparticles with black arrows indicating the nanoparticles (scale bar 100 nm), (e) scheme of the electron transfer conduit obtained for bacterial cells functionalized with Au-nanoparticles. *Source:* Reproduced from Lasala et al. [107] / American Chemical Society / CC BY 4.0.

of bioelectrocatalysis have been discussed. Stressing this last aspect, it is interesting to note the interdisciplinary field of bioelectrocatalysis, which requires knowledge in electrochemistry, biology, material science, and engineering, underlined by the various approaches presented to facilitate the ET process (i.e. the use of redox polymers, nanomaterials, synthetic biology, etc.). The combination of these skills in a highly collaborative environment has enabled the community of scientists and engineers to achieve remarkable results in recent years, both in understanding the fundamental mechanisms governing bioelectrocatalysis and in practical applications of the technology, as extensively discussed in the following chapters. These achievements pave the way for an exciting future in bioelectrocatalysis, positioning it as a key player in achieving a more sustainable society.

Acknowledgments

M. Grattieri would like to acknowledge the funding from Fondazione CON IL SUD, Grant "Brains to South 2018," project number 2018-PDR-00914. S.D. Minteer would like to thank the National Science Foundation, grant number (2406605).

References

1 Hill, H.A.O. and Higgins, I.J. (1981). Bioelectrocatalysis. *Philos. Trans. R. Soc. London, Ser. A Math. Phys. Sci.* 302: 267–273.

2 Riedel, M., Höfs, S., Ruff, A. et al. (2020). A tandem solar biofuel cell: harnessing energy from light and biofuels. *Angew. Chem. Int. Ed.* 60: 2078–2083.

3 Cao, B., Zhao, Z., Peng, L. et al. (2021). Silver nanoparticles boost charge-extraction efficiency in Shewanella microbial fuel cells. *Science* 373: 1336–1340.

4 Bollella, P. and Gorton, L. (2018). Enzyme based amperometric biosensors. *Curr. Opin. Electrochem.* 10: 157–173.

5 Grattieri, M. and Minteer, S.D. (2018). Self-powered biosensors. *ACS Sens.* 3: 44–53.

6 Hara, T.O. and Singh, B. (2021). Electrochemical biosensors for detection of pesticides and heavy metal toxicants in water: recent trends and progress. *ACS ES&T Water* 1: 462–478.

7 Simoska, O., Gaffney, E.M., Minteer, S.D. et al. (2021). Recent trends and advances in microbial electrochemical sensing technologies: an overview. *Curr. Opin. Electrochem.* 30: 100762.

8 Atkinson, J.T., Su, L., Zhang, X. et al. (2022). Real-time bioelectronic sensing of environmental contaminants. *Nature* 611: 548–553.

9 Kalita, N., Gogoi, S., Minteer, S.D., and Goswami, P. (2023). Advances in bioelectrode design for developing electrochemical biosensors. *ACS Meas. Sci. Au* 3: 404–433.

10 Xiao, X., McGourty, K.D., and Magner, E. (2020). Enzymatic biofuel cells for self-powered, controlled drug release. *J. Am. Chem. Soc.* 142: 11602–11609.

11 Xu, Y., De la Paz, E., Paul, A. et al. (2023). In-ear integrated sensor array for the continuous monitoring of brain activity and of lactate in sweat. *Nat. Biomed. Eng.* 7: 1307–1320.

12 Saha, T., Del Caño, R., Mahato, K. et al. (2023). Wearable electrochemical glucose sensors in diabetes management: a comprehensive review. *Chem. Rev.* 123: 7854–7889.

13 Babanova, S., Jones, J., Phadke, S. et al. (2020). Continuous flow, large-scale, microbial fuel cell system for the sustained treatment of swine waste. *Water Environ. Res.* 92: 60–72.

14 Babanova, S., Santoro, C., Jones, J. et al. (2021). Practical demonstration of applicability and efficiency of platinum group metal-free based catalysts in microbial fuel cells for wastewater treatment. *J. Power Sources* 491: 229582.

15 Rossi, R., Hur, A.Y., Page, M.A. et al. (2022). Pilot scale microbial fuel cells using air cathodes for producing electricity while treating wastewater. *Water Res.* 215: 118208.

16 Milton, R.D., Cai, R., Abdellaoui, S. et al. (2017). Bioelectrochemical Haber-Bosch process: an ammonia-producing H_2/N_2 fuel cell. *Angew. Chem. Int. Ed.* 56: 2680–2683.

17 Chen, H., Cai, R., Patel, J. et al. (2019). Upgraded bioelectrocatalytic N_2 fixation: from N_2 to chiral amine intermediates. *J. Am. Chem. Soc.* 141: 4963–4971.

18 Luna-López, G., del Barrio, M., Fize, J. et al. (2023). Photobio-electrocatalytic production of H_2 using fluorine-doped tin oxide (FTO) electrodes covered with a NiO-In_2S_3 p-n junction and NiFeSe hydrogenase. *Bioelectrochemistry* 150: 108361.

19 Dong, F., Lee, Y.S., Gaffney, E.M. et al. (2021). An engineered, non-diazotrophic cyanobacterium and its application in bioelectrochemical nitrogen fixation. *Cell Rep. Phys. Sci.* 2: 100444.

20 Punekar, N.S. (2018). *ENZYMES: Catalysis, Kinetics and Mechanisms*. Singapore: Springer Nature Singapore Pte Ltd.

21 Masa, J., Andronescu, C., and Schuhmann, W. (2020). Electrocatalysis as the Nexus for sustainable renewable energy: the Gordian knot of activity, stability, and selectivity. *Angew. Chem. Int. Ed.* 59: 15298–15312.

22 Ferri, M., Delafontaine, L., Guo, S. et al. (2022). Steering Cu-based CO2RR electrocatalysts' selectivity: effect of hydroxyapatite acid/base moieties in promoting formate production. *ACS Energy Lett.* 7: 2304–2310.

23 Nutting, J.E., Rafiee, M., and Stahl, S.S. (2018). Tetramethylpiperidine N-oxyl (TEMPO), phthalimide N-oxyl (PINO), and related N-oxyl species: electrochemical properties and their use in electrocatalytic reactions. *Chem. Rev.* 118: 4834–4885.

24 Lienemann, M. (2021). Molecular mechanisms of electron transfer employed by native proteins and biological-inorganic hybrid systems. *Comput. Struct. Biotechnol. J.* 19: 206–213.

25 Marcus, R.A. and Sutin, N. (1985). Electron transfer in chemistry and biology. *Biochim. Biophys. Acta* 811: 265–322.
26 Leech, D., Kavanagh, P., and Schuhmann, W. (2012). Enzymatic fuel cells: recent progress. *Electrochim. Acta* 84: 223–234.
27 Hickey, D.P., Milton, R.D., Rasmussen, M. et al. (2015). Fundamentals and applications of bioelectrocatalysis. In: *Electrochemistry* (ed. C. Banks, R. Mortimer, and S. McIntosh), 97, 97–132. Cambridge, UK: Royal Society of Electrochemistry.
28 Gray, H.B. and Winkler, J.R. (2005). Long-range electron transfer. *Proc. Natl. Acad. Sci.* 102: 3534–3539.
29 Chi, Q., Farver, O., and Ulstrup, J. (2005). Long-range protein electron transfer observed at the single-molecule level: in situ mapping of redox-gated tunneling resonance. *Proc. Natl. Acad. Sci.* 102: 16203–16208.
30 Solomon, E.I., Heppner, D.E., Johnston, E.M. et al. (2014). Copper active sites in biology. *Chem. Rev.* 114: 3659–3853.
31 Piontek, K., Antorini, M., and Choinowski, T. (2002). Crystal structure of a Laccase from the FungusTrametes versicolor at 1.90-Å resolution containing a full complement of coppers. *J. Biol. Chem.* 277: 37663–37669.
32 Skálová, T., Dušková, J., Hašek, J. et al. (2010). Structure of laccase fromStreptomyces coelicolorafter soaking with potassium hexacyanoferrate and at an improved resolution of 2.3 Å. *Acta Crystallogr., Sect. F: Struct. Biol. Cryst. Commun.* 67: 27–32.
33 Hecht, H.J., Kalisz, H.M., Hendle, J. et al. (1993). Crystal structure of glucose oxidase from Aspergillus niger refined at 2·3 Å reslution. *J. Mol. Biol.* 229: 153–172.
34 Leskovac, V., Trivic, S., Wohlfahrt, G. et al. (2005). Glucose oxidase from Aspergillus niger: the mechanism of action with molecular oxygen, quinones, and one-electron acceptors. *Int. J. Biochem. Cell Biol.* 37: 731–750.
35 Bartlett, P.N. and Al-Lolage, F.A. (2018). There is no evidence to support literature claims of direct electron transfer (DET) for native glucose oxidase (GOx) at carbon nanotubes or graphene. *J. Electroanal. Chem.* 819: 26–37.
36 Schuster, D.I., Cheng, P., Jarowski, P.D. et al. (2004). Design, synthesis, and photophysical studies of a porphyrin-fullerene dyad with parachute topology; charge recombination in the marcus inverted region. *J. Am. Chem. Soc.* 126: 7257–7270.
37 Eckermann, A.L., Feld, D.J., Shaw, J.A., and Meade, T.J. (2010). Electrochemistry of redox-active self-assembled monolayers. *Coord. Chem. Rev.* 254: 1769–1802.
38 Zakeeruddin, S.M., Fraser, D.M., Nazeeruddin, M.-K., and Grätzel, M. (1992). Towards mediator design: Characterization of tris-(4,4′-substituted-2,2′-bipyridine) complexes of iron(II), ruthenium(II) and osmium(II) as mediators for glucose oxidase of Aspergillus niger and other redox proteins. *J. Electroanal. Chem.* 337: 253–283.
39 Gallaway, J. and Barton, S.C. (2008). Kinetics of redox polymer-Mediated enzyme electrode. *J. Am. Chem. Soc.* 130: 8527–8536.
40 McMillan, D.G.G., Marritt, S.J., Kemp, G.L. et al. (2013). The impact of enzyme orientation and electrode topology on the catalytic activity of adsorbed redox enzymes. *Electrochim. Acta* 110: 79–85.
41 Trammell, S.A., Spano, A., Price, R., and Lebedev, N. (2006). Effect of protein orientation on electron transfer between photosynthetic reaction centers and carbon electrodes. *Biosens. Bioelectron.* 21: 1023–1028.

42 Matanovic, I., Babanova, S., Chavez, M.S., and Atanassov, P. (2016). Protein-support interactions for rationally designed bilirubin oxidase based cathode: a computational study. *J. Phys. Chem. B* 120: 3634–3641.

43 Reginald, S.S., Lee, H., Fazil, N. et al. (2022). Control of carbon monoxide dehydrogenase orientation by site-specific immobilization enables direct electrical contact between enzyme cofactor and solid surface. *Commun. Biol.* 5: 390.

44 Zhao, F., Wang, P., Ruff, A. et al. (2019). A photosystem I monolayer with anisotropic electron flow enables Z-scheme like photosynthetic water splitting. *Energy Environ. Sci.* 12: 3133–3143.

45 Sugimoto, Y., Kitazumi, Y., Tsujimura, S. et al. (2015). Electrostatic interaction between an enzyme and electrodes in the electric double layer examined in a view of direct electron transfer-type bioelectrocatalysis. *Biosens. Bioelectron.* 63: 138–144.

46 Yates, N.D.J., Fascione, M.A., and Parkin, A. (2018). Methodologies for "wiring" redox proteins/enzymes to electrode surfaces, chemistry – a. *Chem. Eur. J.* 24: 12164–12182.

47 Léger, C. and Bertrand, P. (2008). Direct electrochemistry of redox enzymes as a tool for mechanistic studies. *Chem. Rev.* 108: 2379–2438.

48 Liu, Y., Webb, S., Moreno-García, P. et al. (2023). Facile functionalization of carbon electrodes for efficient electroenzymatic hydrogen production. *JACS Au* 3: 124–130.

49 Milton, R.D. and Minteer, S.D. (2017). Direct enzymatic bioelectrocatalysis: differentiating between myth and reality. *J. R. Soc. Interface* 14: 20170253.

50 Cadoux, C. and Milton, R.D. (2020). Recent enzymatic electrochemistry for reductive reactions. *ChemElectroChem* 7: 1974–1986.

51 Davis, V., Heidary, N., Guiet, A. et al. (2023). Immobilization of O2-tolerant [NiFe] hydrogenase from Cupriavidus necator on Tin-rich indium oxide alters the catalytic bias from H_2 oxidation to proton reduction. *ACS Catal.* 13: 6312–6327.

52 Mao, F., Mano, N., and Heller, A. (2003). Long tethers binding redox centers to polymer backbones enhance electron transport in enzyme "wiring" hydrogels. *J. Am. Chem. Soc.* 125: 4951–4957.

53 Mano, N., Yoo, J.E., Tarver, J. et al. (2007). An electron-conducting cross-linked polyaniline-based redox hydrogel formed in one step at pH 7.2, wires glucose oxidase. *J. Am. Chem. Soc. Commun.* 129: 7006–7007.

54 Lisdat, F. (2017). Trends in the layer-by-layer assembly of redox proteins and enzymes in bioelectrochemistry. *Curr. Opin. Electrochem.* 5: 165–172.

55 Calvo, E.J. and Wolosiuk, A. (2005). Wiring enzymes in nanostructures built with electrostatically self-assembled thin films. *ChemPhysChem: Eur. J. Chem. Phys. Phys. Chem.* 6: 43–47.

56 Calvo, E.J., Etchenique, R.A., Pietrasanta, L., and Wolosiuk, A. (2001). Layer-by-layer self-assembly of glucose oxidase and Os(Bpy)2ClPyCH2NH-poly(allylamine) bioelectrode. *Anal. Chem.* 73: 1161–1168.

57 Sokol, K.P., Mersch, D., Hartmann, V. et al. (2016). Rational wiring of photosystem II to hierarchical indium tin oxide electrodes using redox polymers. *Energy Environ. Sci.* 9: 3698–3709.

58 Zhang, L., Vilà, N., Klein, T. et al. (2016). Immobilization of cysteine-tagged proteins on electrode surfaces by Thiol–Ene click chemistry. *ACS Appl. Mater. Interfaces* 8: 17591–17598.

59 Lee, H., Reginald, S.S., Sravan, J.S. et al. (2024). Advanced strategies for enzyme–electrode interfacing in bioelectrocatalytic systems. *Trends Biotechnol.* 43: 1328–1355.

60 Erkmen, O. and Bozoglu, F.T. (2016). Microbial Metabolism of Food Components. In: *Food Microbiology: Principles into Practice* (ed. O. Erkmen and F.T. Bozoglu), 217, 217–227. John Wiley & Sons, Ltd.

61 Macazo, F.C. and Minteer, S.D. (2017). Enzyme cascades in biofuel cells. *Curr. Opin. Electrochem.* 5: 114–120.

62 Bedendi, G., De Moura Torquato, L.D., Webb, S. et al. (2022). Enzymatic and microbial electrochemistry: approaches and methods. *ACS Meas. Sci. Au* 2: 517–541.

63 Pankratova, G. and Gorton, L. (2017). Electrochemical communication between living cells and conductive surfaces. *Curr. Opin. Electrochem.* 5: 193–202.

64 Rosenbaum, M.A. and Henrich, A.W. (2014). Engineering microbial electrocatalysis for chemical and fuel production. *Curr. Opin. Biotechnol.* 29: 93–98.

65 Aiyer, K.S. (2020). *World J. Microbiol. Biotechnol.* 36: 19.

66 Logan, B.E., Rossi, R., Ragab, A., and Saikaly, P.E. (2019). Electroactive microorganisms in bioelectrochemical systems. *Nat. Rev. Microbiol.* 17: 307–319.

67 Childers, S.E., Ciufo, S., and Lovley, D.R. (2002). Geobacter metallireducens accesses insoluble Fe(III) oxide by chemotaxis. *Nature* 416: 767–769.

68 Bond, D.R., Holmes, D.E., Tender, L.M., and Lovley, D.R. (2002). Electrode-reducing microorganisms that harvest energy from marine sediments. *Science* 295: 483–485.

69 Reguera, G., McCarthy, K.D., Mehta, T. et al. (2005). Extracellular electron transfer via microbial nanowires. *Nature* 435: 1098–1101.

70 Marsili, E., Baron, D.B., Shikhare, I.D. et al. (2008). Shewanella secretes flavins that mediate extracellular electron transfer. *PNAS* 105: 3968–3973.

71 El-Naggar, M.Y., Wanger, G., Leung, K.M. et al. (2010). Electrical transport along bacterial nanowires from Shewanella oneidensis MR-1. *PNAS* 107: 18127–18131.

72 Okamoto, A., Hashimoto, K., Nealson, K.H., and Nakamura, R. (2013). Rate enhancement of bacterial extracellular electron transport involves bound flavin semiquinines. *Proc. Natl. Acad. Sci. USA* 110: 7856–7861.

73 Shi, L., Dong, H., Reguera, G. et al. (2016). Extracellular electron transfer mechanisms between microorganisms and minerals. *Nat. Rev. Microbiol.* 14: 651–662.

74 Ishii, S.I., Suzuki, S., Tenney, A. et al. (2018). Comparative metatranscriptomics reveals extracellular electron transfer pathways conferring microbial adaptivity to surface redox potential changes. *ISME J.* 12: 2844–2863.

75 Edwards, M.J., Richardson, D.J., Paquete, C.M., and Clarke, T.A. (2019). Role of multiheme cytochromes involved in extracellular anaerobic respiration in bacteria. *Protein Sci.* 29: 830–842.

76 Heidary, N., Kornienko, N., Kalathil, S. et al. (2020). Disparity of cytochrome utilization in anodic and cathodic extracellular electron transfer pathways of geobacter sulfurreducens biofilms. *J. Am. Chem. Soc.* 142: 5194–5203.

77 Rowe, A.R., Salimijazi, F., Trutschel, L. et al. (2021). Identification of a pathway for electron uptake in Shewanella oneidensis. *Commun. Biol.* 4: 957.

78 Koch, C. and Harnisch, F. (2016). Is there a specific ecological niche for electroactive microorganisms? *ChemElectroChem* 3: 1282–1295.

79 Hasan, K., Grattieri, M., Wang, T. et al. (2017). Enhanced bioelectrocatalysis of Shewanella oneidensis MR-1 by a naphthoquinone redox polymer. *ACS Energy Lett.* 2: 1947–1951.

80 Fang, X., Kalathil, S., Divitini, G. et al. (2020). A three-dimensional hybrid electrode with electroactive microbes for efficient electrogenesis and chemical synthesis. *Proc. Natl. Acad. Sci.* 117: 5074–5080.

81 Zhang, L., Zhang, Y., Liu, Y. et al. (2024). High power density redox-mediated Shewanella microbial flow fuel cells. *Nat. Commun.* 15: 8302.

82 Longatte, G., Sayegh, A., Delacotte, J. et al. (2018). Investigation of photocurrents resulting from a living unicellular algae suspension with quinones over time. *Chem. Sci.* 9: 8271–8281.

83 Grattieri, M., Rhodes, Z., Hickey, D.P. et al. (2019). Understanding biophotocurrent generation in photosynthetic Purple bacteria. *ACS Catal.* 9: 867–873.

84 Zhang, J.Z., Bombelli, P., Sokol, K.P. et al. (2018). Photoelectrochemistry of photosystem II in vitro vs in vivo. *J. Am. Chem. Soc.* 140: 6–9.

85 Clifford, E.R., Bradley, R.W., Wey, L.T. et al. (2021). Phenazines as model low-midpoint potential electron shuttles for photosynthetic bioelectrochemical systems. *Chem. Sci.* 12: 3328–3338.

86 Coman, V., Gustavsson, T., Finkelsteinas, A. et al. (2009). Electrical wiring of live, metabolically enhanced bacillus subtilis cells with flexible osmium-redox polymers. *J. Am. Chem. Soc.* 131: 16171–16176.

87 Hasan, K., Patil, S.A., Leech, D. et al. (2012). Electrochemical communication between microbial cells and electrodes via osmium redox systems. *Biochem. Soc. Trans.* 40: 1330–1335.

88 Hasan, K., Patil, S.A., Gorecki, K. et al. (2013). Electrochemical communication between heterotrophically grown Rhodobacter capsulatus with electrodes mediated by an osmium redox polymer. *Bioelectrochemistry* 93: 30–36.

89 Darus, L., Sadakane, T., Ledezma, P. et al. (2016). Redox-polymers enable uninterrupted day/night photo-driven electricity generation in biophotovoltaic devices. *J. Electrochem. Soc.* 164: H3037–H3040.

90 Yuan, M. and Minteer, S.D. (2019). Redox polymers in electrochemical systems: From methods of mediation to energy storage. *Curr. Opin. Electrochem.* 15: 1–6.

91 Pankratova, G., Pankratov, D., Milton, R.D. et al. (2019). Following nature: bioinspired mediation strategy for gram-positive bacterial cells. *Adv. Energy Mater.* 9: 1900215.

92 Herrero-Medina, Z., Wang, P., Lielpetere, A. et al. (2022). A biophotoelectrode based on boronic acid-modified Chlorella vulgaris cells integrated within a redox polymer. *Bioelectrochemistry* 146: 108128.

93 Weliwatte, N.S., Grattieri, M., Simoska, O. et al. (2021). Unbranched hybrid conducting redox polymers for intact chloroplast-based photobioelectrocatalysis. *Langmuir* 37: 7821–7833.

94 Buscemi, G., Vona, D., Stufano, P. et al. (2022). Bio-inspired redox-adhesive polydopamine matrix for intact bacteria biohybrid photoanodes. *ACS Appl. Mater. Interfaces* 14: 26631–26641.

95 Benjamin Ocheja, O., Wahid, E., Honorio Franco, J. et al. (2024). Polydopamine-immobilized yeast cells for portable electrochemical biosensors applied in environmental copper sensing. *Bioelectrochemistry* 157: 108658.

96 Weliwatte, N.S., Grattieri, M., and Minteer, S.D. (2021). Rational design of artificial redox-mediating systems toward upgrading photobioelectrocatalysis. *Photochem. Photobiol. Sci.* 20: 1333–1356.

97 Wang, R., Li, H., Sun, J. et al. (2021). Nanomaterials facilitating microbial extracellular electron transfer at interfaces. *Adv. Mater.* 33: 2004051.

98 Hsu, L., Deng, P., Zhang, Y. et al. (2018). Nanostructured interfaces for probing and facilitating extracellular electron transfer. *J. Mater. Chem. B* 6: 7144–7158.

99 Pinck, S., Jorand, F.P.A., Xu, M., and Etienne, M. (2019). Protamine promotes direct electron transfer between shewanella oneidensis cells and carbon nanomaterials in bacterial biocomposites. *ChemElectroChem* 6: 2398–2406.

100 Antonucci, A., Reggente, M., Roullier, C. et al. (2022). Carbon nanotube uptake in cyanobacteria for near-infrared imaging and enhanced bioelectricity generation in living photovoltaics. *Nat. Nanotechnol.* 17: 1111–1119.

101 Wang, H., Feng, L., Duan, Z. et al. (2023). Carbon nanotubes modified reticulated vitreous carbon materials enhancing extracellular electron transfer of Shewanella loihica PV-4. *Colloids Surf. A Physicochem. Eng. Asp.* 676: 132126.

102 Shiraki, T., Niidome, Y., Roy, A. et al. (2024). Single-walled carbon nanotubes wrapped with charged polysaccharides enhance extracellular electron transfer. *ACS Appl. Bio Mater.* 7: 5651–5661.

103 Guo, X., Yang, C., Wu, J. et al. (2024). Ultra-small carbon dots boost bioelectricity generation by accelerating extracellular electron transfer. *J. Power Sources* 610: 234711.

104 Chen, Z., Liu, Y., Zhang, C. et al. (2019). Titanium dioxide nanoparticles induced an enhanced and intimately coupled photoelectrochemical-microbial reductive dissolution of As(V) and Fe(III) in flooded arsenic-enriched soils. *ACS Sustain. Chem. Eng.* 7: 13236–13246.

105 Liu, L. and Choi, S. (2021). Enhanced biophotoelectricity generation in cyanobacterial biophotovoltaics with intracellularly biosynthesized gold nanoparticles. *J. Power Sources* 506: 230251.

106 Xia, Q., Liu, R., Chen, X. et al. (2023). in vivo voltammetric imaging of metal nanoparticle-catalyzed single-cell electron transfer by fermi level-responsive graphene. *Research* 6: 0145.

107 Lasala, P., Matteucci, R.M., Volpicella, S.R. et al. (2024). Deciphering the role of inorganic nanoparticles' surface functionalization on biohybrid microbial photoelectrodes. *ACS Appl. Mater. Interfaces* 16: 58598–58608.

108 Liu, X.-W., Chen, J.-J., Huang, Y.-X. et al. (2014). Experimental and theoretical demonstrations for the mechanism behind enhanced microbial electron transfer by CNT network. *Sci. Rep.* 4: 3732.

109 Wu, X., Qiao, Y., Shi, Z. et al. (2018). Hierarchically porous N-doped carbon nanotubes/reduced graphene oxide composite for promoting flavin-based interfacial electron transfer in microbial fuel cells. *ACS Appl. Mater. Interfaces* 10: 11671–11677.

110 Dong, F., Simoska, O., Gaffney, E., and Minteer, S.D. (2022). Applying synthetic biology strategies to bioelectrochemical systems. *Electrochem. Sci. Adv.* 2: e2100197.

111 Bird, L.J., Kundu, B.B., Tschirhart, T. et al. (2021). Engineering wired life: synthetic biology for electroactive bacteria. *ACS Synth. Biol.* 10: 2808–2823.

112 Gescher, J.S., Cordova, C.D., and Spormann, A.M. (2008). Spormann, dissimilatory iron reduction in Escherichia coli: identification of CymA of Shewanella oneidensis and NapC of E. coli as ferric reductases. *Mol. Microbiol.* 68: 706–719.

113 Jensen, H.M., Albers, A.E., Malley, K.R. et al. (2010). Engineering of a synthetic electron conduit in living cells. *PNAS* 107: 19213–19218.

114 TerAvest, M.A., Zajdel, T.J., and Ajo-Franklin, C.M. (2014). The Mtr pathway of Shewanella oneidensis MR-1 couples substrate utilization to current production in Escherichia coli. *ChemElectroChem* 1: 1874–1879.

115 Jensen, H., TerAvest, M., Kokish, M., and Ajo-Franklin, C.M. (2016). CymA and exogenous flavins improve extracellular electron transfer and couple it to cell growth in mtr-expressing Escherichia coli. *ACS Synth. Biol.* 5: 679–688.

116 Goldbeck, C.P., Jensen, H.M., TerAvest, M.A. et al. (2013). Tuning promoter strengths for improved synthesis and function of electron conduits in escherichia coli. *ACS Synth. Biol.* 2: 150–159.

117 Mouhib, M., Reggente, M., Li, L. et al. (2023). Extracellular electron transfer pathways to enhance the electroactivity of modified Escherichia coli. *Joule* 7: 2092–2106.

118 Dong, F., Lee, Y.S., Gaffney, E.M. et al. (2021). Engineering cyanobacterium with transmembrane electron transfer ability for bioelectrochemical nitrogen fixation. *ACS Catal.* 11: 13169–13179.

2

Electron Transfer in Enzymatic Catalysis

Fred Lisdat and Daniel Schäfer

Department of Biosystems Technology, Technical University of Applied Sciences Wildau, Brandenburg, Germany

2.1 Introduction

Enzymes are the natural catalysts that provide a solid basis for life as a sophisticated and advanced phenomenon, clearly distinguished from other complex chemical reactions. These molecules have many fascinating properties that one can use in different fields of application. As the first aspect, the key feature of a specific interaction with a substrate molecule should be emphasized. This can exclusively be one type of molecule or a class of molecules. The basis is a molecular recognition event occurring in the binding pocket of the enzyme, which originates from a steric and chemical fitting of this pocket to the properties and shape of the substrate. One has, however, to state that this interaction is also dynamic, and thus, changes in the 3D structure of the enzyme can also occur. For some substrate–enzyme couples, this leads to the "induced fitting" behavior in which both enzyme and substrate can change during the recognition process. The second aspect is the catalytic rate at which the enzyme converts the substrate. Enzyme action results in a dramatic increase in conversion rate compared to the uncatalyzed process. The acceleration is in the range of 10^3 to 10^{20}, which may illustrate the usefulness in applying enzymes for synthetic purposes. This enormous efficiency in substrate conversion can also be exemplified by the turnover number, which is in the range of 10^2 to 10^7 s^{-1}. Furthermore, it should be mentioned here that the resulting product is a well-defined substance – from the structural and chemical point of view – in contrast to organic reactions where often a mixture of molecules is obtained or racemates of a certain substance.

These advantages are contrasted by the nature of the biocatalyst. They are proteins with a defined 3D structure, and thus, if this structure is affected by the conditions of application, then the activity is strongly influenced up to complete inactivation. This means that conditions need to be carefully controlled in order to ensure a certain operational time of work. Inactivation over time cannot be completely avoided, but can be drastically reduced. This, of course, also depends on the detailed enzyme used for the specific application.

The selective and fast conversion of substances by enzymes has already been exploited in several fields. Analytical applications represent one direction. Thus, there are numerous bioanalytical systems that are based on enzymes. These range from simple optical test systems with all components in solution to test strips with immobilized reagents up to flow systems with an immobilized

Bioelectrocatalysis: From Electron Transfer Processes to Emerging Technological Applications, First Edition.
Edited by Shelley D. Minteer and Matteo Grattieri.

enzyme within columns or cartridges. Enzymes are also valuable tools in immunosensing since antibodies are often coupled to an enzyme to exploit the enzymatic conversion for a sensitive detection of bound antibodies. Biosensors are a special development, as here the biological recognition event is directly coupled to the transducer element. Enzymes have been the first biomolecules for which such a combination has been successfully shown and commercial developments have been started. While there are several transduction principles possible, electrochemical transduction is a clear focus in the application of enzymes for sensing (see Chapters 4 and 5). Another area of application is the use of enzymes for the catalytic generation of a desired product, thus replacing organic synthesis steps. Advantageously, the enzyme is immobilized and can, thus, be recycled after product formation. However, all necessary cofactors and co-substrates have to be provided in solution. A special direction within the field of biocatalysis is bioelectrocatalysis. This is naturally focused on oxidoreductases, and here an electrode governs the enzymatic reaction. This means that electrons that are collected at the electrode originate from the enzymatic substrate conversion, or the substrate conversion is fed by electrons from the electrode. The communication of the electrode with the enzyme reaction can be realized in different ways, as will be explained in this chapter.

A complete bioelectrocatalytic system may be exemplified by the scheme in Figure 2.1, which points to the different key parts for its performance. This means that it is not only the chosen biocatalyst with its specificity and conversion rate that determines the overall performance in producing a desired product, but also how this biocatalyst is coupled to the electrode. This includes not only the electron pathways between the two components but also the amount of enzyme that can be actively coupled to the electrode. Another aspect is the transport of substrate molecules to the biocatalyst (or even several biocatalysts) and the removal of the product. For a very efficient enzyme–electrode coupling, this process can become limiting. The mass transport aspect is not simply related to the substrate molecules that need to reach the catalyst but also to products of the enzyme reaction, which may inhibit the enzyme activity. This might be a typical product inhibition, but also a change in the local environment caused by generated side products. For an efficient system, all aspects have to be considered, and the limiting process needs to be identified for further optimization.

A more recent development can be seen in the use of complete microorganisms as catalysts at the anode or cathode side. Here, not only one (or few) enzyme reactions are occurring, but a whole bundle of reaction pathways exists. This also means that conditions have to be provided

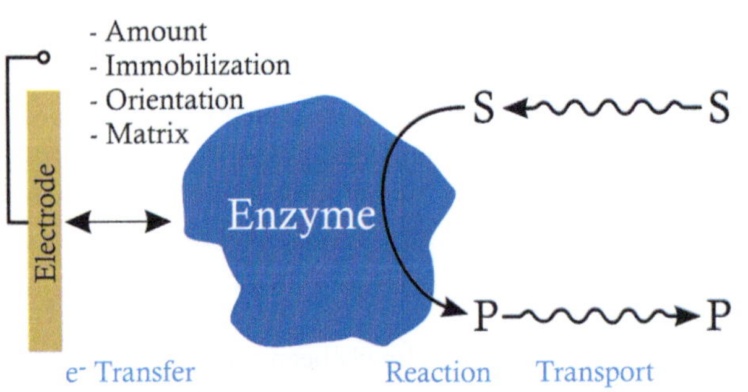

Figure 2.1 Schematic illustration of a biocatalytic system combining a biocatalyst with an electrode for substrate conversion and product generation. Key issues to be considered include electron transfer between the enzyme and the electrode, the biocatalytic substrate conversion at the enzyme, and finally the transport of substrates toward the catalytic site and product transport away. The enzyme can be diffusible in front of the electrode or is mostly immobilized. Consequently, also the amount, the kind of immobilization, additional material used for fixation or embedment, and the orientation of the enzyme play an important role in the overall performance.

to make life possible and that sufficient energy can be used for maintaining a stable metabolism. There are several similarities between bioelectrocatalytic systems based on enzymes or microorganisms. This applies to the principles of electron transfer and the possibilities of enhancing the electron transfer rate, but it also includes general rules for an electrode design with high surface area and sufficiently fast mass transport. Despite the similarities, there are many details in which the application of microorganisms differs from isolated enzymes. The case of microbial electrocatalysis will be discussed in Chapter 3, and the focus of this chapter will be on the use of enzymes as biocatalysts.

2.2 Overview of Approaches

In bioelectrocatalysis, an enzyme reaction is coupled to an electrode. The fundamental principles for establishing communication between the biocatalyst and the electrode are similar in bioanalysis and biocatalysis; however, not all approaches are feasible when an electrode shall be used for biosynthetic purposes. This short chapter will first introduce the different principles and then focus on the suitability for application.

The first kind, or first generation, of enzyme electrodes is based on the electrochemical conversion of substances that participate in the enzyme reaction. This excludes the enzyme and the substrate, but includes educts, side products, and cofactors. It means that a signal is only produced at the electrode when the enzyme successfully converts a certain substrate. Obviously, this is mainly attractive for bioanalytical applications. Here, the electrode cannot be used to influence the enzymatic reaction, but can be applied to obtain a signal from this conversion. Figure 2.2a illustrates the basic idea. Since this approach is mainly relevant for analytical applications and not for bioelectrocatalytic production processes, it will not be treated in more detail here [1].

In the second kind, or second generation, of enzyme electrodes, a shuttle molecule is applied, acting between the electrode and the enzyme. This means the molecule reacts, on the one hand, with the electrode and, on the other hand, with the biocatalyst. These shuttle molecules are termed as mediators, and most often they replace a natural co-substrate of the enzyme. The approach to search for nonnatural reaction partners has already been used in biochemistry since for many optical enzyme activity assays, colored substances were found to be efficient reaction partners of enzymes. For mediated enzyme electrodes, the property of color is not important, but the redox behavior is essential since the mediator molecule needs to be subsequently converted at the electrode. The approach is interesting for both bioanalysis and biocatalysis. The electrode actively takes part in the enzymatic cycle and the electrode potential becomes a tool to influence the conversion at the enzyme. Figure 2.2b gives a schematic illustration – both for cathodic (i) and anodic (ii) reactions. A substance that is intended to work as a mediator needs to fulfill several requirements. This will be described in more detail in Section 2.3.

In the third kind, or third generation, of enzyme electrodes, a direct electron transfer takes place between the electrode and the enzyme molecule. In this way, no shuttle molecules are necessary, and the electrode takes the role of one natural reaction partner of the enzyme – either the role of the electron donor or the role of the electron acceptor. Again, the electrode directly participates in the enzyme reaction and the electrode potential can be used to control the enzymatic conversion. Consequently, this approach is interesting for bioanalysis and biocatalysis as well. A direct electron transfer, however, is not trivial – several demands have to be met so that this process becomes efficient, and not all enzymes can be applied here. This approach is described in more detail in Section 2.4.

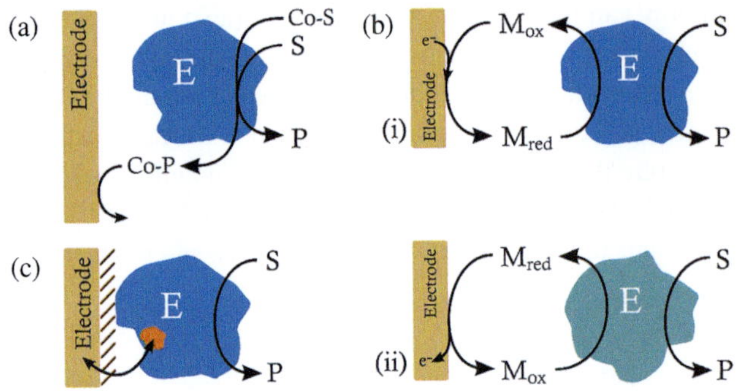

Figure 2.2 Schematic illustration of the different approaches in connecting enzyme reactions with electrodes. (a) Electrodes of the first generation use products or educts of the enzymatic conversion and detect them by electrochemical oxidation or reduction at the electrode and thus, generate a current signal when the enzyme is working. (b) For enzyme electrodes of the second generation, a shuttle molecule, which is called mediator, transfers electrons between the enzyme and the electrode. Here, the current flow can be cathodic (illustrated under [i]) or anodic (illustrated under [ii]). (c) The third generation of enzyme electrodes is based on a direct exchange of electrons between the electrode and an enzyme.

In the literature, another term is often used as an alternative approach to establish communication between an enzyme and an electrode material – a wired electron transfer. The basic idea is the connection of the catalytic center of the enzyme with the electrode by a kind of molecular wire. This is a material of small dimensions and of sufficient conductivity that is able to come close to the catalytic center and allow electron transfer reactions to occur. When one is analyzing what kind of materials are used here, it becomes obvious that they follow either the idea of mediators or the idea of enlarging the electrode surface by conductive substructures and thus, allowing direct electron transfer to take place. An example of the first direction is the use of redox polymers, and an example of the latter is the use of carbon nanotubes. In principle, this is not an alternative fourth approach, but the concept and the advancements in synthesis and materials have resulted in several attractive systems applied in bioelectrocatalysis.

So far, the term enzyme electrodes has been used a little abstractly since the situation of the enzyme has not been specified. In general, two situations can be distinguished: (i) the enzyme is in solution around the electrode but confined in a defined space by means of membranes which allow substrate and co-substrate transport but hinder enzyme molecules from leaving or (ii) the enzyme is immobilized in front of the electrode. The confinement by means of membranes is mainly based on size exclusion, while for enzyme immobilization, a huge variety of approaches have been developed. While this chapter cannot describe the immobilization strategies in detail, some of the most frequently used ones are mentioned below:

- Protein entrapment or encapsulation into a polymeric network
- Formation of chelates between the enzyme and groups on the electrode surface
- Adsorption of the enzyme onto the electrode surface
- Covalent linkage of the enzyme to groups on the electrode (or groups which have been introduced to the surface)
- Layer-by-layer deposition of enzymes by means of polyelectrolytes or nanoparticles of defined charge (also called polyelectrolyte adsorption technique)
- Bioaffinity binding by exploiting a specific interaction between two biomolecules and coupling one of these biomolecules to the enzyme (e.g. biotinylated enzyme and avidin)

2.3 Mediated Electron Transfer

2.3.1 Concept

As explained in Section 2.2, enzymes can communicate with an electrode by means of mediators. In order to construct an efficient system, several requirements have to be fulfilled:

1) Since the mediator must react efficiently with the electrode in order to generate the desired reaction partner for the enzyme, its redox behavior should be of reversible or quasi-reversible nature. This means a fast heterogeneous electron transfer has to take place here. It should be mentioned that not only the molecule itself determines this reaction rate, but also the electrode material, and to some extent, the solution composition. To quantify the behavior of molecules, the heterogeneous electron transfer rate constant k_s can be used. The determination of this characteristic parameter will be explained in Section 2.3.2.
2) The mediator also needs to react with the enzyme. This reaction must be thermodynamically feasible, but also has to be fast in order not to limit the whole process.
   For the first prerequisite, the consideration of redox potentials is essential. When the mediator should be able to take electrons from the enzyme, then its redox potential must be larger than the redox potential of the catalytic site within the enzyme from which electrons have to be taken. When the mediator shall deliver electrons for the enzymatic conversion, then the redox potential of the mediator must be lower than that of the catalytic center that has to be provided with electrons. Regarding the kinetics of this reaction, the rate constant is a helpful tool to compare different suitable molecules. The reaction rate of such molecules with the enzyme can be determined by different means. When optical properties can be exploited, then simple UV/VIS measurements are feasible. However, electrochemical methods are also well-suited to solve this problem. Since the reaction rate of the mediator with the electrode has to be determined electrochemically, only some additional experiments are required in order to elucidate the reaction rate with the enzyme. This is an inherent advantage of the electrochemical methods. The details will be explained in Section 2.3.3.
   It has to be stressed here that the mediator replaces a natural co-substrate of the enzyme. So, in the presence of a mediator, enzyme, and substrate, a defined reaction pathway from/to the electrode exists. However, sometimes the absence of the natural co-substrate is not easy to achieve (e.g. oxygen); consequently, the reaction rate of the mediator with the enzyme must be much higher than that with the natural substrate. This can be achieved by a high rate constant and/or by higher concentrations of the mediator.

These are the two basic requirements that have to be met by the chosen mediator. In addition, it is desirable to work with mediators that allow to operate at rather low applied potential at the electrode in order to avoid undesired side reactions there. It also needs to be ensured that the mediator does not show any spontaneous reaction with components occurring in solution (e.g. oxygen). Only in this way, defined reaction pathways can be created and defined reaction products are obtained. When working with mediators, their pH behavior has to be considered too. Some mediators are highly dependent on pH since protons are involved in the reaction. Here, it should be clarified in which pH range operation is feasible – from the redox potential of the mediator molecule (which becomes pH dependent), via its reaction rate with the electrode and the reaction rate with the enzyme. Other requirements arise from economic or sustainable considerations. It would, thus, be preferable when the stability of the mediator is good in operation as well as in storage, and it is not a toxic compound.

Mainly based on their applications in sensors, many substances have been studied as artificial reaction partners of enzymes in combination with electrodes. Thus, a wide variety of substances with different chemical nature and very different redox potentials are available [2]. Examples of mediators with a rather negative redox potential are 2,8-diaminoacridine (−0.731 V vs. SHE) or methyl viologen (−0.44 V vs. SHE), with medium values: Meldola's blue (0.067 V vs. SHE) or 2,6-dichloroindophenol DCIP (0.217 vs. SHE) and with comparatively high redox potential: ferrocene (0.476 V vs. SHE) or $[Os(CN)_6]^{4-}$ (0.6 V vs. SHE).

2.3.2 Reaction of the Mediator with the Electrode

In order to evaluate the reaction capability of a given substance with an electrode, the heterogeneous electron transfer rate constant k_s is a suitable parameter. There exist several methods for its quantitative determination. The focus in this chapter will be on voltammetric methods. This is also because a simple cyclic voltammogram can already give some first ideas on the redox properties of the potential mediator and the reversibility of its electrochemical behavior. By changing the scan rate during voltammetry, the kinetics can be analyzed not only qualitatively but also quantitatively.

It has to be mentioned here again that the electrode material (metal electrode, different types of carbon materials, or even metal oxide electrodes) has a significant influence on the reaction rate of a given redox molecule. This offers some flexibility since one may simply replace the electrode if the reaction appears to be rather slow. Furthermore, the solution composition (mainly the type of buffer and the kind of added salts) can influence the electrochemical conversion. Mostly of moderate magnitude, sometimes this effect becomes more pronounced.

For the analysis of the kinetic mediator behavior, two situations can be distinguished: (i) the mediator is in solution and diffuses freely or (ii) the mediator is immobilized on the electrode surface. The two situations are treated separately in the following chapters.

2.3.2.1 Mediator in Solution

When a potential mediator is added to a solution in an electrochemical cell, cyclic voltammetry (CV) experiments at varying scan rates (v) can provide qualitative and quantitative information on its redox behavior. From a CV measurement, several parameters can be elucidated such as the anodic peak current I_{pox}, the cathodic peak current I_{pred}, the anodic peak potential E_{pox}, the cathodic peak potential E_{pred}, the formal potential $E_f = ([E_{pox} + E_{pred}]/2)$, and the peak separation $\Delta E_p = (E_{pox} - E_{pred})$.

For a reversible system where the heterogeneous electron transfer rate is not limiting (but diffusion of the molecules – systems under mass transport limitation), the following criteria should be fulfilled:

- $I_{pox}/I_{pred} = 1$
- $E_f \neq f(v) = E^\circ$
- $\Delta E_p = 0.059\,V/n$ (298 K)
- I_p follows the Randles–Sevcik equation

Here, E° represents the redox potential of the redox couple and n is the number of transferred electrons.

Experimentally, a large window of scan rates can be tested with the mediator molecules. However, care has to be taken because of increasing noise, limited current collection rate, uncompensated IR drop, and also because the charging current is linearly increasing with scan rate. So, the applicable scan rate might be practically limited. When the heterogeneous electron transfer rate constant shall be elucidated, then a scan rate region needs to be reached, where the peak position changes with the applied scan rate (i.e. the peaks are moving with increasing scan rate – the anodic peak toward positive potentials, the cathodic toward more negative potentials). In this region, the scan rate is not anymore slower than the electron transfer process to be characterized and thus, only here quantitative information on the rate constant can be obtained. This behavior has been analyzed by Nicholson and Shain [3] who give an equation for the irreversible case:

$$I_p = 0.227 \cdot n \cdot F \cdot A \cdot c \cdot k_s \cdot e^{-[(\alpha \cdot n \cdot F/R \cdot T) \cdot (E_p - E^\circ)]} \tag{2.1}$$

where I_p is the peak current, A is the electrode area in electrolyte contact, k_s is the heterogeneous electron transfer rate constant, α (alpha) is the transfer coefficient, E_p is the peak potential, and E° is the redox potential.

For the analysis of the rate constant, this equation means that ln I_p can be plotted vs. the distance of the peak from the redox potential ($E_p - E^\circ$). This can be done separately for the oxidation and the reduction peak. In the region where the peak position is changing with the scan rate, experimental data points should be on a line. By a regression analysis, the transfer coefficient (α) can be obtained from the slope and the k_s value from the intersection with the logarithmic current axis. E° can be determined from measurements at low scan rate (no change of peak position) and plotting E_f as a function of the scan rate v. When there are only experimental variations in the values obtained, then the medium value can be used as E°; otherwise, an extrapolation to the scan rate "0" can be performed. When a large peak separation (~200mV) is not reached experimentally, then another kind of data evaluation has to be used. Here, the analysis is based on the following function Ψ [4, 5]:

$$\Psi = \frac{\gamma^\alpha \cdot k_s}{\sqrt{\frac{\pi \cdot D_O \cdot n \cdot F \cdot v}{R \cdot T}}} \tag{2.2}$$

$$\gamma = \sqrt{\frac{D_O}{D_R}} \tag{2.3}$$

where D_O and D_R are the diffusion coefficients of oxidized and reduced species, v is the scan rate, and k_s is the heterogeneous electron transfer rate constant.

This function depends on the peak separation observed experimentally and, consequently, the value of k_s can be calculated. The logarithmic function of Ψ vs. ΔE_p is given in the literature [4, 6]. Often, it is reasonable to assume that D_O and D_R are rather equal; otherwise, these values can also be determined experimentally. The method is applicable in the range of alpha values $0.3 < \alpha < 0.7$ and is well suited for quasi-reversible systems where it might be difficult to reach such high scan rates to obtain a sufficiently high peak separation (>200 mV) as described in the previous case.

When the mediator behaves reversibly, then $k_s > 10^{-1}$ cm s^{-1}, while if a quasi-reversible process occurs, k_s is in the range of $10^{-1} > k_s > 10^{-5}$ cm s^{-1}. A quasi-reversible behavior can often be observed and would still be suited for application. In the abovementioned reversibility criteria, this leads to some deviations ($I_{pox}/I_{pred} \approx 1$, $\Delta E_p > 0.059$ V/n and E_f can become a function of scan rate). As a general rule, the electrochemical conversion of the mediator should be fast in order not to limit the substrate conversion at the enzyme.

An alternative method that can be used for the rate constant determination is electrochemical impedance spectroscopy [7, 8]. Here, an ac signal of small amplitude is applied and the resulting ac current is detected. The quotient of both is the impedance, which is a complex value. Since a reliable model for an electrode in contact with an electrolyte solution containing redox molecules is available, the charge transfer resistance of the redox couple under investigation can be elucidated from

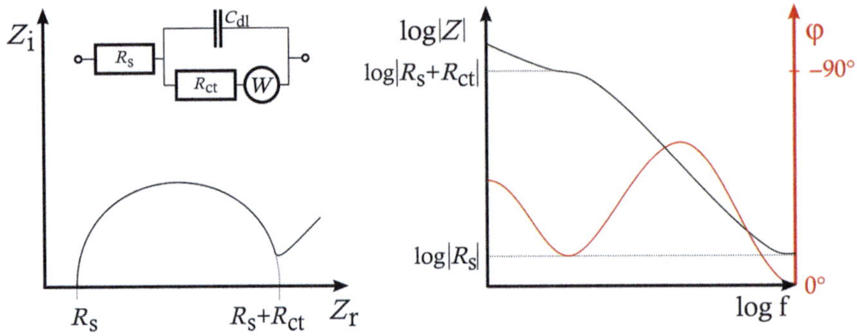

Figure 2.3 An equivalent circuit representing an electrode in contact with an electrolyte solution containing redox species that can be oxidized/reduced at the electrode (Randles model). The model contains R_s as the resistance of the electrolyte solution, C_{dl} as the double layer capacity, R_{ct} as the charge transfer resistance of the redox couple at the electrode, and W as the Warburg impedance originated from the diffusion of redox species toward the electrode surface. Typical Nyquist plot (left) and Bode plot (right) of an impedance measurement at such an electrode. The solution resistance and the charge transfer resistance can be seen directly in the plot. More reliable data for R_{ct} are obtained from a fit of the experimental curve using the Randles circuit.

the impedance measurement. Figure 2.3 shows the Randles equivalent circuit that can describe an uncovered electrode rather well. Experimental data can be fitted according to the model and the charge transfer resistance R_{ct} can be deduced. However, care has to be taken when the roughness of the electrode is rather high as the capacitance (resembling the double layer capacity) has probably to be replaced by a constant phase element. The charge transfer resistance represents the resistance of the electron transfer process at the interface (from the redox species to the electrode and vice versa) and is correlated to the exchange current density, which depends on the concentration of the redox species and the rate constant. For the measurement, this means that impedance needs to be analyzed under equilibrium conditions (at the redox potential of the investigated couple and in solutions with oxidized and reduced species with the same concentration). If the exchange rate is not very high, the equilibrium may not be obtained easily – here a CV experiment before the impedance measurement can be helpful (finishing at $E°$). Calculation would be as follows:

$$R_{ct} = \frac{R \cdot T}{n^2 \cdot F^2 \cdot A \cdot c \cdot k_s} \tag{2.4}$$

2.3.2.2 Mediator Immobilized

When molecules are immobilized, diffusion does not need to be considered. This will, however, only influence the evaluation of the data, but not the experiments. So, also, in this case, CV at different scan rates has to be performed, and the region of interest is provided by scan rates where peak position becomes a function of the scan rate (peaks are “moving”). Laviron has formulated here a comprehensive theory that allows the data evaluation even when only small peak separations can be achieved experimentally. Thus, two different procedures have to be used for the determination of k_s [9]:

1) $\Delta E_p > 200\,\text{mV}$
   When such large peak separations can be obtained experimentally, a rather simple plot can help to elucidate the rate constant. A graph of $E_p = f(\log v)$ results in two curves – one for the anodic and one for the cathodic peak. The behavior is schematically illustrated in Figure 2.4.

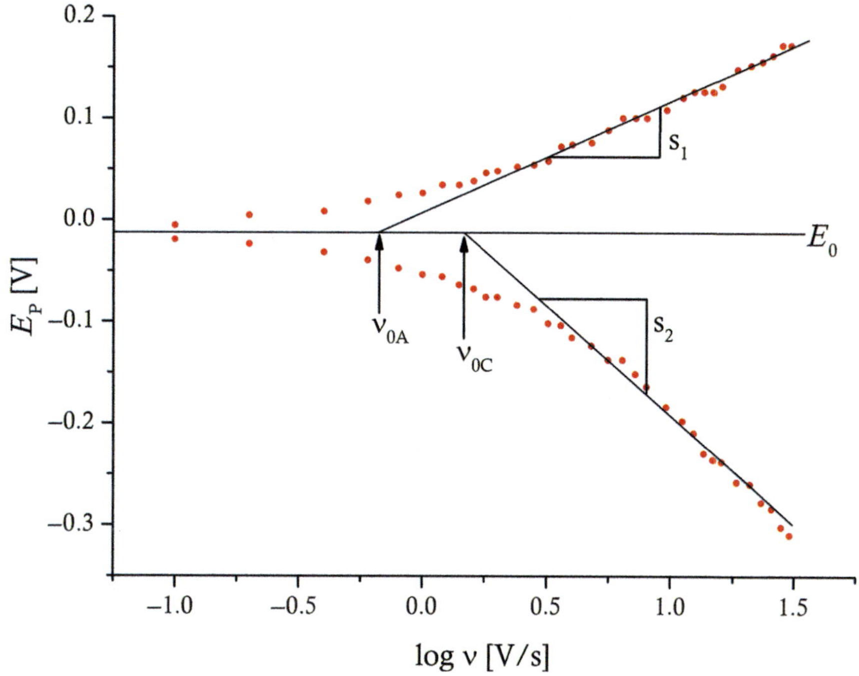

Figure 2.4 Schematic illustration of the determination of the heterogeneous electron transfer constant k_s, with immobilized redox species (mediator or enzyme). The peak potentials E_p for oxidation and reduction in the cyclic voltammetric measurements are plotted vs. the scan rate (log v). $E°$ can be obtained from measurements and midpoint potential analysis at low scan rate. For the kinetic analysis, however, higher scan rates have to be used in order to be in a range where the peak position becomes a function of the scan rate. The scan rate range, in detail, depends on the rate of the heterogeneous electron transfer of the system under study. When peak separations of more than 200 mV can be achieved, then linear functions can be evaluated from the shift of both the anodic and cathodic peak vs. log v. This is indicated by the two lines in the graph. The slope of this function contains the value of α, which is the transfer coefficient. This value gives information about the kinetics of oxidation and reduction, which is rather similar for the two processes when $\alpha = 0.5$ and differs when the value deviates from this value. The slope s_1 of the linear function for the anodic peak is 2.3 RT/$(1-\alpha)nF$ and the slope for the cathodic peak s_2 is −2.3 RT/αnF. In the example given, α is smaller than 0.5 – already indicated by different shift of the peak potential with increasing scan rate. The heterogeneous electron transfer rate constant k_s can be derived from the two scan rates v_{0C} and v_{0A}, which can be obtained from the intersection of the $E°$ curve and the graphs elucidated for the cathodic and anodic peak potential shift, respectively, by applying Eq. (2.5).

The shape of the two curves resembles a "trumpet" because at higher scan rates, both peaks significantly change the position, which does not appear at small scan rates. From the linear part of both curves (in the higher scan rate range), the transfer coefficient α (alpha) can be calculated since the slope will be −2.3 RT/αzF for the cathodic and 2.3 RT/$(1-\alpha)zF$ for the anodic peak. From the interception of the linear part of these curves with $E°$, one can get two defined scan rates for the anodic and cathodic peak position curves (v_{0C} and v_{0A}), which allow the calculation of the rate constant:

$$k_s = \frac{\alpha \cdot n \cdot F \cdot v_{0C}}{R \cdot T} = \frac{(1-\alpha) \cdot n \cdot F \cdot v_{0A}}{R \cdot T} \tag{2.5}$$

2) $\Delta E_p < 200\,\text{mV}$
In case such a large separation of both peaks cannot be achieved during the experiments, another procedure can be applied. First, the transfer coefficient α is determined by plotting $(E_{pred} - E°) / (E_{pox} - E°) = f\,(z\Delta E_p)$ and comparing the obtained curve with normalized curves which are given by Laviron [9].
In the second step, the value of m, which is defined as

$$m = (\text{RT/F})\,k_s/nv \tag{2.6}$$

can be elucidated from the experimentally observed peak separation by a normalized graph given in the same paper. Here, n is the number of transferred electrons and v is the scan rate. Since for every experiment the applied scan rate is known, the value of k_s can be obtained. Finally, the obtained k_s values are analyzed with respect to the scan rate dependence since several experiments at different scan rates have to be performed.

2.3.2.3 Mediators Operating in a Volume in Front of the Electrode

An attractive method has been developed to provide a high concentration of the mediator and at the same time to have the mediator immobilized – redox polymers. Here, the redox active group is fixed to a side chain of a non-redox-active polymer backbone. A large amount of redox molecules is meanwhile available, which can be integrated into such polymers [10]. Thus, a large potential window can be covered by redox polymers. Examples of redox groups applied are viologens, redox dyes, quinones, osmium complexes, iron complexes, and other metal complexes. Electron transfer occurs here by a hopping mechanism from one redox center to the next. The polymer acts not only as an electronic connector to the electrode, but at the same time as immobilization matrix of the biocatalyst and, thus, needs to provide a protein-friendly environment. During the polymer synthesis, side chains can also be introduced that harbor a coupling group. This way, the enzyme can be covalently bound to the polymer and is not only embedded in between the polymer chains. The approach has revolutionized the field of bioelectrochemistry and has been pioneered by the work of Adam Heller in the field of glucose sensing [11–13]. The method is rather versatile since the polymer can be designed according to the idea of a toolbox and several aspects can easily be varied:

- The type and number of monomers (degree of hydrophilicity, coupling possibilities)
- Length of the side chain with the redox site (influences the reaction capability with the enzyme but also the inter redox site electron transfer rate – self-exchange rate)
- Type of redox molecule fixed in the side chain
- For metal complexes – the ligands around the metal (which determine the actual redox potential)
- Crosslinking possibility

This approach also reflects electron transfer steps in biological systems where electrons are transported over larger distances by hopping from one redox center to another. It has also been demonstrated that proteins can be arranged in such a way that this kind of electron transfer becomes possible in an artificial system. This has been exemplified for the mitochondrial heme protein cytochrome c [14, 15]. Multilayers of this redox protein can easily be constructed using the layer-by-layer technique and applying together with this positively charged protein a negatively charged polymer or negatively charged nanoparticles, but also another charged biomolecule such as DNA. In analogy to redox polymers, electrons are transported by a hopping mechanism between the heme centers throughout the assembly. This has been used to integrate different enzymes and, thus, to establish bioelectrocatalytic systems governed by the presence of the enzymatic substrate. Examples are bilirubin oxidase and oxygen [16], sulfite oxidase and

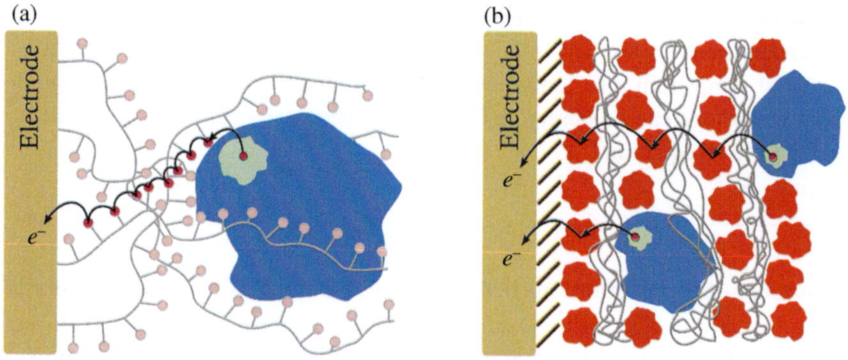

Figure 2.5 Schematic illustration of electron pathways for electrode systems combining a biocatalyst and a redox polymer consisting of a backbone chain to which redox moieties are coupled via a flexible linker. Electrons can be transported by a hopping mechanism from one redox site to the next. Enzymes can also be integrated into redox protein-based multilayer assemblies. Also, here, electron transfer occurs via multiple steps from one redox protein to another. Note that in these assemblies, the redox protein is immobilized, but not covalently fixed in order to ensure some rotational flexibility.

sulfite [17], cellobiose dehydrogenase and lactose [18], or fructose dehydrogenase and fructose [19]. However, even larger biocatalytic entities such as bacteria can be connected to electrodes using this approach [20]. Figure 2.5 illustrates the redox polymer and the redox protein-based approaches in a schematic way.

Electron transfer in these electron hopping systems does not only include the reaction of the selected redox molecule with the enzyme and its reaction with the electrode but also several electron transfer steps between the redox centers. Compared to the size of the molecules, rather large distances appear in the immobilization matrix and, thus, this charge transport can limit the whole process. The electron self-exchange rate constant serves as a characteristic parameter for this process [21–23]. It is defined as:

$$D_e = k_{ex} \cdot \delta^2 \cdot C/6 \quad (2.7)$$

where D_e is the apparent electron diffusion coefficient, δ is the average redox site distance, and C is the concentration of the redox sites in the matrix.

Within these systems, electrons move to areas where the redox sites are oxidized and thus, they behave similarly to a diffusible redox molecule in a concentration gradient in front of an electrode. It has to be stressed here that not the redox sites are diffusing in this model (although they have some local mobility), but only the electrons, which move from one redox site to the next one. δ needs to be calculated according to the size of redox molecules and the average distance of these sites within the immobilization matrix (redox polymer or redox protein multilayer) and C is the concentration within this matrix. For the determination of C, the overall amount can be determined from the peak area in a CV experiment at small scan rates. With the information on the layer thickness and the covered area, the concentration is accessible. The value to be determined experimentally is the electron diffusion coefficient D_e. Here, different experimental techniques have been reported.

In a rather simple approach, CV experiments can be used since the peak current behaves according to the Randles–Sevcik equation known from the behavior of diffusible redox species in solution.

$$I_p = 2.69 \cdot 10^5 \cdot \sqrt[3]{n} \cdot A \cdot \sqrt[2]{D} \cdot c \cdot \sqrt[2]{v} \quad (2.8)$$

where n is the number of transferred electrons per molecule, D is the diffusion coefficient (here D_e), c is the concentration, and v is the scan rate.

Consequently, D_e can be determined from experiments at different scan rates (without catalysis). As an alternative to voltammetry, potential jump experiments can be used. Here, the current is analyzed after a potential jump from a region where no redox conversion occurs to a potential where mass transport limitations can be observed. The current response in such an experiment can be described by the Cottrell equation [5]:

$$i = \frac{n \cdot F \cdot A \cdot c_j^0 \cdot \sqrt{D_j}}{\sqrt{\pi \cdot t}} \tag{2.9}$$

where c is the concentration, D is the diffusion coefficient, and A is the electrode area in contact with the electrolyte.

Care has to be taken here because of instrument limitations in current measurement and selection of a proper time frame since a potential jump will also result in double layer charging. However, the charging current rapidly decays with time and, consequently, a time > about five times the time constant $t = C_{dl}{}^*R$ (where R is the uncompensated cell resistance) should be evaluated.

2.3.3 Reaction of the Mediator with the Enzyme

The fast reaction at the electrode surface is one of the major requirements for a mediator, and the fast reaction with the enzyme is the other very important demand. This reaction capability of a mediator can also be analyzed by CV [24]. Here, the idea of a catalytic current is exploited, which can be expressed by the following equations:

$$\boldsymbol{O} + \boldsymbol{e}^- \overset{k_s}{\longleftrightarrow} \boldsymbol{R} \tag{2.10}$$

a) $\boldsymbol{R} + E_{ox} \xrightarrow{k_1} O + E$

$E + S \longrightarrow E_{ox} + P$

b) $\boldsymbol{O} + E_{red} \xrightarrow{k_1} R + E$

$E + S \longrightarrow E_{red} + P$

where O and R are the oxidized and reduced mediator species, respectively, k_s is the heterogeneous electron transfer rate constant, and k_1 is the rate constant for the reaction of the substrate oxidized enzyme (i) or the substrate reduced enzyme (ii) with the mediator.

This means that the enzymatic reaction reverses the process at the electrode. Such a catalytic behavior can happen in both directions – the enzyme reaction can supply electrons to the mediator or the enzyme reaction consumes electrons from the mediator. This will, however, significantly change the cyclic voltammogram. For case (i), a catalytic reduction current is obtained, and no, or a diminished, oxidation process occurs at the electrode (since mediator oxidation is catalyzed by the enzyme). For case (ii), this means that a catalytic oxidation current is obtained, and no, or a diminished, reduction process occurs at the electrode. This is illustrated in Figure 2.6 with the example of ferrocene carboxylic acid (FCA) as mediator, glucose oxidase as enzyme, and glucose as substrate.

The development of a catalytic current can be used to obtain information about the rate constant k_1 of the reaction between the enzyme and the mediator (when the substrate is not limiting – meaning under substrate saturation conditions). This situation was also analyzed by Nicholson/Shain, resulting in the following procedure for the rate constant determination [3]:

1) Determination of the peak current I_D without catalytic reaction (diffusion-controlled current without substrate added).

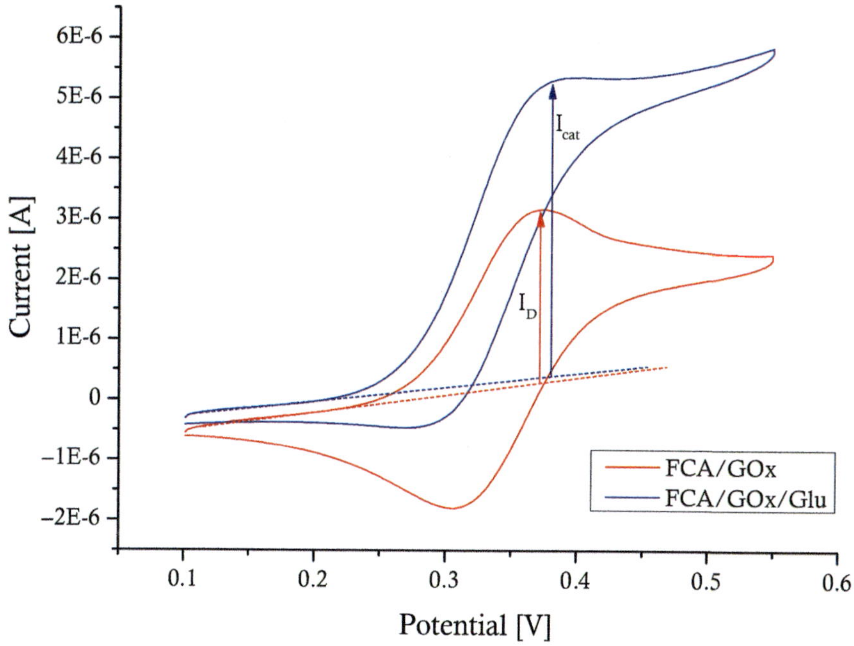

Figure 2.6 Development of a catalytic current at a gold electrode with a free diffusing mediator (here ferrocene carboxylic acid – FCA) and glucose oxidase (GOx) as the reaction partner. Without the substrate, the quasi-reversible electrochemistry of FCA can be seen and the diffusion-limited current I_D can be elucidated from the peak. After the addition of glucose, the enzyme will be reduced and thus, can act as a reaction partner for the oxidized FCA. This biochemical reaction reverses the electrochemical conversion of FCA, resulting in an enhanced oxidation current – here termed as I_{cat}. In the back scan, no reduction peak is detectable since the enzyme is reducing the mediator molecules efficiently (100 mM phosphate buffer, pH 7, 0.5 mM FCA, glucose oxidase (20 U), and 25 mM glucose (Glu), scan rate 20 mV s^{-1}).

2) Determination of the catalytic current I_{cat} (reaction-controlled current in the presence of substrate).
3) Determination of the ratio of I_D/I_{cat} for different scan rates. It should be noticed that although different scan rates also have to be measured here, the scan rate range greatly differs from the measurements for k_s analysis. At high scan rates, the conversion, according to Eq. (2.10), will be fast and, thus, the biochemical reaction will have no influence on the CV. In order to see (and analyze) clear catalytic currents, one has to move to low scan rates.
4) For every ratio I_D/I_{cat}, a kinetic parameter has to be determined according to ([3]; Table XII). The kinetic parameter is defined as:

$$k_{cat}/a = k_{cat}\,\mathrm{RT}/zFv \tag{2.11}$$

5) The plot of the kinetic parameter as a function of inverse scan rate: $k_{cat}/a = f(1/v)$ will result in a linear graph with the slope of k_{cat}.
6) This procedure has to be repeated with different amounts of enzyme since k_{cat} contains not only the rate constant with the mediator k_1, but also the enzyme concentration. Finally, k_{cat} is plotted against the enzyme concentration $k_{cat} = f(E)$, resulting in a linear graph with the slope of k_1 – the rate constant of the reaction of substrate-saturated enzyme with the mediator [24].

In addition to electrochemical techniques, spectroscopic techniques allow the evaluation of the reaction capability of an enzyme with a certain redox substance and its quantification in the form of the reaction rate constant – provided that the redox molecules have a clear optical signature that differs in the two redox states. This is similar to the biochemical characterization of enzymes and, thus, not the focus of this chapter.

2.4 Direct Electron Transfer

Many oxidoreductases have an inherent cofactor or prosthetic group. This can be a metal compound or a metal-free compound. In both cases, the redox properties are defined not only by the compound, but also by the specific protein environment in which they are embedded. Consequently, the redox potential of such a redox site can significantly differ from that of the free form. Other oxidoreductases do not have their own redox center, and cofactors have to dock to the protein in a similar way as the substrate. For electrodes based on direct electron transfer, the first group of oxidoreductases is in focus [25, 26]. The electrode has here to replace one of the natural substrates, and supply electrons for the reaction, or withdraw electrons that have been transferred to the catalytic center during the substrate conversion. This type of electron transfer resembles the biological situation when two proteins come close together to form a charge transfer complex and electrons are transferred from one protein to another. This so-called homogeneous electron transfer has several similarities to the heterogeneous electron transfer where one partner is replaced by the electrode.

The possibility for such an electron transfer was not yet in focus in the 1960s, when enzymes have first been coupled with electrodes, but in the 1980s, pioneering work by several groups has shown the feasibility and also the benefits (mostly for analytical applications). Naturally, small redox proteins such as cytochromes [27–29], ferredoxins [30, 31], or azurin [30, 32–34] have been in the focus, but also more complex enzymes, such as laccase [35–41], peroxidase [26, 42–47], hydrogenases [48–51], or cyt P450 [52] have been of considerable interest. Interestingly, all types of electrodes were studied here – from metal electrodes with several modifications – most successfully based on self-assembled monolayers (SAMs) or chemisorption layers – over different carbon electrodes (with and without pretreatment) to metal oxide electrodes such as indium tin oxide. A DET could not be achieved for all enzymes tested and rather different rates were found for the electron transfer. This points to a model that can be used in order to rationalize the experimental findings and may also help in designing electrodes and proteins for an efficient direct electron transfer. Here, the Marcus theory with all its subsequent refinements is well recognized as such a tool. As also discussed in Chapter 1, the rate constant for the electron transfer process is in focus. The general structure follows the models in chemical kinetics comprising a pre-exponential factor and an exponential term [53]:

$$k_s = A \cdot e^{\frac{-(\Delta G + \lambda)^2}{4 \cdot \lambda \cdot R \cdot T}} \tag{2.12}$$

The exponential factor contains two energies. The free energy change ΔG can be attributed to the overpotential applied at the electrode with respect to the equilibrium potential and the reorganization energy λ with its two components λ_{in} (inner shell reorganization energy) and λ_{out} (outer shell reorganization energy). This means that the rate of electron transfer is exponentially increasing when the electrode is biased more positive (for oxidation) or more negative (for reduction) than the redox potential $E°$ of the species under investigation. The reorganization energy reflects the changes inside the molecule and also in its surrounding, which are accompanied by the electron transfer step. For many proteins, the changes are rather obvious since properties in the oxidized or reduced state are not identical. At $E = E°$, the rate constant can be given as follows:

$$k_s = A \cdot e^{\frac{-\lambda}{4 \cdot R \cdot T}} \tag{2.13}$$

This equation underlines the role of the reorganization energy λ as a kinetic barrier for the entire process. This is one of the starting points of protein engineering with the aim of creating mutants with a lower reorganization energy for faster electron transfer. In addition, the solution composition can also influence the reorganization energy (more specifically λ_{out}). Several factors are compiled within the pre-exponential factor A since it is a function of λ_{out}, the temperature, the dielectric properties of the medium between the redox center and the electrode (ε), the conductivity of the medium, the distance d of the redox center and the electrode surface, and ν_f – the nuclear frequency of the formation of a productive electron transfer complex. In consequence, the solution composition shows some influence on the electron transfer kinetics. This effect is a little underestimated in the literature, but besides the rather obvious pH influence, the kind of ions and their concentration can have pronounced effects on the protein behavior on the surface [54, 55]. More importantly, however, is the role of the distance of the redox center in the molecule and the electrode surface. This can be given as follows:

$$A \sim e^{\beta \cdot (d - d_{min})} \tag{2.14}$$

As introduced in Chapter 1, direct electron transfer can only take place when both partners are in close contact. Usually, it is considered that for distances of more than 20 Å there is no efficient electron transfer, although this is also dependent on the decay factor β that reflects the properties of the medium between the electrode and the redox center. When changing the distance, the rate constant decreases exponentially. This has also been verified experimentally using SAM layers of different thickness to which redox proteins have been attached [56, 57]. As a further consequence out of this, the orientation of the protein molecule in front of the electrode becomes an important factor for the observation and the efficiency of DET. When natural enzymes are used, the modification of the electrode surface is the main approach to achieve a "productive" orientation in which DET becomes feasible. Deglycosylation and truncation of the protein can also be a valuable way for reaching a lower distance. In the area of engineering proteins for DET, the achievement of a defined orientation of all proteins is consequently one of the major routes (see Section 2.8). All this follows directly from the quantities which are contained within the pre-exponential factor.

When proteins are in solution, then the surface modification can significantly contribute to achieve a higher frequency of productive interaction with the electrode. In protein electrochemistry, such an electrode-attached molecule is termed as "promotor." It interacts, on the one hand, with the chosen electrode material (most often via a chemisorption bound) and, on the other hand, with a certain region of the enzyme molecule [58–61]. For enzyme electrodes, where the enzyme shall be attached to the electrode, a layer of such promotor molecules is very valuable for protein fixation. The immobilization protocol, however, needs to result in a preferential if not uniform orientation of the protein molecules with a rather short electron transfer distance between the redox center and the electrode. Numerous approaches have been reported here, and some of the most successful routes are summarized below:

- Use of surface properties of the enzyme such as accumulated charges in a certain area or the presence of hydrophobic areas and modification of the electrode surface in a complementary way (e.g. by chemisorption layers with different head groups)
- Covalent coupling of surface groups near the redox center to existing or introduced groups on the electrode
- Binding of the protein using affinity tags (e.g. histidines to Ni-complexes fixed onto the electrode)
- Reconstitution of the holoenzyme at the electrode where the prosthetic group has been fixed by a linker molecule

- Exploitation of a preferential protein adsorption to certain surfaces (such as edge plane or basal plane of graphite)
- Use of nanomaterials such as carbon nanotubes or metal nanoparticles
- Application of electrodes with a 3D structure to exploit "curvature effects" when the 3D structuring comes into the size range of the enzyme and, thus, a higher probability for electron transfer occurs [62].

During the voltammetric investigation of enzymes showing DET, distinct redox peaks may appear. In this case, it is rather simple to study which redox center of the enzyme is in contact with the electrode (when the enzyme contains several centers). The influence of the immobilization procedure on the redox properties can also be analyzed in this situation. For example, covalent immobilization can lead to a shift of the redox potential to slightly more negative values. However, the Faradaic current is sometimes so weak, in that a clear analysis is not feasible. In this case, the addition of substrate can help clarify whether a direct electron transfer has been achieved and a catalytic current occurs in the CV. In contrast, it may also happen that no catalytic current occurs although the redox center can be addressed by the electrode. Here, the accessibility of the catalytic center for the substrate might be hampered by the chosen immobilization procedure.

The structure of enzymes can be quite complex. As illustrated in Figure 2.7, an enzyme may have only one redox center. An oriented immobilization is important to ensure a sufficiently short electron transfer distance, but at the same time, substrate molecules also need to reach the catalytic active site. Here, a linker structure can be beneficial in anchoring the protein to the surface [63]. Some enzyme molecules contain several redox centers with different functions: a catalytic site where substrate conversion takes place and one or more other redox centers that are devoted to the transfer of electrons only. Some enzymes even comprise a second catalytic site. Here, a 3D structure is extremely helpful in designing surfaces for a DET-based electrode. If an in-built electron transfer center (or even a whole electron transfer subunit) is present, this can advantageously be used for the electrode interaction. Examples are cellobiose dehydrogenase (with a cyt b subunit) [64] or fructose dehydrogenase (with a subunit containing three hemes) [65]. Further examples are fungal PQQ-dependent pyranose dehydrogenase, bacterial FAD-dependent GDH [66], or PQQ-dependent alcohol and aldehyde dehydrogenase [67]. When enzymes contain several active sites, this is originated from the catalysis of one substrate conversion at one site and the reaction of a second substrate (or cofactor) at the second catalytic center. For the DET, the question arises, which conversion shall be addressed by such an electrode and, thus, which substrate needs to be replaced by the electrode. Some enzymes also catalyze reactions that can go in both directions. Here, not only the catalytic centers are of interest but also potential redox centers that are connecting both catalytic centers. An example of such a complex enzyme is tungsten-dependent formate dehydrogenase [68].

For some enzyme groups, recent reviews summarize the state-of-the-art in this field: dehydrogenases [69], nitrogenases [70], iron-sulfur cluster containing proteins [71], heme enzymes [46], fructose dehydrogenase (FDH) [72], PQQ-dependent glucose dehydrogenase (PQQ-GDH) [73], FAD-GDH [74], cellobiose dehydrogenase (CDH) [64], cytochrome P450 [75], and hydrogenases [76, 77]. A different approach has been developed for a subgroup of dehydrogenases, which do not contain a prosthetic group but are cofactor-dependent. If NAD^+ or NADH are the cofactors that need to be present to enable a catalytic reaction to occur, then these cofactors can be coupled to an electrode in an electroactive way. This often means a surface modification is necessary since electrochemistry of these cofactors is mostly ill-defined at bare electrodes [78]. Subsequently, the biospecific interaction with the cofactor-dependent enzyme can be used for an oriented immobilization, and thus an electrode preparation enabling a bioelectrocatalytic conversion of the substrate [79, 80].

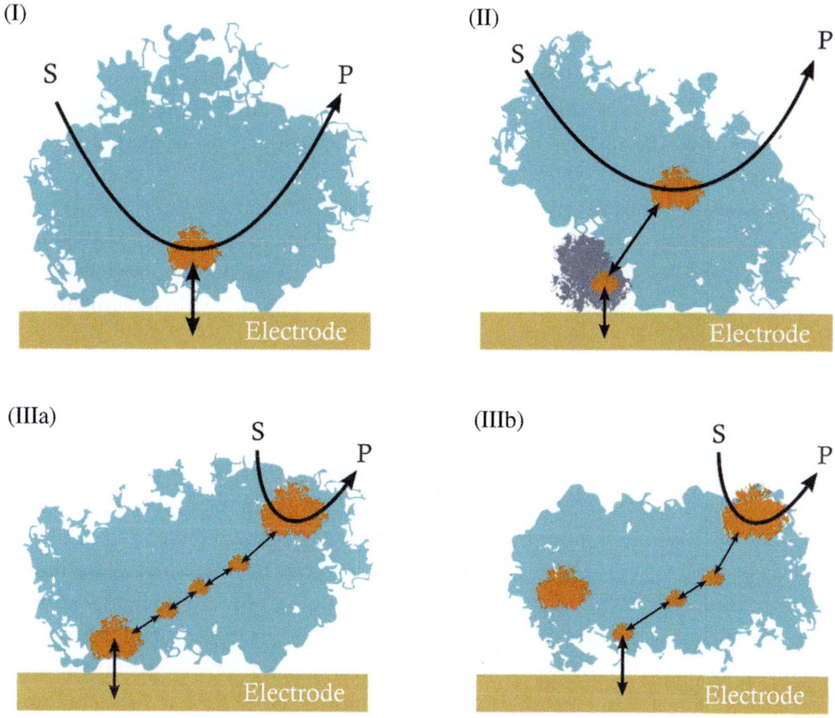

Figure 2.7 Schematic illustration of different types of redox enzymes and their interaction with electrodes. (I) The enzyme may have only one redox center and thus, connection to the electrode and accessibility for the substrate have to be ensured. (II) The enzyme may have an additional subunit which is able to perform electron transfer with the catalytic center on the one hand and the electrode on the other hand. Here, the electrode is often replacing a natural acceptor. (III) In this group, the enzyme has two catalytic sites for two different substrates (or substrate and cofactor) and an electron transfer relay of several redox centers might be present in order to connect these two sites. (a) Here, one reaction partner of the enzyme is replaced by the electrode and thus, one center needs to be in direct contact with the electrode. (b) When one redox center from the electron transfer relay is close enough to the surface, this can also be used for connecting a catalytic site with the electrode.

A special group of enzymes are membrane proteins with a relatively large hydrophobic patch on the surface. Here, other types of surface modifications are often necessary, reflecting the situation of these proteins in a biological membrane [81–85]. Lipid bilayers or advanced versions, such as tethered lipid bilayers, have been developed for this purpose. Not only sophisticated methods such as Langmuir–Blodgett or Langmuir–Schäfer techniques are applicable, but also a comparatively simple spreading of vesicles can be used for surface preparation. Other approaches use surfactants in order to immobilize proteins. Interestingly, this approach has also been shown to be valuable for proteins which are not embedded in a membrane in the biological context. An example here is myoglobin [86]. Furthermore, it needs to be added that for some membrane proteins, it seems to be sufficient to provide a minimal amount of detergents in order to couple them to electrodes. Thus, even for large protein complexes such as photosystem II [87] and photosystem I [88, 89], direct electron transfer could be demonstrated.

In order to analyze the kinetics of the heterogeneous electron transfer of proteins at electrodes, in principle, the same methods are suitable as described for the heterogeneous electron transfer of mediators (see Section 2.3). Here, the difficulty is in the occurrence of clear voltammetric peaks in a broader scan rate range for evaluation of the peak shift. If the heterogeneous electron transfer is not fast enough, this restricts the dynamic range for the substrate as saturation of the current signal is already reached at lower concentrations compared to the enzyme kinetics in solution.

At this point, it must be added that not only electrochemical methods can be used to clarify the mechanisms of protein–electrode interaction including electron transfer. A number of different techniques have been developed over the years. These methods are based on structural investigations of the surface (e.g. scanning electron microscopy SEM, atomic force microscopy AFM), on the analysis of the protein binding to the surface (e.g. quartz crystal microbalance QCM [90–92], surface plasmon resonance SPR [93–95]), or spectroscopic methods supporting the development of mechanistic models [96, 97]. Spectroelectrochemical studies are particularly valuable due to the coupled detection of current signals and optical features [98].

2.5 Conducting Polymers

Conducting polymers are a special group of macromolecules with the inherent property of electronic conductivity. In contrast to redox polymers, which are based on the attachment of redox molecules to a polymeric backbone, here the polymer itself enables electron transport. Examples for this type of polymers are polypyrrole, polyaniline, polyacetylene, or polythiophene [99–102]. The structure of some of these conducting polymers is shown in Figure 2.8. The most important structural feature is a conjugated Π electron system which occurs in situations where single and double bonds are alternating. When these polymers are compared to the classical series of materials – metals, semiconductors, and insulators – then often the closeness to semiconductors is emphasized [103]. However, the transport of electrons in the polymeric network is more sophisticated than in a classical semiconductor. This can already be recognized by simple temperature-dependent measurements of the polymer conductivity.

Obviously, a polymeric network is not a very ordered structure. This starts from the macromolecules themselves, as they usually have different chain lengths and may even possess different defects along the chains. Interaction between the chains can mostly take place via Π–Π interactions. This can result in crystalline domains, but not throughout the whole polymer, and also amorphous regions are contained. The absence of long-distance-order results in localization of the charge wave-function, and charge transport can be understood as quantum mechanical jumps from one site to the other. Different refinements of the electron transfer theory have been developed [104, 105].

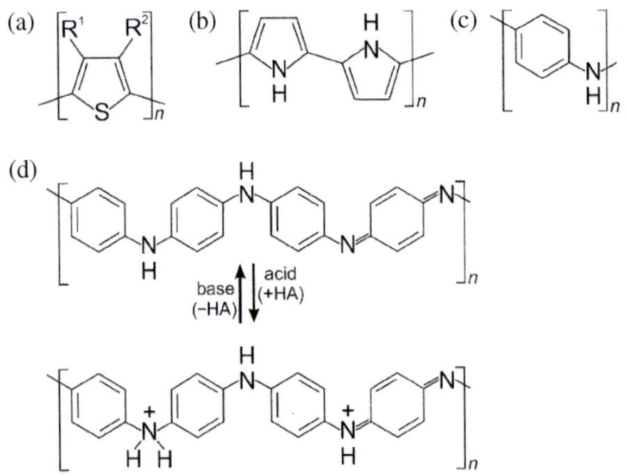

Figure 2.8 Chemical formula for some of the most often used conducting polymers: (a) polythiophene, (b) polypyrrole, and (c) polyaniline. For polyanilines, the formula of the most conducting form is additionally given, which is the mixed redox state in its protonated form (d).

In order to achieve high conductivity, conducting polymers need, however, "doping." This can be seen in analogy to doping in semiconductors, but it differs in the kind of dopant and the quantities used for the doping procedure. Whereas semiconductor doping is in the range of ppm, the doping in conducting polymers is in the range of mol%. The doping of conducting polymers does not result in a change of electron and hole concentrations in the conduction or valence band, but introduces charges and thus, lowers the energy barrier for electron transport. The doping can be achieved by different methods such as by ions, charged groups in the polymer, redox molecules, or by charge injection (e.g. from an electrode). Conductivity raises with the doping level, but this increase levels off at higher dopant concentrations. This is mainly originated from the fact that high dopant concentrations lead to higher disorder that suppresses carrier mobility. Mostly, the concept of polarons is used to describe the charge carrier situation within a conducting polymer. The doping procedure results in distortions of the local lattice structure forming a polaron that can extend over several monomeric units. The transport of this polaron can occur along the polymeric chain but also between chains. In general, ordered structures will show a higher mobility compared to disordered regions. It should also be mentioned that some polymers can reach such high conductivities as in metals. This is mainly achieved by doping or by a higher degree of order within the polymer [104].

The charge transport within these polymers is, however, also influenced by ions in solution [106]. Although conducting polymers can be studied in the dry state, this is not the relevant situation in bioelectrochemistry. Here, the state of the polymer might be described as hydrated or swollen or even appear as a hydrogel. In these situations, one can also observe ionic transport within the polymers. Increasing hydration increases the ionic conductivity significantly since water–ion interaction becomes dominating over ion–polymer interaction. In the swollen state, water and ions are in equilibrium with the surrounding electrolyte. The degree of swelling can be influenced by the electronic situation (e.g. by charge injection from an electrode) and points to a coupling of ion movement and electronic state of the polymer. Particularly, high ionic conductance can be achieved with protons, for example, through a hydrogen-bonded network within the polymer [107]. When hydrogels are considered, the ion transport can be considered as decoupled from the polymer dynamics, as there are continuous liquid pathways available. Consequently, conducting polymers for application in bioelectrocatalysis can often be considered as mixed electronic-ionic conductors.

Although charge transport within conducting polymers is not simple, they have gained considerable attention in combining enzymes with electrodes. This is mainly connected to the electronic conductivity, but is also originated from the rather high biocompatibility of these materials and the ease in processability [100]. The latter aspect is attributed to the preparation of conducting polymers and their surface application. For the preparation, different methods can be used:

- Chemical synthesis (Different protocols are available for the different classes of conducting polymers, which apply different starters for the polymer chain reaction.)
- Electrochemical synthesis (Polymerization can also be initiated by means of an electrode. Here, monomer oxidation is usually applied to generate a radical cation as a starter for the chain reaction. The tremendous advantage is that this type of generation takes place at the surface where the polymer shall operate in a bioelectrocatalytic system. Furthermore, since the polymeric chains are conductive, the thickness of these films can easily be controlled by the current flow during the electropolymerization. Simple CV and constant potential approaches can be applied, but pulse methods have been found particularly valuable for polymer preparations avoiding any overoxidation.)
- Biocatalytic synthesis (Enzymes can also be used to prepare conducting polymers as long as the monomers are acceptable reaction partners. One example is peroxidase, which can start the

monomer oxidation in the presence of hydrogen peroxide [108–110]. Other enzymes that have been applied are tyrosinase [111] or laccase [112]. There are also examples known where whole microorganisms have been used for the polymer preparation [113, 114].)

Enzyme immobilization can be achieved in different ways. The enzyme might be embedded within the polymeric structure or attached to the polymer (covalently or noncovalently).

But how can operation be understood when conducting polymers are applied on electrode surfaces to which enzyme reactions shall be coupled?

A) The original idea seems rather simple – when electronic conductivity can be ensured within the polymer, this material may act as an extension of the electrode and would allow a closer contact with the catalytic site of an enzyme due to the molecular flexibility of the polymeric chains. This means that the polymer acts here as an electronic wire from the electrode to the biocatalyst. The approach follows the concept of a DET system and, thus, electron flow should be possible around the redox potential of the protein site in contact with the conducting polymer (with a certain overpotential). The idea can be realized, but it turned out that not so many enzymes can be connected in this way with electrodes. Examples are polypyrrole and quinohemoprotein alcohol dehydrogenase [115], sulfonated polyanilines and PQQ-GDH [116] or xanthine dehydrogenase XDH [117], polythiophene and PQQ-GDH [118], and PEDOT:PSS and cyt c [119].
B) From the mechanistic point of view, the situation becomes more complicated because conducting polymers can also show defined redox reactions, resulting in clear voltammetric peaks when applied to electrodes. Conductivity is often found to be different in the different redox states. The enzyme reaction may require a certain redox state of the polymer in order to observe electron transfer between these partners. This, however, is very similar to mediators where bioelectrocatalysis starts when the mediator is being oxidized (for an oxidase) or being reduced (for a reductase). When this is the case for an enzyme reaction with a conducting polymer at the potential of the polymer redox couple, then the mechanism would also have to be termed as MET-based. Examples have been shown with some PQQ-dependent enzymes and sulfonated polyaniline [120].
C) Operation of an enzyme electrode with a conducting polymer can, however, also be based on product formation. Conducting polymers can exhibit fast reactions with side products of enzymatic reactions (e.g. hydrogen peroxide) and thus, a significant current signal appears when substrate is present in solution. This is valuable in sensing, but of less importance when the enzyme reactions shall be controlled by the electrode potential.

Based on conducting polymers, different developments can be observed within the last 20 years. In one direction, the conducting polymeric backbone is combined with the construction principle of redox polymers, that is, the attachment of redox moieties in introduced side chains. Here, a mixed conductivity – hopping of electrons between the redox sites and electronic conductivity in the backbone – can be achieved [121–125]. This approach has been valuable not only for enzymes, but also for more complex biological entities such as microorganisms or photo-active cells. Another – very large – directions deal with the combination of conducting polymers with the so-called "nanofillers", which are different nanomaterials such as carbon black, carbon nanotubes, fullerenes, graphene, or metal nanoparticles [126]. These particles are modifying the properties of the conducting polymer but can also interact with enzyme molecules as will be explained in the next section. Some composites even combine several of these components. Often, "synergistic" effects are reported, but mostly it is hard to recognize the mechanism behind and the role of the individual component.

2.6 Application of Nanoparticles

Nanostructures and nanoparticles are considered as a powerful tool in the concept of creating a "molecular wire" between the redox center of a biocatalyst and an electrode. The nanostructures are intended to provide electronic conductivity, and due to their size in the range of biomolecules, they show a better interaction with them compared to flat large electrodes. The approach is mainly devoted to a DET of the enzyme. Consequently, the application of nanomaterials is directed to an enlargement of the electrode surface and to provide more anchor points for protein immobilization. It should be emphasized that the orientation of the enzyme can also be improved this way. The key advantages can be summarized as follows:

- Nanostructures can be prepared in different sizes
- Additionally, the shape of nanoparticles can often be changed
- Surface modification is feasible and thus, properties can be adjusted to the enzyme
- Different protein coupling strategies have been developed from simple adsorption, charge-determined interaction, and covalent linkage or bioaffinity binding

A "molecular wire" par excellence are carbon nanotubes. Since their commercial availability, they have seen a considerable application in many fields, including bioelectrocatalysis [127, 128]. For this direction of application, the conducting types of these nanostructures are particularly interesting. They can be divided into single-walled carbon nanotubes (SWCNTs) and multiwalled carbon nanotubes (MWCNTs) according to the number of graphitic sidewalls they possess. They have a very good electronic conductivity and, in addition, several other remarkable properties such as an extremely high elastic modulus. With a size range of few up to several 10 nm in diameter, they are well suited for interaction with biomolecules, including enzymes. They can significantly increase the active electrode area and can harbor a large amount of biomolecules compared to a flat electrode. In bioelectrocatalysis, the main focus is on an enhanced DET of enzymes since the distance between a redox center in the enzyme and the conducting nanotubes can be much smaller compared to a large macroscopic electrode. In addition, several molecules can be nicely converted at the CNTs, allowing the sensitive determination, which can be used for co-substrate or product detection according to Figure 2.2a. Examples are NADH or hydrogen peroxide oxidation. They can, however, also be combined with mediators such as redox polymers as they can provide fast electron transport over longer distances, which is here combined with a fast reaction of the enzyme with the redox polymer.

An important aspect of the DET-based approaches in using CNTs is the possibility of surface modification. This can be accomplished by several methods of partial oxidation of the side walls. Here, however, care has to be taken that the introduced defects do not harm the electronic conductivity to a larger extent. In this way, new functional groups can be introduced that are useful for adsorption, layer-by-layer procedures, or covalent binding. Alternatively, π–π stacking interactions can be applied for a reliable adjustment of the surface properties with respect to the properties of the chosen enzyme. Molecules with condensed aromatic ring structures are used for this purpose. Few examples are given in Figure 2.9. This way, easy covalent fixation can also be achieved, for example, when activated esters are introduced by the π stacking compound [129–131].

Alternatively, metal nanoparticles have been used for the efficient connection of enzymes with electrodes. Gold nanomaterials are clearly the most commonly used nanostructures [132]. By applying them on macroscopic electrodes also here, the increase in surface area has been in focus and could be demonstrated in the application with several enzymes. Most investigations have concentrated on gold nanoparticles in the size range of few nanometers up to 100 nm. This

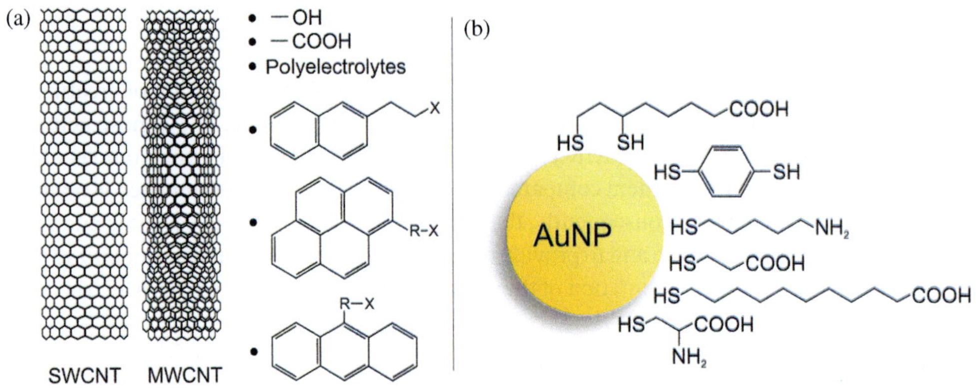

Figure 2.9 Nanostructures which are often used in bioelectrocatalytic systems (a) carbon nanotubes with SWCNTs – single-walled carbon nanotubes and MWCNTs – multiwalled carbon nanotubes and potential modifications of the surface by chemical treatment (introduction of hydroxy and carboxy groups), by the application of polyelectrolytes or by modification with π stacking compounds such as derivatives of naphthalene, anthracene, or pyrene (X represents a functional group which is intended to be introduced to the surface) and (b) gold nanoparticles and their surface modification by thiol compounds of different length and head groups.

type of nanostructure can also be easily modified. In that case, chemisorption helps immobilize particularly thiol compounds on the nanoparticles (see Figure 2.8). Since the sulfur-containing compounds may carry different head groups, the charge and other chemical properties can be adapted to the protein properties. This also includes a covalent coupling of enzymes to the nanoparticle surface.

The application of nanoparticles is generally aimed at increasing the catalytic current in bioelectrocatalysis. This is mostly based on a current increase per geometric area. When the real area is analyzed and efficiency is compared, then rather similar values can often be found.

Examples for such investigations are studies on bilirubin oxidase [133, 134], fructose dehydrogenase [133], laccase [135], PQQ-dependent dehydrogenases [136], cellobiose dehydrogenase [137], or peroxidase [138]. It should be mentioned here that efficient DET could not always be observed. Here, the details of the immobilization procedure also play an important role [67, 135, 139]. However, since synthesis protocols of nanostructures have developed further, even smaller nanoclusters can be prepared in a controlled way. Nanoclusters are particles with a diameter < 2 nm and, thus, consist of 15–200 atoms. With these very small particles, some additional effects can be observed – improvements in the kinetics of the enzyme reaction and sometimes also improvements with respect to overpotentials. This means that the mechanisms of action must be different compared to larger particles and it can be explained by the more molecule-like electrochemical behavior of such nanoclusters. As a consequence, defined electron transfer steps can be observed from an electrode toward nanoclusters. Thus, they behave more similar to mediator molecules [140]. Examples for such investigations are electrode interactions with partially deglycosylated glucose oxidase [141], fructose dehydrogenase [142], galactose oxidase [143], or bilirubin oxidase [144].

This may again illustrate that the term "wired electron transfer" does not explain the mechanism behind the current flow obtained. As for the metal nanoparticles, it seems that more research is necessary to clarify the transition from a behavior determined by an extension of the electrode area to a behavior dominated by the redox chemistry of a discrete redox object [145]. It has to be added here that mediators can also be prepared in the form of nanoparticles and applied together with the enzyme to electrodes [146].

In addition to the application of nanoparticles on the electrode surface, the nanostructuring of the electrode material itself has also been given special attention. This is particularly relevant for carbon [147–149] metal oxides [150] and metal electrodes [147, 151, 152]. Interestingly, there have been some reports avoiding any promotor layer on metal electrodes by applying a nanostructuring protocol. Examples are laccase (from *Trametes hirsuta* [153]), bilirubin oxidase (*Myrothecium verrucaria* [154]), or hemoglobin [155].

2.7 Mass Transport Limited Systems

As indicated in Figure 2.1, there are several partial processes occurring at an enzyme electrode under conversion conditions. When electron transfer and the reaction rate of the enzyme are fast, then the transport of the enzyme substrate to the converting electrode can influence or even determine the current flow. Diffusion is the process that brings the substrate molecule close to the converting enzyme (even under convection conditions). According to the kinetics of diffusion (Fick's first law), the thickness of the diffusion layer is the key parameter influencing the rate of this transport process because the surface concentration can only be zero as the lowest value, a certain solution concentration has been chosen according to the properties of the enzyme (high enough for substrate saturation, but not too high to avoid substrate inhibition) and the diffusion coefficient is a constant for a certain substance. The diffusion layer thickness can be changed by changing the stirring or flow conditions in the solution. This has led to the development of rotating disks experiments since here the rotation speed can be directly correlated to the thickness of this layer. Consequently, by increasing the rotation speed, the diffusion can be drastically enhanced and thus, information on the enzyme immobilized at the electrode surface can be obtained [5, 156].

In a very simple approach, the enzyme reaction in DET mode at the electrode appears in two steps:

$$E + S \xrightarrow{k_1} \in + P$$

$$\in \xrightarrow{k_s} E \pm e^-$$

where E is the enzyme in the resting state, $\in$ is the enzyme after reaction with a substrate molecule, S is the substrate and P is the product, k_1 is the reaction rate constant of E with S, and k_s is the rate constant for the heterogeneous electron transfer at the electrode.

The current at a rotating disk electrode can be described as

$$\frac{1}{I} = \frac{1}{I_L} + \frac{1}{I_k} \tag{2.15}$$

where I_L is the total mass transfer limited current and I_k is the current in the absence of any mass transport effects (reaction rate limited), as given by

$$I_L = 0.62 \cdot n \cdot F \cdot A \cdot D^{2/3} \cdot \omega^{1/2} \cdot \nu^{-1/6} \cdot c_b \tag{2.16}$$

$$\frac{1}{I_k} = \left(\frac{1}{n \cdot F \cdot A \cdot \Gamma}\right) \cdot \left(\frac{1}{k_1 \cdot c_b} + \frac{1}{k_s}\right) \tag{2.17}$$

where ω is the angular frequency, Γ is the surface concentration of the enzyme, D is the diffusion coefficient of the substrate, and c_b is the substrate concentration.

For evaluation of the behavior, the current is usually plotted in Koutecky–Levich coordinates ($1/I$ vs. $\omega^{-1/2}$) with different substrate concentrations. Here, one can clearly see in which region the reaction is controlled by diffusion (I changes with rotation speed) and which region is not (since the surface reaction controls the current). According to Eq. (2.16), the slope of the plot should be dependent on the substrate concentration and the number of transferred electrons. The kinetically limited current I_k can be obtained from the intercept of these curves. By plotting it vs. the substrate concentration, the resulting slope is dependent on the rate of the reaction (k_1), whereas the intercept is proportional to the heterogeneous electron transfer rate constant k_s. For the evaluation of these rate constant, one needs the surface concentration of the enzyme. This value is not always easy to obtain. For enzymes, which exhibit a clear DET, one may integrate the area under CV peaks in the absence of substrate, but not always clear peaks can be obtained. Thus, other methods, for example, based on a complete removal of the enzyme and determination of inherent cofactors, have to be applied.

The potential of this type of measurement is great since the entire system can be analyzed and even more complex reaction schemes are accessible. Furthermore, experiments under DET conditions can be compared with experiments in the presence of mediators (MET). This may help elucidate, for example, which fraction of the immobilized enzyme is capable of DET and which fraction can only be connected to the electrode by means of mediators [156].

In this paragraph, a rather simple behavior has been assumed. A more detailed description can be found in the literature [157–160].

2.8 Protein Engineering

Protein engineering has become a powerful tool for creating new molecules with properties different from those of naturally occurring proteins. It has also helped to understand the mechanisms behind the catalytic activity and structure–function relationships in these complex biomolecules. The achievements in the last couple of decades are a good example of interdisciplinarity, as enzymology, molecular biology, structural biology, and bioinformatics have contributed to the successful development. Although enzymes have evolved over thousands of years, they operate in biology in living organisms. Technical applications of enzymes by humans can significantly differ in the environment and conditions, but also in the purpose, these biocatalysts are applied for. This has clearly been one driving force for the generation of new molecules with some new functions better adapted to the intended technical functionality. The main directions of protein engineering are directed toward protein molecules with:

- Modified substrate spectrum (mostly a strict selectivity vs. a selected substrate)
- Improved stability (this includes thermal stability, but can also be directed to stability against variations in solution composition such as pH or organic solvent content)
- Enhanced catalytic activity, that is, an enhanced substrate conversion rate
- Higher expression level in protein production organisms
- Improved electron transfer with external reaction partners, within the enzyme (internal electron transfer) and also in the interaction with electrodes (heterogeneous electron transfer)

Different general approaches can be identified in this area. One can create new protein molecules through principles of directed evolution or by rational design [161, 162]. In addition, combined approaches have been developed. In directed evolutionary approaches, the idea is to produce many variants of a given protein by random variation in the amino acids and to select among these mutants the molecules of the desired function. Obviously, here phenotype–genotype information needs to be combined and a suitable screening method for a desired property is required. This makes this approach rather labor-intensive, but it does not require any specific knowledge on the enzyme and its structure or mechanism. In order to reduce the number of variants to be produced and studied, some methods analyze related proteins and their sequence. Ancestral sequence reconstruction or consensus design are examples of this direction. In contrast to this, rational design starts from a known protein and its structure and function. Thus, only few variants are produced with specific predicted changes. Historically, this has started in the region of the catalytic center of the enzyme (where most knowledge has been accumulated) but has expanded considerably to other regions since more and more tools are available which help in the prediction of properties. There is a variety of computational methods available that support the researcher in the design of new mutants. Meanwhile, the number of tools is rather overwhelming, and one needs to find the right tool for one's own question [163]. One may distinguish between tools identifying "hotspots" and tools, which are predicting the effects of mutations. Here, the focus might be rather different (e.g. rates, allosteric sites, protein interaction, electrostatics), which explains the diversity. Within this aspect, protein dynamics plays an increasing role and is not reflected in the classical 3D model derived from the analysis of X-ray diffraction patterns of protein crystals. Particularly, motions occurring in the microsecond time scale seem to be relevant for the catalytic reaction [164]. A recent development in computational models can be seen in the application of machine learning tools since they can help in protein structure predictions [165]. When predictions are becoming more and more accurate, this can provide more efficient ways to new proteins.

It should also be mentioned here that combinations of the two basic approaches – directed evolution and rational design – can also be recognized and this is sometimes termed as "semi-rational design". Here, libraries of mutants are prepared by amino acid variations in well-selected areas of the protein.

Since stability is a major issue in the practical application of many enzymes, proteins appearing in thermostable organisms can be considered as a good starting point for protein engineering since these molecules are already optimized for operation under rather harsh conditions. Most of the protein engineering work is based on the 20 amino acids, which are the building blocks of natural proteins. The introduction of unnatural amino acids can be considered as an alternative direction of research.

Driven by an extensive use of enzymes in analytical sciences, but also in the food industry, the main activities in protein engineering have focused on sensitivity (catalytic rate), selectivity, and stability. Besides these aspects, the value of protein engineering has also been recognized as a tool in clarifying electron transfer pathways within proteins, between proteins and of proteins with electrodes. These developments have started in the late 1980s and have naturally focused first on small redox proteins [166]. In order to shine light on structure–function relationships in ET, three directions have mainly been established here:

- Mutations in the vicinity of the catalytic center in order to change the redox potential and the reorganization energy
- Modifications in the region between redox centers since the intervening medium strongly influences the β value in the distance dependence of the electron transfer rate (Eq. 2.14)
- Changes on the protein surface to study electron transfer reactions of introduced surface groups with the catalytic redox center and to modify interaction with reaction partners

Heme proteins such as cytochrome *c* have been studied here in order to clarify the interplay between environment of the heme group, redox potential, and stability. In addition, small blue copper proteins such as azurin have been investigated. Often, negatively charged axial ligands have been found to stabilize the oxidized species, whereas hydrophobic residues near the redox site tend to stabilize the reduced state. Site-specific surface modifications have been introduced to couple

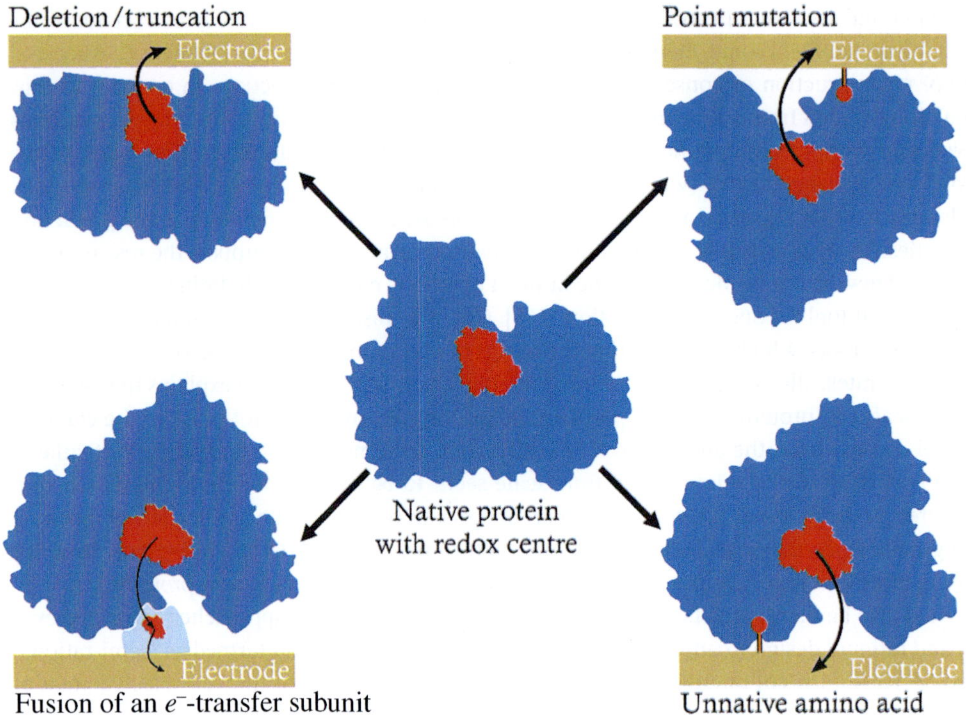

Figure 2.10 Illustration of different approaches used in protein engineering in order to improve the electrode protein interaction for efficient electron transfer. Native proteins may be truncated or deglycosylated in order to minimize the distance of the redox center to the electrode. Specific point mutations can be introduced and applied for a more oriented coupling of the enzyme to the electrode. This same idea can also be realized by introduction of an unnatural amino acid which may give more advanced coupling schemes. It is also feasible to design a fusion protein by introduction of an electron transfer relay subunit from one well-studied multisite protein to a protein with only one redox center.

redox species to a certain location on the protein surface and study the electron transfer capability with the redox center inside the protein. The polypeptide structure was found to modulate the electronic coupling between the redox sites and, thus, influences the electron transfer rate [166, 167].

Since these early years of protein engineering, a continuous development can be seen up to the present time [168–170]. As indicated before, this is accompanied by a better understanding of protein structure and factors influencing electron transfer reactions and by advanced methods in preparing variants of native proteins. Furthermore, the interaction of a redox protein with an electrode surface has become stronger in the focus since developments on enzyme electrodes have demonstrated the advantages of direct electron transfer systems. Thus, nowadays, one may distinguish the following directions in producing protein variants for improved electron transfer. They are additionally illustrated schematically in Figure 2.10:

1) In order to increase the accessibility of the redox center for electron transfer reactions, the protein can be truncated. This means that a part of the protein is removed, which can be done on the DNA level, but also in some cases by specific proteases at the prepared protein. The removal of glycosylations from the protein surface may also be considered as an approach within this group.

2) To establish a new feature in an existing enzyme molecule, electron transfer domains or whole subunits known from other proteins can be introduced in the form of a fusion protein. This is directed to the creation of a multistep electron transfer cascade from the redox center of the chosen enzyme toward the electrode and follows, thus, the natural examples when considering such important physiological redox processes as respiration or photosynthesis.
3) Another research direction is aimed at the introduction of point mutations at selected places within the protein. The concept is intended to improve the orientation of the enzyme molecule with respect to the electrode surface. This can result in a more uniform situation of the molecules and shorter electron transfer distances. These point mutations can also be accompanied by mutations within the region between the redox center and the electrode or near the redox center in order to change the properties for a more efficient electron transfer (mainly devoted to changes in reorganization energy λ and β factor in the distance dependence or changes in the redox potential and thus, the driving force for reactions).
4) One may also define a fourth direction since successful developments have resulted in the possibility to introduce artificial amino acids into the protein structures. This can be considered as its own direction since here much more effort is necessary to ensure a sufficient expression efficiency in preparing these protein variants. In addition, different chemical coupling strategies can be applied compared to the natural amino acids.

Despite the increasing technological application of protein variants, protein engineering still remains a valuable tool in clarifying electron transfer pathways and mechanisms in enzymes.

2.8.1 Truncation

If protein truncations are to be used for variant production, care has to be taken in the selection of the region to be removed. These are often regions at the C- or N-terminus of the polypeptide chain. Besides the effect of a better availability of the redox center for the electrode, the downsizing of the protein may allow a higher protein density on the surface. Both effects will result in higher catalytic currents when no mass transport limitations occur.

Examples of this approach are cytochrome *c* [171], microperoxidase (which is a truncated cytochrome *c*), or laccase [172]. For the enzyme FAD-GDH from *Burkholderia cepacia*, truncation has been used to clarify the electron transfer pathway from the catalytic FAD domain via an Fe S cluster and two hemes toward the electrode [74]. Also, for fructose dehydrogenase from *Gluconobacter japonicus*, truncation was helpful to elucidate the role of the three hemes in the heme subunit of the enzyme and prepare a variant with higher DET activity [173, 174].

Carbohydrates on the protein surface exhibit a strong insulating effect. Thus, the removal of glycans from the protein surface has been found to profoundly increase the electronic communication of enzymes with electrodes. This has been studied with several enzymes, such as glucose oxidase [175] or cellobiose dehydrogenase [176, 177]. The approach might also be interesting when redox polymers are applied, as shown for FAD-dependent glucose dehydrogenase [178].

2.8.2 Fusion Proteins

The idea behind the use of fusion proteins is the introduction of a new functionality into the enzyme which is not present in the native molecule. Different approaches can be distinguished here. On the one hand, electron transfer domains of other proteins can be introduced and, on the other hand, peptide sequences can be added which show a specific interaction with surface materials and thus, help for an oriented immobilization of the enzyme molecule [169, 179].

Cytochromes are small redox proteins serving as electron carriers in biological systems. They have often been used as electron transfer relays since the heme centers are well accessible for electron transfer reactions. Examples are fusion constructs based on cytochrome *c* (and FAD GDH [180]), flavocytochrome *b*2 (and lactate dehydrogenase [181]), and cytochrome *b*562 (and FAD-GDH [182]).

The introduction of affinity tags is an approach to change the orientation of the enzyme on the electrode surface and narrow the distribution of possible protein orientations. For DET systems, this approach is only applicable when the distance of the redox center to the electrode under these optimal immobilization conditions is suitable for direct electron exchange. Such an optimal orientation might be obtained by point mutations (see subchapter below), but also by introduction of an affinity tag. Peptides have proven to be valuable molecules despite their small size compared to whole proteins. The value comes mostly from a specific affinity which is very helpful in interaction studies. Recognition is, however, not limited to discrete molecules, but is also feasible at surfaces. This means that surface binding becomes a rather selective process. When such a binding peptide sequence is attached to an enzyme, this will help in an oriented immobilization at a specific electrode material. Examples are gold binding peptides (fused e.g. to FAD-GDH [183]), Met-rich motifs, and other peptides which are binding to CNTs (e.g. copper efflux oxidase [184], P450 monooxygenase [185]). Another type of affinity tag originates from affinity chromatography used in the purification of recombinant proteins – the histidine tag. This tag of six histidines can easily bind via chelate formation to surface-fixed transition metal ions (usually via nitrilotriacetic acid derivatives and nickel ions). Examples are a PQQ-dependent aldehyde dehydrogenase [186] or multicopper oxidase from *Pyrobaculum aerophilum* [187].

2.8.3 Point Mutations

In order to ensure a better interaction of the enzyme with the electrode as a reaction partner, several ideas have been realized based on point mutations. Cysteine has often been considered as new junction of the protein to a gold electrode. The amino acid can be introduced as a point mutation or as a tag at the C- or N-terminus. In case there are already several cysteines on the surface of the native protein, the reduction in this number can also be an approach. For immobilization of the modified enzyme, sulfur-gold chemisorption can be exploited or specific reactions of the thiol group such as with maleimide groups.

An improved coupling of the protein variant with electrodes has been demonstrated for several examples, such as cytochrome *c* [188], the P450cam enzyme [189], superoxide dismutase [190], glucose oxidase [141], cellobiose dehydrogenase [191], or bilirubin oxidase [192].

Alternatively, histidine can also be an intended connection site. Here, chelate formation of histidines with metal ions can be exploited for site-specific immobilization. Examples are the ferredoxin $NADP^+$ reductase [193] or cytochrome *c* [194].

Cellobiose dehydrogenases represent another interesting group of enzymes that can be isolated from different organisms. The enzyme consists of two subunits – the FAD containing one is responsible for sugar oxidation, whereas the cytochrome *b*-type subunit is responsible for electron transfer. Both domains are coupled by a flexible linker, and protein engineering has helped to elucidate factors that determine the interaction between both subunits [195]. This internal electron transfer has been found rather limiting in several cases.

Another background for point mutations is intended changes in the redox properties. For example, the multicopper enzymes laccase [196], bilirubin oxidase [197], and copper efflux oxidase [198] have been studied with respect to changes of the redox potential of the copper sites and, consequently, the catalytic rate for substrate conversion and oxygen reduction.

2.8.4 Unnatural Amino Acids

These are amino acids with the usual amino- and carboxy-groups, but a side chain which does not occur in any of the 20 native amino acids present in proteins. This allows the introduction of completely new chemical functionalities at specific positions in a protein. This can have rather different practical implications. One direction is still the modification of the properties of the catalytic center of the enzyme [199]. With respect to an improved electron transfer, the main idea is that a specific functional group is introduced which helps in the oriented immobilization of the enzyme variant on the electrode surface [179]. Clearly, this new functionality should not influence the catalytic cycle. "Click" chemistry approaches are in focus for the immobilization of these modified proteins. The approach differs in the protein expression systems used and the chemical reactions which occur but follows the idea of a site-directed mutagenesis in a rather rational design. As for point mutations, the selection of the site where the new functionality is introduced is of special importance.

2.9 Conclusions and Outlook

Electron transfer is one of the key issues in constructing bioelectrocatalytic systems. This process should not determine the overall performance and, thus, considerations have to be devoted to the efficiency of this part of the electrode system. Particularly valuable are methods allowing the determination of the heterogeneous electron transfer rate constant in order to evaluate the quality of the established contact of the redox enzyme or the redox mediator and the electrode. Whereas mediator-based approaches often provide higher current values, direct electron transfer-based approaches avoid any additional component in the system, which can also directly react with substances from the solution. DET-based electrodes can operate close to the redox potential of the chosen enzyme but have inherent limitations in the amount of enzyme molecules, which can directly exchange electrons with the electrode material. Here, 3D electrodes can bring new perspectives as, for example, highly doped metal oxide electrodes prepared by template-based approaches. Electron transfer cannot only be considered from the side of the electrode (material, preparation, modification) and suitable redox molecules or nanostructures, but also from the side of the biomolecules. Here, the advancements in protein engineering can be expected to open new possibilities.

References

1 Lisdat, F. and Scheller, F. (2000). Electrodes. In: *Biosensors for Environmental Monitoring*. Harwood Academic Publishers.

2 Chaubey, A. and Malhotra, B. (2002). Mediated biosensors. *Biosens. Bioelectron.* 17 (6–7): 441–456. https://doi.org/10.1016/S0956-5663(01)00313-X.

3 Nicholson Richard, S. and Shain, I. (1964). Theory of stationary electrode polarography - single scan and cyclic methods applied to reversible, irreversible and kinetic systems. *Anal. Chem.* 36 (4): 706–723.

4 Nicholson, R.S. (1965). The theory and application of cyclic voltammetry for measurement of electrode reaction kinetics. *Anal. Chem.* 37 (11): 1351–1355.

5 Bard, A.J. and Faulkner, L.R. (2001). *Electrochemical Methods - Fundamentals and Applications*, 2e. New York: John Wiley & Sons, Ltd.

6 Lavagnini, I., Antiochia, R., and Magno, F. (2004). An extended method for the practical evaluation of the standard rate constant from cyclic voltammetric data. *Electroanalysis* 16 (6): 505–506. https://doi.org/10.1002/elan.200302851.

7 Lazanas, A. and Prodromidis, M. (2023). Electrochemical impedance spectroscopy-a tutorial. *ACS Meas. Sci. Au* 3 (3): 162–193. https://doi.org/10.1021/acsmeasuresciau.2c00070.

8 Paenke, O., Balkenhohl, T., Kafka, J. et al. (2008). Impedance spectroscopy and biosensing. In: *Biosensing for the 21st Century*, vol. 109 (ed. R. Renneberg and F. Lisdat), 195–237. https://doi.org/10.1007/10_2007_081.

9 Laviron, E. (1979). General expression of the linear potential sweep voltammogram in the case of diffusionless electrochemical systems. *J. Electroanal. Chem.* 101 (1): 19–28. https://doi.org/10.1016/S0022-0728(79)80075-3.

10 Ruff, A. (2017). Redox polymers in bioelectrochemistry: common playgrounds and novel concepts. *Curr. Opin. Electrochem.* 5 (1): 66–73. https://doi.org/10.1016/j.coelec.2017.06.007.

11 Heller, A. (1990). Electrical wiring of redox enzymes. *Acc. Chem. Res.* 23 (5): 128–134. https://doi.org/10.1021/ar00173a002.

12 Heller, A. (2006). Electron-conducting redox hydrogels: design, characteristics and synthesis. *Curr. Opin. Chem. Biol.* 10 (6): 664–672. https://doi.org/10.1016/j.cbpa.2006.09.018.

13 Forster, R. and Vos, J. (1990). Synthesis, characterization, and properties of a series of osmium-containing and ruthenium-containing metallopolymers. *Macromolecules* 23 (20): 4372–4377. https://doi.org/10.1021/ma00222a008.

14 Lisdat, F., Dronov, R., Möhwald, H. et al. (2009). Self-assembly of electro-active protein architectures on electrodes for the construction of biomimetic signal chains. *Chem. Commun.* 3: 274–283. https://doi.org/10.1039/b813559b.

15 Feifel, S., Kapp, A., and Lisdat, F. (2014). Protein multilayer architectures on electrodes for analyte detection. In: *Biosensors Based on Aptamers and Enzymes*, vol. 140 (ed. M. Gu and H. Kim), 253–298. https://doi.org/10.1007/10_2013_236.

16 Dronov, R., Kurth, D., Moehwald, H. et al. (2008). Communication in a protein stack:: electron transfer between cytochrome c and bilirubin oxidase within a polyelectrolyte multilayer. *Angew. Chem. Int. Ed.* 47 (16): 3000–3003. https://doi.org/10.1002/anie.200704049.

17 Dronov, R., Kurth, D., Möhwald, H. et al. (2008). Layer-by-layer arrangement by protein-protein interaction of sulfite oxidase and cytochrome c catalyzing oxidation of sulfite. *J. Am. Chem. Soc.* 130 (4): 1122. https://doi.org/10.1021/ja0768690.

18 Feifel, S., Kapp, A., Ludwig, R., and Lisdat, F. (2014). Nanobiomolecular multiprotein clusters on electrodes for the formation of a switchable cascadic reaction scheme. *Angew. Chem. Int. Ed.* 53 (22): 5676–5679. https://doi.org/10.1002/anie.201310437.

19 Wettstein, C., Kano, K., Schäfer, D. et al. (2016). Interaction of flavin-dependent fructose dehydrogenase with cytochrome c as basis for the construction of biomacromolecular architectures on electrodes. *Anal. Chem.* 88 (12): 6382–6389. https://doi.org/10.1021/acs.analchem.6b00815.

20 Pinck, S., Etienne, M., Dossot, M., and Jorand, F. (2017). A rapid and simple protocol to prepare a living biocomposite that mimics electroactive biofilms. *Bioelectrochemistry* 118: 131–138. https://doi.org/10.1016/j.bioelechem.2017.07.010.

21 Dahms, H. (1968). Electronic conduction in aqueous solution. *J. Phys. Chem.* 72: 362–364.

22 Ruff, L. and Friedrich, V. (1971). Transfer diffusion I theoretical. *J. Phys. Chem.* 75: 3297–3302.

23 Dalton, E. and Murray, R. (1991). Viologen(2+/1+) and viologen(1+/0) electron-self-exchange reactions in a redox polymer. *J. Phys. Chem.* 95 (16): 6383–6389. https://doi.org/10.1021/j100169a056.

24 Cass, A.E.G. (1990). *Biosensors - A Practical Approach; A Practical Approach Series*. IRL Press.

25 Bollella, P. and Katz, E. (2020). Enzyme-based biosensors: tackling electron transfer issues. *Sensors* 20 (12): https://doi.org/10.3390/s20123517.

26 Schachinger, F., Chang, H., Scheiblbrandner, S., and Ludwig, R. (2021). Amperometric biosensors based on direct electron transfer enzymes. *Molecules* 26 (15): https://doi.org/10.3390/molecules26154525.

27 Reed, D. and Hawkridge, F. (1987). Direct electron-transfer reactions of cytochrome-c at silver electrodes. *Anal. Chem.* 59 (19): 2334–2339. https://doi.org/10.1021/ac00146a005.

28 Senaratne, V. and Bowden, E. (1988). The electrode-reaction of euglena-gracilis cytochrome c-552 at edge-oriented pyrolytic-graphite. *Biochem. Biophys. Res. Commun.* 157 (3): 1021–1026. https://doi.org/10.1016/S0006-291X(88)80976-8.

29 Bagby, S., Barker, P., Guo, L., and Hill, H. (1990). Direct electrochemistry of protein protein complexes involving cytochrome-c, cytochrome-b5 and plastocyanin. *Biochemistry* 29 (13): 3213–3219. https://doi.org/10.1021/bi00465a010.

30 Armstrong, F., Hill, H., Oliver, B., and Walton, N. (1984). Direct electrochemistry of redox proteins at pyrolytic-graphite electrodes. *J. Am. Chem. Soc.* 106 (4): 921–923. https://doi.org/10.1021/ja00316a015.

31 Zhang, J., Christensen, H., Ooi, B., and Ulstrup, J. (2004). In situ STM imaging and direct electrochemistry of pyrococcus furiosus ferredoxin assembled on thiolate-modified Au(111) surfaces. *Langmuir* 20 (23): 10200–10207. https://doi.org/10.1021/la048853i.

32 Sakurai, T., Nose, F., Fujiki, T., and Suzuki, S. (1996). Reduction and oxidation processes of blue copper proteins, azurin, pseudoazurin, umecyanin, stellacyanin, plantacyanin, and plastocyanin approached by cyclic and potential step voltammetries. *Bull. Chem. Soc. Jpn.* 69 (10): 2855–2862. https://doi.org/10.1246/bcsj.69.2855.

33 Chi, Q., Zhang, J., Nielsen, J. et al. (2000). Molecular monolayers and interfacial electron transfer of *Pseudomonas aeruginosa* azurin on Au(111). *J. Am. Chem. Soc.* 122 (17): 4047–4055. https://doi.org/10.1021/ja993174t.

34 Khoshtariya, D., Dolidze, T., Shushanyan, M. et al. (2010). Fundamental signatures of short- and long-range electron transfer for the blue copper protein azurin at Au/SAM junctions. *Proc. Natl. Acad. Sci. USA* 107 (7): 2757–2762. https://doi.org/10.1073/pnas.0910837107.

35 Yaropolov, A., Kharybin, A., Emneus, J. et al. (1996). Electrochemical properties of some copper-containing oxidases. *Bioelectrochem. Bioenerg.* 40 (1): 49–57. https://doi.org/10.1016/0302-4598(96)01919-8.

36 Johnson, D., Thompson, J., Brinkmann, S. et al. (2003). Electrochemical characterization of purified rhus vernicifera laccase:: voltammetric evidence for a sequential four-electron transfer. *Biochemistry* 42 (34): 10229–10237. https://doi.org/10.1021/bi034268p.

37 Shleev, S., Jarosz-Wilkolazka, A., Khalunina, A. et al. (2005). Direct electron transfer reactions of laccases from different origins on carbon electrodes. *Bioelectrochemistry* 67 (1): 115–124. https://doi.org/10.1016/j.bioelechem.2005.02.004.

38 Nazaruk, E., Michota, A., Bukowska, J. et al. (2007). Properties of native and hydrophobic laccases immobilized in the liquid-crystalline cubic phase on electrodes. *J. Biol. Inorg. Chem.* 12 (3): 335–344. https://doi.org/10.1007/s00775-006-0193-7.

39 Blanford, C., Foster, C., Heath, R., and Armstrong, F. (2008). Efficient electrocatalytic oxygen reduction by the "blue" copper oxidase, laccase, directly attached to chemically modified carbons. *Faraday Discuss.* 140: 319–335. https://doi.org/10.1039/b808939f.

40 Le Goff, A., Holzinger, M., and Cosnier, S. (2015). Recent progress in oxygen-reducing laccase biocathodes for enzymatic biofuel cells. *Cell. Mol. Life Sci.* 72 (5): 941–952. https://doi.org/10.1007/s00018-014-1828-4.

41 Mano, N. and de Poulpiquet, A. (2018). O_2 reduction in enzymatic biofuel cells. *Chem. Rev.* 118 (5): 2392–2468. https://doi.org/10.1021/acs.chemrev.7b00220.

42 Varfolomeev, S., Kurochkin, I., and Yaropolov, A. (1996). Direct electron transfer effect biosensors. *Biosens. Bioelectron.* 11 (9): 863–871. https://doi.org/10.1016/0956-5663(96)89435-8.

43 Ghindilis, A., Atanasov, P., and Wilkins, E. (1997). Enzyme-catalyzed direct electron transfer: fundamentals and analytical applications. *Electroanalysis* 9 (9): 661–674. https://doi.org/10.1002/elan.1140090902.

44 Ferapontova, E. (2004). Direct peroxidase bioelectrocatalysis on a variety of electrode materials. *Electroanalysis* 16 (13–14): 1101–1112. https://doi.org/10.1002/elan.200403003.

45 Wu, Y. and Hu, S. (2007). Biosensors based on direct electron transfer in redox proteins. *Microchim. Acta* 159 (1–2): 1–17. https://doi.org/10.1007/s00604-007-0749-4.

46 Zuccarello, L., Barbosa, C., Todorovic, S., and Silveira, C. (2021). Electrocatalysis by heme enzymes-applications in biosensing. *Catalysis* 11 (2): https://doi.org/10.3390/catal11020218.

47 Andreu, R., Ferapontova, E., Gorton, L., and Calvente, J. (2007). Direct electron transfer kinetics in horseradish peroxidase electrocatalysis. *J. Phys. Chem. B* 111 (2): 469–477. https://doi.org/10.1021/jp064277i.

48 Smith, E., Ensign, S., Ludden, P., and Feinberg, B. (1992). Direct electrochemical studies of hydrogenase and co dehydrogenase. *Biochem. J.* 285: 181–185. https://doi.org/10.1042/bj2850181.

49 Guiral-Brugna, M., Giudici-Orticoni, M., Bruschi, M., and Bianco, P. (2001). Electrocatalysis of the hydrogen production by [Fe] hydrogenase from desulfovibrio vulgaris hildenborough. *J. Electroanal. Chem.* 510 (1–2): 136–143. https://doi.org/10.1016/S0022-0728(01)00502-2.

50 Léger, C., Jones, A., Albracht, S., and Armstrong, F. (2002). Effect of a dispersion of interfacial electron transfer rates on steady state catalytic electron transport in [NiFe]-hydrogenase and other enzymes. *J. Phys. Chem. B* 106 (50): 13058–13063. https://doi.org/10.1021/jp0265687.

51 Luo, X., Brugna, M., Tron-Infossi, P. et al. (2009). Immobilization of the hyperthermophilic hydrogenase from aquifex aeolicus bacterium onto gold and carbon nanotube electrodes for efficient H_2 oxidation. *J. Biol. Inorg. Chem.* 14 (8): 1275–1288. https://doi.org/10.1007/s00775-009-0572-y.

52 Bistolas, N., Wollenberger, U., Jung, C., and Scheller, F. (2005). Cytochrome P450 biosensors - a review. *Biosens. Bioelectron.* 20 (12): 2408–2423. https://doi.org/10.1016/j.bios.2004.11.023.

53 Fedurco, M. (2000). Redox reactions of heme-containing metalloproteins:: dynamic effects of self-assembled monolayers on thermodynamics and kinetics of cytochrome c electron-transfer reactions. *Coord. Chem. Rev.* 209: 263–331. https://doi.org/10.1016/S0010-8545(00)00292-7.

54 Battistuzzi, G., Borsari, M., and Sola, M. (2001). Medium and temperature effects on the redox chemistry of cytochrome c. *Eur. J. Inorg. Chem.* 12: 2989–3004.

55 Carucci, C., Salis, A., and Magner, E. (2017). Electrolyte effects on enzyme electrochemistry. *Curr. Opin. Electrochem.* 5 (1): 158–164. https://doi.org/10.1016/j.coelec.2017.08.011.

56 Niki, K., Sprinkle, J., and Margoliash, E. (2002). Intermolecular biological electron transfer: an electrochemical approach. *Bioelectrochemistry* 55 (1–2): 37–40. https://doi.org/10.1016/S1567-5394(01)00157-8.

57 Feng, Z., Imabayashi, S., Kakiuchi, T., and Niki, K. (1997). Long-range electron-transfer reaction rates to cytochrome c across long- and short-chain alkanethiol self-assembled monolayers: electroreflectance studies. *J. Chem. Soc., Faraday Trans.* 93 (7): 1367–1370.

58 Armstrong, F.A. and Wilson, G.S. (2000). Recent developments in faradaic bioelectrochemistry. *Electrochim. Acta* 45 (15–16): 2623–2645. https://doi.org/10.1016/S0013-4686(00)00342-X.

59 Niki, K., Yagi, T., Inokuchi, H., and Kimura, K. (1979). Electrochemical behavior of cytochrome-c3 of desulfovibrio-vulgaris, strain miyazaki, on the mercury-electrode. *J. Am. Chem. Soc.* 101 (12): 3335–3340. https://doi.org/10.1021/ja00506a033.

60 Eddowes, M. and Hill, H. (1977). Novel method for investigation of electrochemistry of metalloproteins - cytochrome-c. *J. Chem. Soc., Chem. Commun.* 21: 771–772.

61 Yeh, P. and Kuwana, T. (1977). Reversible electrode-reaction of cytochrome-c. *Chem. Lett.* 10: 1145–1148. https://doi.org/10.1246/cl.1977.1145.

62 Kano, K. (2022). Fundamental insight into redox enzyme-based bioelectrocatalysis. *Biosci. Biotech. Bioch.* 86 (2): 141–156. https://doi.org/10.1093/bbb/zbab197.

63 Hitaishi, V., Clement, R., Bourassin, N. et al. (2018). Controlling redox enzyme orientation at planar electrodes. *Catalysis* 8 (5): https://doi.org/10.3390/catal8050192.

64 Scheiblbrandner, S. and Ludwig, R. (2020). Cellobiose dehydrogenase: bioelectrochemical insights and applications. *Bioelectrochemistry* 131. https://doi.org/10.1016/j.bioelechem.2019.107345.

65 Hibino, Y., Kawai, S., Kitazumi, Y. et al. (2019). Protein-engineering improvement of direct electron transfer-type bioelectrocatalytic properties of d-fructose dehydrogenase. *Electrochemistry* 87 (1): 47–51. https://doi.org/10.5796/electrochemistry.18-00068.

66 Takeda, K. and Nakamura, N. (2021). Direct electron transfer process of pyrroloquinoline quinone-dependent and flavin adenine dinucleotide-dependent dehydrogenases: fundamentals and applications. *Curr. Opin. Electrochem.* 29. https://doi.org/10.1016/j.coelec.2021.100747.

67 Neto, S., Hickey, D., Milton, R. et al. (2015). High current density PQQ-dependent alcohol and aldehyde dehydrogenase bioanodes. *Biosens. Bioelectron.* 72: 247–254. https://doi.org/10.1016/j.bios.2015.05.011.

68 Yoshikawa, T., Makino, F., Miyata, T. et al. (2022). Multiple electron transfer pathways of tungsten-containing formate dehydrogenase in direct electron transfer-type bioelectrocatalysis. *Chem. Commun.* 58 (45): 6478–6481. https://doi.org/10.1039/d2cc01541b.

69 Bollella, P., Gorton, L., and Antiochia, R. (2018). Direct electron transfer of dehydrogenases for development of 3rd generation biosensors and enzymatic fuel cells. *Sensors* 18 (5): https://doi.org/10.3390/s18051319.

70 Milton, R. and Minteer, S. (2019). Nitrogenase bioelectrochemistry for synthesis applications. *Acc. Chem. Res.* 52 (12): 3351–3360. https://doi.org/10.1021/acs.accounts.9b00494.

71 Zanello, P. (2019). Structure and electrochemistry of proteins harboring iron-sulfur clusters of different nuclearities. Part IV. Canonical, non-canonical and hybrid iron-sulfur proteins. *J. Struct. Biol.* 205 (2): 103–120. https://doi.org/10.1016/j.jsb.2019.01.003.

72 Adachi, T., Kaida, Y., Kitazumi, Y. et al. (2019). Bioelectrocatalytic performance of d-fructose dehydrogenase. *Bioelectrochemistry* 129: 1–9. https://doi.org/10.1016/j.bioelechem.2019.04.024.

73 Lisdat, F. (2020). PQQ-GDH - structure, function and application in bioelectrochemistry. *Bioelectrochemistry* 134. https://doi.org/10.1016/j.bioelechem.2020.107496.

74 Okuda-Shimazaki, J., Loew, N., Hirose, N. et al. (2018). Construction and characterization of flavin adenine dinucleotide glucose dehydrogenase complex harboring a truncated electron transfer subunit. *Electrochim. Acta* 277: 276–286. https://doi.org/10.1016/j.electacta.2018.04.060.

75 Kumar, N., He, J., and Rusling, J. (2023). Electrochemical transformations catalyzed by cytochrome p450s and peroxidases. *Chem. Soc. Rev.* 52 (15): 5135–5171. https://doi.org/10.1039/d3cs00461a.

76 Sensi, M., del Barrio, M., Baffert, C. et al. (2017). New perspectives in hydrogenase direct electrochemistry. *Curr. Opin. Electrochem.* 5 (1): 135–145. https://doi.org/10.1016/j.coelec.2017.08.005.

77 Armstrong, F., Evans, R., Hexter, S. et al. (2016). Guiding principles of hydrogenase catalysis instigated and clarified by protein film electrochemistry. *Acc. Chem. Res.* 49 (5): 884–892. https://doi.org/10.1021/acs.accounts.6b00027.

78 Munteanu, F., Mano, N., Kuhn, A., and Gorton, L. (2002). Mediator-modified electrodes for catalytic NADH oxidation: high rate constants at interesting overpotentials. *Bioelectrochemistry* 56 (1–2): 67–72. https://doi.org/10.1016/S1567-5394(02)00009-9.

79 Willner, I., Katz, E., and Willner, B. (1997). Electrical contact of redox enzyme layers associated with electrodes: routes to amperometric biosensors. *Electroanalysis* 9 (13): 965–977. https://doi.org/10.1002/elan.1140091302.

80 Bardea, A., Katz, E., Buckmann, A., and Willner, I. (1997). NAD(+)-dependent enzyme electrodes: electrical contact of cofactor-dependent enzymes and electrodes. *J. Am. Chem. Soc.* 119 (39): 9114–9119. https://doi.org/10.1021/ja971192+.

81 Makarchuk, I., Seica, A., Melin, F., and Hellwig, P. (2021). Probing the reaction of membrane proteins via infrared spectroscopies, plasmonics, and electrochemistry. *Curr. Opin. Electrochem.* 30. https://doi.org/10.1016/j.coelec.2021.100770.

82 Braasch-Turi, M., Koehn, J., and Crans, D. (2022). Chemistry of lipoquinones: properties, synthesis, and membrane location of ubiquinones, plastoquinones, and menaquinones. *Int. J. Mol. Sci.* 23 (21): https://doi.org/10.3390/ijms232112856.

83 Jeuken, L. (2009). Electrodes for integral membrane enzymes. *Nat. Prod. Rep.* 26 (10): 1234–1240. https://doi.org/10.1039/b903252e.

84 Naumann, R., Nowak, C., and Knoll, W. (2011). Proteins in biomimetic membranes: promises and facts. *Soft Matter* 7 (20): 9535–9548. https://doi.org/10.1039/c1sm05626c.

85 Zhang, H., Catania, R., and Jeuken, L. (2020). Membrane protein modified electrodes in bioelectrocatalysis. *Catalysis* 10 (12): https://doi.org/10.3390/catal10121427.

86 Rusling, J. (1998). Enzyme bioelectrochemistry in cast biomembrane-like films. *Acc. Chem. Res.* 31 (6): 363–369. https://doi.org/10.1021/ar970254y.

87 Kato, M., Cardona, T., Rutherford, A., and Reisner, E. (2012). Photoelectrochemical water oxidation with photosystem II integrated in a mesoporous; indium tin oxide electrode. *J. Am. Chem. Soc.* 134 (20): 8332–8335. https://doi.org/10.1021/ja301488d.

88 Morlock, S., Subramanian, S., Zouni, A., and Lisdat, F. (2022). Bio-inorganic hybrid structures for direct electron transfer to photosystem I in photobioelectrodes. *Biosens. Bioelectron.* 214. https://doi.org/10.1016/j.bios.2022.114495.

89 Morlock, S., Subramanian, S., Zouni, A., and Lisdat, F. (2023). Closing the green gap of photosystem I with synthetic fluorophores for enhanced photocurrent generation in photobiocathodes. *Chem. Sci.* 14 (7): 1696–1708. https://doi.org/10.1039/d2sc05324a.

90 He, P., Hu, N., and Rusling, J. (2004). Driving forces for layer-by-layer self-assembly of films of SiO_2 nanoparticles and heme proteins. *Langmuir* 20 (3): 722–729. https://doi.org/10.1021/la035006r.

91 Nakano, K., Yoshitake, T., Yamashita, Y., and Bowden, E. (2007). Cytochrome c self-assembly on alkanethiol monolayer electrodes as characterized by AFM, IR, QCM, and direct electrochemistry. *Langmuir* 23 (11): 6270–6275. https://doi.org/10.1021/la063697w.

92 Easley, A., Ma, T., Eneh, C. et al. (2022). A practical guide to quartz crystal microbalance with dissipation monitoring of thin polymer films. *J. Polym. Sci.* 60 (7): 1090–1107. https://doi.org/10.1002/pol.20210324.

93 Chen, X., Ferrigno, R., Yang, J., and Whitesides, G. (2002). Redox properties of cytochrome c adsorbed on self-assembled monolayers:: a probe for protein conformation and orientation. *Langmuir* 18 (18): 7009–7015. https://doi.org/10.1021/la0204794.

94 Lojou, E. and Bianco, P. (2006). Assemblies of dendrimers and proteins on carbon and gold electrodes. *Bioelectrochemistry* 69 (2): 237–247. https://doi.org/10.1016/j.bioelechem.2006.03.028.

95 Geilfuss, D., Boukherroub, R., Dostalek, J. et al. (2022). Can classical surface plasmon resonance advance via the coupling to other analytical approaches? *Front. Anal. Sci.* 2. https://doi.org/10.3389/frans.2022.1091869.

96 de Souza, J., Macedo, L., Hassan, A. et al. (2021). In situ and operando techniques for investigating electron transfer in biological systems. *ChemElectroChem* 8 (3): 431–446. https://doi.org/10.1002/celc.202001327.

97 Kornienko, N., Ly, K., Robinson, W. et al. (2019). Advancing techniques for investigating the enzyme-electrode interface. *Acc. Chem. Res.* 52 (5): 1439–1448. https://doi.org/10.1021/acs.accounts.9b00087.

98 Wilson, G. (2022). Spectroelectrochemistry of proteins. *Electroanalysis* 34 (12): 1834–1841. https://doi.org/10.1002/elan.202100535.

99 Ates, M. (2013). A review study of (bio)sensor systems based on conducting polymers. *Mater. Sci. Eng., C-Mater. Biol. Appl.* 33 (4): 1853–1859. https://doi.org/10.1016/j.msec.2013.01.035.

100 Ramanavicius, S. and Ramanavicius, A. (2021). Charge transfer and biocompatibility aspects in conducting polymer-based enzymatic biosensors and biofuel cells. *Nanomaterials* 11 (2): https://doi.org/10.3390/nano11020371.

101 Kaloni, T., Giesbrecht, P., Schreckenbach, G., and Freund, M. (2017). Polythiophene: from fundamental perspectives to applications. *Chem. Mater.* 29 (24): 10248–10283. https://doi.org/10.1021/acs.chemmater.7b03035.

102 Molapo, K., Ndangili, P., Ajayi, R. et al. (2012). Electronics of conjugated polymers (I): polyaniline. *Int. J. Electrochem. Sci.* 7 (12): 11859–11875.

103 Liu, Y., Gao, S., Zhang, X. et al. (2022). Probing the nature of charge carriers in one-dimensional conjugated polymers: a review of the theoretical models, experimental trends, and thermoelectric applications. *J. Mater. Chem. C* 11 (1): 12–47. https://doi.org/10.1039/d2tc03574j.

104 Gueye, M., Carella, A., Faure-Vincent, J. et al. (2020). Progress in understanding structure and transport properties of PEDOT-based materials: a critical review. *Prog. Mater. Sci.* 108. https://doi.org/10.1016/j.pmatsci.2019.100616.

105 Kang, S. and Snyder, G. (2017). Charge-transport model for conducting polymers. *Nat. Mater.* 16 (2): 252–257. https://doi.org/10.1038/NMAT4784.

106 Paulsen, B., Fabiano, S., and Rivnay, J. (2021; Clarke, D., Ed.; 2021; Vol. 51). Mixed ionic-electronic transport in polymers. *Annu. Rev. Mater. Res.* 51: 73–99. https://doi.org/10.1146/annurev-matsci-080619-101319.

107 Liu, J., Davis, N., Liu, D., and Hammond, P. (2012). Highly transparent mixed electron and proton conducting polymer membranes. *J. Mater. Chem.* 22 (31): 15534–15539. https://doi.org/10.1039/c2jm32296j.

108 Curvetto, N., Figlas, D., Brandolin, A. et al. (2006). Efficiency of enzymatic and non-enzymatic catalysts in the synthesis of insoluble polyphenol and conductive polyaniline in water. *Biochem. Eng. J.* 29 (3): 191–203. https://doi.org/10.1016/j.bej.2005.11.016.

109 Datta, B. and Schuster, G. (2008). DNA-directed synthesis of aniline and 4-aminobiphenyl, oligomers: programmed transfer of sequence information to a conjoined polymer nanowire. *J. Am. Chem. Soc.* 130 (10): 2965–2973. https://doi.org/10.1021/ja0726106.

110 Liu, W., Kumar, J., Tripathy, S. et al. (1999). Enzymatically synthesized conducting polyaniline. *J. Am. Chem. Soc.* 121 (1): 71–78. https://doi.org/10.1021/ja982270b.

111 Dai, M., Huang, T., Chao, L. et al. (2016). Tyrosinase-catalyzed polymerization of L-DOPA (versus L-tyrosine and dopamine) to generate melanin-like biomaterials for immobilization of

enzymes and amperometric biosensing. *RSC Adv.* 6 (21): 17016–17022. https://doi.org/10.1039/c5ra27478h.

112 Kashima, K., Fujisaki, T., Serrano-Luginbühl, S. et al. (2018). How experimental details matter. The case of a laccase-catalysed oligomerisation reaction. *RSC Adv.* 8 (58): 33229–33242. https://doi.org/10.1039/c8ra05731a.

113 Apetrei, R., Carac, G., Bahrim, G. et al. (2018). Modification of aspergillus niger by conducting polymer, polypyrrole, and the evaluation of electrochemical properties of modified cells. *Bioelectrochemistry* 121: 46–55. https://doi.org/10.1016/j.bioelechem.2018.01.001.

114 Kisieliute, A., Popov, A., Apetrei, R. et al. (2019). Towards microbial biofuel cells: improvement of charge transfer by self-modification of microoganisms with conducting polymer - polypyrrole. *Chem. Eng. J.* 356: 1014–1021. https://doi.org/10.1016/j.cej.2018.09.026.

115 Ramanavicius, A., Habermüller, K., Csöregi, E. et al. (1999). Polypyrrole entrapped quinohemoprotein alcohol dehydrogenase.: evidence for direct electron transfer via conducting-polymer chains. *Anal. Chem.* 71 (16): 3581–3586. https://doi.org/10.1021/ac981201c.

116 Schubart, I.W., Göbel, G., and Lisdat, F. (2012). A pyrroloquinolinequinone-dependent glucose dehydrogenase (PQQ-GDH)-electrode with direct electron transfer based on polyaniline modified carbon nanotubes for biofuel cell application. *Electrochim. Acta* 82: 224–232. https://doi.org/10.1016/j.electacta.2012.03.128.

117 Sarauli, D., Borowski, A., Peters, K. et al. (2016). Investigation of the pH-dependent impact of sulfonated polyaniline on bioelectrocatalytic activity of xanthine dehydrogenase. *ACS Catal.* 6 (10): 7152–7159. https://doi.org/10.1021/acscatal.6b02011.

118 Fusco, G., Göbel, G., Zanoni, R. et al. (2018). Aqueous polythiophene electrosynthesis: a new route to an efficient electrode coupling of PQQ-dependent glucose dehydrogenase for sensing and bioenergetic applications. *Biosens. Bioelectron.* 112: 8–17. https://doi.org/10.1016/j.bios.2018.04.014.

119 López-Bemabeu, S., Huerta, F., Morallón, E., and Montilla, F. (2017). Direct electron transfer to cytochrome c induced by a conducting polymer. *J. Phys. Chem. C* 121 (29): 15870–15879. https://doi.org/10.1021/acs.jpcc.7b05204.

120 Xu, S. and Minteer, S. (2014). Pyrroloquinoline quinone-dependent enzymatic bioanode: incorporation of the substituted polyaniline conducting polymer as a mediator. *ACS Catal.* 4 (7): 2241–2248. https://doi.org/10.1021/cs500442b.

121 Pickup, P. (1999). Conjugated metallopolymers. Redox polymers with interacting metal based redox sites. *J. Mater. Chem.* 9 (8): 1641–1653. https://doi.org/10.1039/a902244i.

122 Gracia, R. and Mecerreyes, D. (2013). Polymers with redox properties: materials for batteries, biosensors and more. *Polym. Chem.* 4 (7): 2206–2214. https://doi.org/10.1039/c3py21118e.

123 Holliday, B. and Swager, T. (2005). Conducting metallopolymers: the roles of molecular architecture and redox matching. *Chem. Commun.* 1: 23–36. https://doi.org/10.1039/b408479a.

124 Foulds, N. and Lowe, C. (1988). Immobilization of glucose-oxidase in ferrocene-modified pyrrole polymers. *Anal. Chem.* 60 (22): 2473–2478. https://doi.org/10.1021/ac00173a008.

125 Karlsson, C., Huang, H., Stromme, M. et al. (2015). Ion- and electron transport in pyrrole/quinone conducting redox polymers investigated by in situ conductivity methods. *Electrochim. Acta* 179: 336–342. https://doi.org/10.1016/j.electacta.2015.02.193.

126 Nasajpour-Esfahani, N., Dastan, D., Alizadeh, A. et al. (2023). A critical review on intrinsic conducting polymers and their applications. *J. Ind. Eng. Chem.* 125: 14–37. https://doi.org/10.1016/j.jiec.2023.05.013.

127 Wang, J. (2005). Carbon-nanotube based electrochemical biosensors: a review. *Electroanalysis* 17 (1): 7–14. https://doi.org/10.1002/elan.200403113.

128 Gupta, S., Murthy, C., and Prabha, C. (2018). Recent advances in carbon nanotube based electrochemical biosensors. *Int. J. Biol. Macromol.* 108: 687–703. https://doi.org/10.1016/j.ijbiomac.2017.12.038.

129 Yang, C., Denno, M., Pyakurel, P., and Venton, B. (2015). Recent trends in carbon nanomaterial-based electrochemical sensors for biomolecules: a review. *Anal. Chim. Acta* 887: 17–37. https://doi.org/10.1016/j.aca.2015.05.049.

130 Ding, S., Cargill, A., Medintz, I., and Claussen, J. (2015). Increasing the activity of immobilized enzymes with nanoparticle conjugation. *Curr. Opin. Biotechnol.* 34: 242–250. https://doi.org/10.1016/j.copbio.2015.04.005.

131 Bollella, P. and Katz, E. (2020). Bioelectrocatalysis at carbon nanotubes. In: *Methods in Enzymology: Nanoarmoring of Enzymes with Carbon Nanotubes and Magnetic Nanoparticles*, vol. 630 (ed. C. Kumar, A. Pyle, and D. Christianson), 215–247. https://doi.org/10.1016/bs.mie.2019.10.012.

132 Pingarrón, J., Yáñez-Sedeño, P., and González-Cortés, A. (2008). Gold nanoparticle-based electrochemical biosensors. *Electrochim. Acta* 53 (19): 5848–5866. https://doi.org/10.1016/j.electacta.2008.03.005.

133 Suzuki, M., Murata, K., Nakamura, N., and Ohno, H. (2012). The effect of particle size on the direct electron transfer reactions of metalloproteins using au nanoparticle-modified electrodes. *Electrochemistry* 80 (5): 337–339. https://doi.org/10.5796/electrochemistry.80.337.

134 Pankratov, D., Sundberg, R., Suyatin, D. et al. (2014). The influence of nanoparticles on enzymatic bioelectrocatalysis. *RSC Adv.* 4 (72): 38164–38168. https://doi.org/10.1039/c4ra08107b.

135 Dagys, M., Lamberg, P., Shleev, S. et al. (2014). Comparison of bioelectrocatalysis at trichaptum abietinum and trametes hirsuta laccase modified electrodes. *Electrochim. Acta* 130: 141–147. https://doi.org/10.1016/j.electacta.2014.03.014.

136 Treu, B., Arechederra, R., and Minteer, S. (2009). Bioelectrocatalysis of ethanol via PQQ-dependent dehydrogenases utilizing carbon nanomaterial supports. *J. Nanosci. Nanotechnol.* 9 (4): 2374–2380. https://doi.org/10.1166/jnn.2009.SE33.

137 Bozorgzadeh, S., Hamidi, H., Ortiz, R. et al. (2015). Direct electron transfer of phanerochaete chrysosporium cellobiose dehydrogenase at platinum and palladium nanoparticles decorated carbon nanotubes modified electrodes. *Phys. Chem. Chem. Phys.* 17 (37): 24157–24165. https://doi.org/10.1039/c5cp03812j.

138 Taurino, I., Sanzò, G., Antiochia, R. et al. (2016). Recent advances in third generation biosensors based on Au and Pt nanostructured electrodes. *TrAC, Trends Anal. Chem.* 79: 151–159. https://doi.org/10.1016/j.trac.2016.01.020.

139 Diaconu, M., Chira, A., and Radu, L. (2014). Modulating indium doped tin oxide electrode properties for laccase electron transfer enhancement. *Thin Solid Films* 565: 84–88. https://doi.org/10.1016/j.tsf.2014.06.055.

140 Quinn, B., Liljeroth, P., Ruiz, V. et al. (2003). Electrochemical resolution of 15 oxidation states for monolayer protected gold nanoparticles. *J. Am. Chem. Soc.* 125 (22): 6644–6645. https://doi.org/10.1021/ja0349305.

141 Holland, J., Lau, C., Brozik, S. et al. (2011). Engineering of glucose oxidase for direct electron transfer via site-specific gold nanoparticle conjugation. *J. Am. Chem. Soc.* 133 (48): 19262–19265. https://doi.org/10.1021/ja2071237.

142 Kizling, M., Dzwonek, M., Wieckowska, A., and Bilewicz, R. (2018). Size does mattermediation of electron transfer by gold clusters in bioelectrocatalysis. *ChemCatChem* 10 (9): 1988–1992. https://doi.org/10.1002/cctc.201800032.

143 Abad, J., Gass, M., Bleloch, A., and Schiffrin, D. (2009). Direct electron transfer to a metalloenzyme redox center coordinated to a monolayer-protected cluster. *J. Am. Chem. Soc.* 131 (29): 10229–10236. https://doi.org/10.1021/ja9026693.

144 Chakraborty, S., Babanova, S., Rocha, R. et al. (2015). A hybrid DNA-templated gold nanocluster for enhanced enzymatic reduction of oxygen. *J. Am. Chem. Soc.* 137 (36): 11678–11687. https://doi.org/10.1021/jacs.5b05338.

145 Kizling, M., Dzwonek, M., Wieckowska, A., and Bilewicz, R. (2018). Gold nanoparticles in bioelectrocatalysis - the role of nanoparticle size. *Curr. Opin. Electrochem.* 12: 113–120. https://doi.org/10.1016/j.coelec.2018.05.021.

146 Arduini, F., Micheli, L., Moscone, D. et al. (2016). Electrochemical biosensors based on nanomodified screen-printed electrodes: recent applications in clinical analysis. *TrAC, Trends Anal. Chem.* 79: 114–126. https://doi.org/10.1016/j.trac.2016.01.032.

147 Adachi, T., Kitazumi, Y., Shirai, O., and Kano, K. (2020). Direct electron transfer-type bioelectrocatalysis of redox enzymes at nanostructured electrodes. *Catalysis* 10 (2): https://doi.org/10.3390/catal10020236.

148 Baker, S., Colavita, P., Tse, K., and Hamers, R. (2006). Functionalized vertically aligned carbon nanofibers as scaffolds for immobilization and electrochemical detection of redox-active proteins. *Chem. Mater.* 18 (18): 4415–4422. https://doi.org/10.1021/cm0609000.

149 Zhang, L. (2008). Direct electrochemistry of cytochrome c at ordered macroporous active carbon electrode. *Biosens. Bioelectron.* 23 (11): 1610–1615. https://doi.org/10.1016/j.bios.2008.01.022.

150 Yang, C., Xu, C., and Wang, X. (2012). ZnO/Cu nanocomposite: a platform for direct electrochemistry of enzymes and biosensing applications. *Langmuir* 28 (9): 4580–4585. https://doi.org/10.1021/la2044202.

151 Xu, J., Shang, F., Luong, J. et al. (2010). Direct electrochemistry of horseradish peroxidase immobilized on a monolayer modified nanowire array electrode. *Biosens. Bioelectron.* 25 (6): 1313–1318. https://doi.org/10.1016/j.bios.2009.10.018.

152 Xiao, X., Si, P., and Magner, E. (2016). An overview of dealloyed nanoporous gold in bioelectrochemistry. *Bioelectrochemistry* 109: 117–126. https://doi.org/10.1016/j.bioelechem.2015.12.008.

153 Salaj-Kosla, U., Pöller, S., Schuhmann, W. et al. (2013). Direct electron transfer of trametes hirsuta laccase adsorbed at unmodified nanoporous gold electrodes. *Bioelectrochemistry* 91: 15–20. https://doi.org/10.1016/j.bioelechem.2012.11.001.

154 Salaj-Kosla, U., Pöller, S., Beyl, Y. et al. (2012). Direct electron transfer of bilirubin oxidase (myrothecium verrucaria) at an unmodified nanoporous gold biocathode. *Electrochem. Commun.* 16 (1): 92–95. https://doi.org/10.1016/j.elecom.2011.12.007.

155 Guo, H., Liu, D., Yu, X., and Xia, X. (2009). Direct electrochemistry and electrocatalysis of hemoglobin on nanostructured gold colloid-silk fibroin modified glassy carbon electrode. *Sens. Actuators, B-Chem.* 139 (2): 598–603. https://doi.org/10.1016/j.snb.2009.03.046.

156 Ruzgas, T., Gorton, L., Emneus, J., and Markovarga, G. (1995). Kinetic-models of horseradish-peroxidase action on a graphite electrode. *J. Electroanal. Chem.* 391 (1–2): 41–49. https://doi.org/10.1016/0022-0728(95)03930-F.

157 Léger, C. and Bertrand, P. (2008). Direct electrochemistry of redox enzymes as a tool for mechanistic studies. *Chem. Rev.* 108 (7): 2379–2438. https://doi.org/10.1021/cr0680742.

158 Bartlett, P. and PRATT, K. (1995). Theoretical treatment of diffusion and kinetics in amperometric immobilized enzyme electrodes. 1. Redox mediator entrapped within the film. *J. Electroanal. Chem.* 397 (1–2): 61–78. https://doi.org/10.1016/0022-0728(95)04236-7.

159 Calvo, E., Danilowicz, C., and Diaz, L. (1993). Enzyme catalysis at hydrogel-modified electrodes with redox polymer mediator. *J. Chem. Soc., Faraday Trans.* 89 (2): 377–384. https://doi.org/10.1039/ft9938900377.

160 Britz, D. and Strutwolf, J. (2013). Digital simulation of chronoamperometry at an electrode within a hemispherical polymer drop containing an enzyme: comparison of a hemispherical with a flat disk electrode. *Biosens. Bioelectron.* 50: 269–277. https://doi.org/10.1016/j.bios.2013.06.059.

161 Kapoor, S., Rafiq, A., and Sharma, S. (2017). Protein engineering and its applications in food industry. *Crit. Rev. Food Sci. Nutr.* 57 (11): 2321–2329. https://doi.org/10.1080/10408398.2014.1000481.

162 Wilding, M., Hong, N., Spence, M. et al. (2019). Protein engineering: the potential of remote mutations. *Biochem. Soc. Trans.* 47: 701–711. https://doi.org/10.1042/BST20180614.

163 Sequeiros-Borja, C., Surpeta, B., and Brezovsky, J. (2021). Recent advances in user-friendly computational tools to engineer protein function. *Brief. Bioinform.* 22 (3): https://doi.org/10.1093/bib/bbaa150.

164 Lemay-St-Denis, C., Doucet, N., and Pelletier, J. (2022). Integrating dynamics into enzyme engineering. *Protein Eng. Des. Sel.* 35: https://doi.org/10.1093/protein/gzac015.

165 Qiu, Y. and Wei, G. (2023). Artificial intelligence-aided protein engineering: from topological data analysis to deep protein language models. *Brief. Bioinform.* 24 (5): https://doi.org/10.1093/bib/bbad289.

166 Wuttke, D. and Gray, H. (1993). Protein engineering as a tool for understanding electron-transfer. *Curr. Opin. Struct. Biol.* 3 (4): 555–563. https://doi.org/10.1016/0959-440X(93)90083-W.

167 Gilardi, G. and Fantuzzi, A. (2001). Manipulating redox systems: application to nanotechnology. *Trends Biotechnol.* 19 (11): 468–476. https://doi.org/10.1016/S0167-7799(01)01769-3.

168 Güven, G., Prodanovic, R., and Schwaneberg, U. (2010). Protein engineering an option for enzymatic biofuel cell design. *Electroanalysis* 22 (7–8): 765. https://doi.org/10.1002/elan.200980017.

169 Akram, M., Rehman, J., and Hall, E. Engineered proteins for bioelectrochemistry. *Annu. Rev. Anal. Chem.* 7; Cooks, R., Pemberton, J., Eds.; 2014; Vol. 7: 257–274. https://doi.org/10.1146/annurev-anchem-071213-020143.

170 Zhu, Z., Song, H., Wang, Y., and Zhang, Y. (2022). Protein engineering for electrochemical biosensors. *Curr. Opin. Biotechnol.* 76. https://doi.org/10.1016/j.copbio.2022.102751.

171 Ranieri, A., Borsari, M., Casalini, S. et al. (2021). How to turn an electron transfer protein into a redox enzyme for biosensing. *Molecules* 26 (16): https://doi.org/10.3390/molecules26164950.

172 Gelo-Pujic, M., Kim, H., Butlin, N., and Palmore, G. (1999). Electrochemical studies of a truncated laccase produced in pichia pastoris. *Appl. Environ. Microbiol.* 65 (12): 5515–5521.

173 Hibino, Y., Kawai, S., Kitazumi, Y. et al. (2017). Construction of a protein-engineered variant of D-fructose dehydrogenase for direct electron transfer-type bioelectrocatalysis. *Electrochem. Commun.* 77: 112–115. https://doi.org/10.1016/j.elecom.2017.03.005.

174 Kaida, Y., Hibino, Y., Kitazumi, Y. et al. (2020). Discussion on direct electron transfer-type bioelectrocatalysis of downsized and axial-ligand exchanged variants of D-fructose dehydrogenase. *Electrochemistry* 88 (3): 195–199. https://doi.org/10.5796/electrochemistry.20-00029.

175 Courjean, O., Gao, F., and Mano, N. (2009). Deglycosylation of glucose oxidase for direct and efficient glucose electrooxidation on a glassy carbon electrode. *Angew. Chem. Int. Ed.* 48 (32): 5897–5899. https://doi.org/10.1002/anie.200902191.

176 Karnpakdee, K., Kracher, D., and Ludwig, R. (2024). Electron transfer of cellobiose dehydrogenase in polyethyleneimine films. *ChemElectroChem* 11 (8): https://doi.org/10.1002/celc.202300679.

177 Ortiz, R., Matsumura, H., Tasca, F. et al. (2012). Effect of deglycosylation of cellobiose dehydrogenases on the enhancement of direct electron transfer with electrodes. *Anal. Chem.* 84 (23): 10315–10323. https://doi.org/10.1021/ac3022899.

178 Zafar, M.N., Beden, N., Leech, D. et al. (2012). Characterization of different FAD-dependent glucose dehydrogenases for possible use in glucose-based biosensors and biofuel cells. *Anal. Bioanal. Chem.* 402 (6): 2069–2077. https://doi.org/10.1007/s00216-011-5650-7.

179 Algov, I. and Alfonta, L. (2021). Use of protein engineering to elucidate electron transfer pathways between proteins and electrodes. *ACS Meas. Sci. Au* 2 (2): 78–90. https://doi.org/10.1021/acsmeasuresciau.1c00038.

180 Algov, I., Grushka, J., Zarivach, R., and Alfonta, L. (2017). Highly efficient flavin-adenine dinucleotide glucose dehydrogenase fused to a minimal cytochrome C domain. *J. Am. Chem. Soc.* 139 (48): 17217–17220. https://doi.org/10.1021/jacs.7b07011.

181 Hiraka, K., Tsugawa, W., Asano, R. et al. (2021). Rational design of direct electron transfer type L-lactate dehydrogenase for the development of multiplexed biosensor. *Biosens. Bioelectron.* 176: https://doi.org/10.1016/j.bios.2020.112933.

182 Yanase, T., Okuda-Shimazaki, J., Mori, K. et al. (2020). Creation of a novel DET type FAD glucose dehydrogenase harboring *Escherichia coli* derived cytochrome b562 as an electron transfer domain. *Biochem. Biophys. Res. Commun.* 530 (1): 82–86. https://doi.org/10.1016/j.bbrc.2020.06.132.

183 Lee, Y., Baek, S., Lee, H. et al. (2018). Construction of uniform monolayer- and orientation-tunable enzyme electrode by a synthetic glucose dehydrogenase without electron-transfer subunit via optimized site-specific gold-binding peptide capable of direct electron transfer. *ACS Appl. Mater. Interfaces* 10 (34): 28615–28626. https://doi.org/10.1021/acsami.8b08876.

184 Cui, H., Zhang, L., Söder, D. et al. (2021). Rapid and oriented immobilization of laccases on electrodes via a methionine-rich peptide. *ACS Catal.* 11 (4): 2445–2453. https://doi.org/10.1021/acscatal.0c05490.

185 Llaudet, E., Darimont, D., Samba, R. et al. (2016). Expanding an efficient, electrically driven and CNT-tagged P450 system into the third dimension: a nanowired CNT-containing and enzyme-stabilising 3D sol-gel electrode. *ChemBioChem* 17 (14): 1367–1373. https://doi.org/10.1002/cbic.201600173.

186 Xu, S. and Minteer, S. (2013). Investigating the impact of multi-heme pyrroloquinoline quinone-aldehyde dehydrogenase orientation on direct bioelectrocatalysis via site specific enzyme immobilization. *ACS Catal.* 3 (8): 1756–1763. https://doi.org/10.1021/cs400316b.

187 Takamura, E., Nakamura, T., Sakamoto, H. et al. (2019). Effects of multicopper oxidase orientation in multiwalled carbon nanotube biocathodes on direct electron transfer. *Biotechnol. Appl. Biochem.* 66 (2): 137–141. https://doi.org/10.1002/bab.1710.

188 Schröper, F., Baumann, A., Offenhäusser, A., and Mayer, D. (2010). Bidirectional immobilization of affinity-tagged cytochrome C on electrode surfaces. *Chem. Commun.* 46 (29): 5295–5297. https://doi.org/10.1039/c0cc00850h.

189 Davis, J., Djuricic, D., Lo, K. et al. (2000). A scanning tunnelling study of immobilised cytochrome P450cam. *Faraday Discuss.* 116: 15–22. https://doi.org/10.1039/b001372m.

190 Beissenhirtz, M., Scheller, F., Viezzoli, M., and Lisdat, F. (2006). Engineered superoxide dismutase monomers for superoxide biosensor applications. *Anal. Chem.* 78 (3): 928–935. https://doi.org/10.1021/ac051465g.

191 Ma, S., Laurent, C., Meneghello, M. et al. (2019). Direct electron-transfer anisotropy of a site-specifically immobilized cellobiose dehydrogenase. *ACS Catal.* 9 (8): 7607–7615. https://doi.org/10.1021/acscatal.9b02014.

192 Al-Lolage, F., Bartlett, P., Gounel, S. et al. (2019). Site-directed immobilization of bilirubin oxidase for electrocatalytic oxygen reduction. *ACS Catal.* 9 (3): 2068–2078. https://doi.org/10.1021/acscatal.8b04340.

193 Madoz-Gúrpide, J., Abad, J., Fernández-Recio, J. et al. (2000). Modulation of electroenzymatic NADPH oxidation through oriented immobilization of ferredoxin: NADP+ reductase onto modified gold electrodes. *J. Am. Chem. Soc.* 122 (40): 9808–9817. https://doi.org/10.1021/ja001365m.

194 Schröper, F., Baumann, A., Offenhäusser, A., and Mayer, D. (2012). Direct electrochemistry of novel affinity-tag immobilized recombinant horse heart cytochrome C. *Biosens. Bioelectron.* 34 (1): 171–177. https://doi.org/10.1016/j.bios.2012.01.039.

195 Kadek, A., Kavan, D., Marcoux, J. et al. (2017). Interdomain electron transfer in cellobiose dehydrogenase is governed by surface electrostatics. *BBA-Gen. Subjects* 1861 (2): 157–167. https://doi.org/10.1016/j.bbagen.2016.11.016.

196 Durao, P., Bento, I., Fernandes, A. et al. (2006). Perturbations of the T1 copper site in the CotA laccase from bacillus subtilis:: structural, biochemical, enzymatic and stability studies. *J. Biol. Inorg. Chem.* 11 (4): 514–526. https://doi.org/10.1007/s00775-006-0102-0.

197 Kallio, J., Auer, S., Jänis, J. et al. (2009). Structure-function studies of a melanocarpus albomyces laccase suggest a pathway for oxidation of phenolic compounds. *J. Mol. Biol.* 392 (4): 895–909. https://doi.org/10.1016/j.jmb.2009.06.053.

198 Miura, Y., Tsujimura, S., Kurose, S. et al. (2009). Direct electrochemistry of CueO and its mutants at residues to and near type I Cu for oxygen-reducing biocathode. *Fuel Cells* 9 (1): 70–78. https://doi.org/10.1002/fuce.200800027.

199 Hosseinzadeh, P. and Lu, Y. (2016). Design and fine-tuning redox potentials of metalloproteins involved in electron transfer in bioenergetics. *BBA-Bioenergetics* 1857 (5): 557–581. https://doi.org/10.1016/j.bbabio.2015.08.006.

3

Extracellular Electron Transfer in Microbial Bioelectrocatalysis

César I. Torres[1,2], Christine Lewis[1], Juan F. Ortiz Medina[1], and Jesús A. Pérez García[1]

[1]Biodesign Swette Center for Environmental Biotechnology, Arizona State University, Tempe, AZ, USA
[2]School for Engineering of Matter, Transport, and Energy, Arizona State University, Tempe, AZ, USA

3.1 Introduction

Microorganisms are known to carry out complex metabolic and catabolic reactions that are of interest in various industries [1–5]. From the oxidation of complex organic molecules and pollutants, the synthesis of complex organics, to redox reactions involving nitrogen, sulfur, and many metals of interest, microorganisms hold the key to perform complex redox chemistries that are difficult to catalyze otherwise. The field of microbial electrochemistry has flourished in the last several decades based on the premise of connecting these complex, multienzymatic reactions to an electrode that either collects or delivers the electrons required [4, 6–8]. Yet, to make this connection possible, electrons need to flow in or out of the cell bodies through a process coined extracellular electron transfer (EET) [9–11]. In the field of microbial electrochemistry, EET is a crucial and unique component of every cell/electrode interaction and the focus of this chapter.

In 1911, Potter mimicked a photobiological reactor system and demonstrated cellular energy flow by constructing one of the first microbial fuel cells (MFCs) driven by the yeast *Saccharomyces cerevisiae* [12]. This work provided the evidence that electrodes paired with cellular metabolic processes were possible. After a long hiatus, photosynthetic research with isolated redox proteins gained popularity in the 1980s by artificially shuttling electrons within electrochemical photobioreactors to better understand photosynthesis [13]. It is not surprising that the 1990s mark this rise in interest because important discoveries have begun to unfold. New organisms capable of metal oxide respiration have been discovered and studied, including *Geobacter* and *Shewanella* [14, 15].

3.1.1 Electrons Through Insulating Cell Bodies

It is important to note that EET is not a natural process in most microorganisms. While most microorganisms rely on redox reactions and potential gradients to generate energy for growth, these gradients are handled from within the cell via specialized proteins and potential gradients that can generate enough energy for growth. Electrons, typically in the form of nicotinamide adenine dinucleotide (NADH), are transported along the electron transport chain, reaching an electron acceptor inside the cell, such as oxygen. The cell uses the potential gradient between electron donor and acceptor to create a proton-motive force. This proton translocation is converted back to chemical energy in the form of adenosine triphosphate (ATP), which the cell uses extensively to

Bioelectrocatalysis: From Electron Transfer Processes to Emerging Technological Applications, First Edition.
Edited by Shelley D. Minteer and Matteo Grattieri.

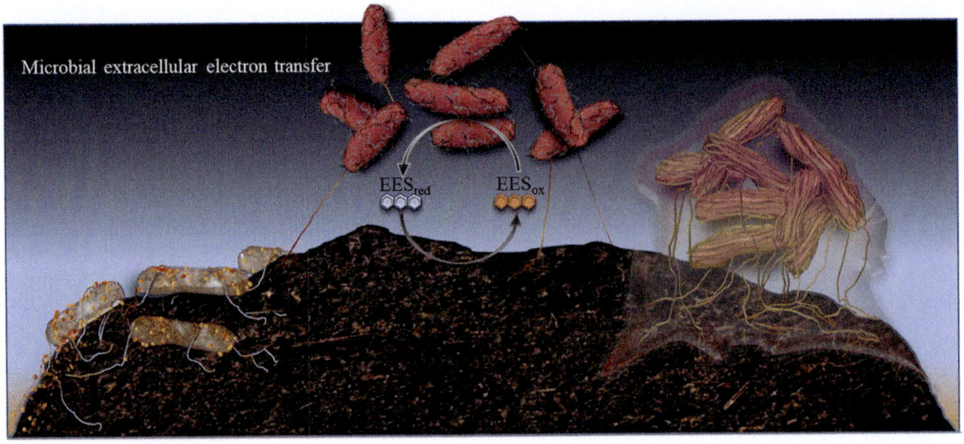

Figure 3.1 Extracellular electron transfer mechanisms. Direct contact of cells (left), electron shuttles or mediators (center), and a solid conductive network through microbial nanowires (right).

catalyze energy intensive reactions. This ATP generation method requires ionically impermeable membranes (phospholipids in most cases). Microbial cell bodies are electrically and ionically insulating, giving rise to the challenge of achieving EET. How can cells move electrons across insulating, lipidic membranes?

Some microorganisms have evolved EET approaches to achieve specific types of respiratory demands [11]. Membrane-anchored cytochromes initiate the transport of electrons outside the cell. In some cases, such as *Geobacter* and *Shewanella* species, electron acceptors can be metal oxides that cannot be transported into the cell; thus, electrons must be transported out of the cell at a high rate to meet the cell's respiratory demands [16]. In other cases, such as *Pseudomonas* species, electrons could be transported out of the cell in order to meet the electron acceptor outside of a biofilm or floc [17]. These interactions give rise to the evolution of natural and efficient EET mechanisms.

The early discoveries with *Geobacter* and *Shewanella* were not only groundbreaking, but they also opened mysteries behind how electrons moved. And so began the great debates – or dare we say – some grand scientific arguments. Were specialized proteins performing EET? Are electrons escorted chemically? If so – how? Were there electrochemically conducting nanowires? How are cytochromes involved in EET? Are all microorganisms using the same EET method? These questions, arguments, and the ensuing research gave rise to the general EET theories, as shown in Figure 3.1.

Three main EET theories are typically discussed in both natural and electrochemical systems: direct contact of cells, electron shuttles, and solid conductive matrix utilizing conductive appendages. It is important to understand these mechanisms as they can limit the performance of bioelectrochemical systems, and the approaches to alleviate these limitations are different for each mechanism [16, 18, 19]. In direct contact, it is assumed that outer membrane proteins capable of electron transfer can touch the conductive surface. As we outline further in Section 3.3, the outlook for direct contact is more complicated than has been traditionally explained. Direct contact alone leads to limited current densities due to the limited number of cells attaching [18]. As such, direct contact is only briefly discussed in this chapter. Electron shuttles (or mediators) are based on a small molecule that can be transported in and out of the cell and serves as a shuttle of electrons between the surface/electrode and the cell's electron transfer proteins [11]. These small molecules are synthesized by microbes to perform EET or can be added externally to help mediate bioelectrochemical connections in bacteria that do not perform EET naturally. Electron shuttles are transported by diffusion to the electrode surface, a process that could limit current production in these systems [20]. Ensuring efficient transport and decreasing transport distances are approaches that have improved current densities in mediated systems.

The discovery of microbial conductive nanowires, initially in *Geobacter sulfurreducens* [21], gave rise to a better understanding of EET in biofilms through a possible solid conductive matrix. This approach allows biofilms to grow on an electrode, with the cells far from the electrode surface participating in electrical current production. Yet, these cells must be attached and connected to the nanowire network to produce current. Despite being almost 20 years since the discovery of this mechanism [21], our knowledge on microbial nanowires is still continuously evolving and limited mostly to *G. sulfurreducens* [22] and a few extra characterizations [23]. Being composed of cytochromes, microbial nanowires are expected to be limited by turnover rates of these proteins or by the potential window they operate at. Yet, microorganisms using this EET mechanism are known to produce the highest current densities to date in microbial electrochemistry (~2 mA cm^{-2} at flat electrodes) due to the higher biomass that allows to be active.

Recent discoveries have expanded our vision of EET as a specialized metabolic approach. First, the discovery of multiple organisms that produce significant current densities on electrodes, both anodic and cathodic, has given a new perspective on the phylogeny of EET [24–27]. In fact, recent studies performing comparative genomics have developed a library of microorganisms that most likely have cytochrome nanowires [23]. Thus, we are still learning about novel microorganisms that interact naturally with an electrode through science and discovery. Second, the concept of direct interspecies electron transfer, in which microorganisms exchange electrons through conductive nanowires, has given rise to a new paradigm for EET utilization beyond the use of solid electron donors/acceptors. EET mostly between *Geobacter* species [28–30] and methanogenic archaea [28, 30–32] has been proposed to enhance interactions and flow of electrons to methane gas.

As reviewed before [18], current densities in microbial electrochemical technologies (METs) are high enough to make their biofilms among the most metabolically active ever recorded. This has led to multiple efforts to use METs for organic removal in wastewater industries, as discussed in Chapter 7. Yet, these rates are still considered low when compared to other bioenergy/bioprocessing technologies. Thus, there is an intrinsic need to understand how to overcome limitations in electrical current generation by electroactive bacteria [33, 34]. As we discuss in this chapter, EET is often limiting current generation, but this is not always the case.

In this chapter, we will expand on the concepts of EET based on model organisms and their knowledge starting with a section on understanding the electrochemical responses of electroactive bacteria. For natural electron shuttles, studies using *Pseudomonas aeruginosa* and *Shewanella oneidensis* are presented, providing an outlook on how these organisms connect to electrodes. Approaches where external electron shuttles are added to facilitate the connection of microorganisms to electrodes are also discussed. For solid conduction, the focus is on *G. sulfurreducens*, the most studied organism in microbial electrochemistry. In all cases, the focus is on our current understanding of how the mechanism works and the underlying equations used to predict their behavior.

The concept of EET has been reviewed extensively in the literature. As the knowledge on EET processes has evolved significantly over the last few years, we caution the reader to interpret each study based on the year it was published. In some cases, the reader is directed to other review publications that provide a different perspective or more extensive discussion on a topic related to EET [6, 9–11, 16, 18, 19, 35–42]. This chapter aims to introduce EET mechanisms and provide a kinetic perspective for the correct interpretation of electrochemical data in microbial electrochemistry.

3.2 Electrochemical Responses of Electroactive Bacteria

3.2.1 Extracellular Electron Shuttle Transport Under Substrate Excess Conditions

Figure 3.2 introduces the extracellular electron shuttle (EES) for microbial electrochemical transport in a cathodic system. An anodic model would be similar, except the redox loop would be in reverse. Modeling EES kinetics relies predominately on diffusive transport whereby

$$\frac{i}{A_e} = \frac{(nFDC_{ox})}{L} \tag{3.1}$$

where i is the current, A_e is the electroactive area, n is the number of electrons, F is Faraday's constant, D is the diffusion coefficient of the EES, C_{ox} is the concentration level of the oxidized EES (vs. reduced), and L is the distance of the diffusion boundary layer. Glasser et al. [36] theorized that certain thresholds can be met with the correct diffusion coefficients, EES concentrations, and individual cell concentrations.

Most EES systems aim to increase transport by decreasing limitations. Figure 3.2 demonstrates limitations at (1) the electrode, (2) and (6) the diffusion boundary layer, (3) and (5) the cell boundary layer, and (4) the cellular redox exchange point. In order to avoid limitation at the electrode (1), r_{cat} can be virtually instantaneous by setting an overpotential [43]. The (L) diffusion boundary layer (2) and (6) can be decreased by increased agitation levels. The diffusion coefficient for each EES can be determined by running different cyclic voltammetry (CV) scan rates and consequently applying the *Randles–Ševčík* equation [44–46]. Advances have been made in extending surface area, catalysis, and adding polymers [47, 48]. Additionally, Guzman et al. found increased current density by

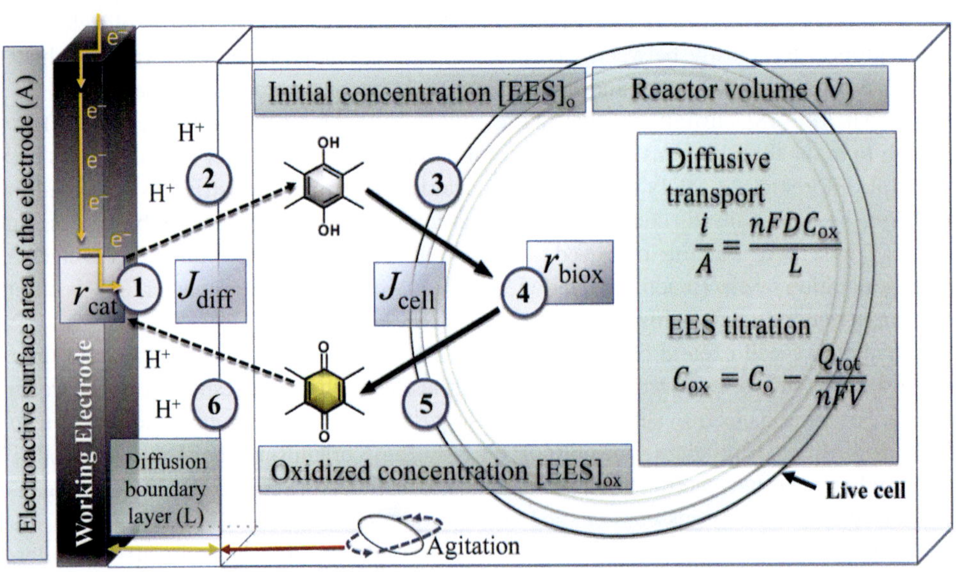

Figure 3.2 Example of a cathodic transport of EES, where J_{Diff} is the electron flux across the diffusion barrier, r_{cat} is the EES's reduction rate at the cathode, J_{cell} is the electron flux across barriers involved with the cell, r_{biox} is the rate at the cell, e^- are electrons, and H^+ are available protons. Diffusive transport parameters include current (i), electroactive area (A_e), diffusion boundary layer (L), number of electrons exchanged (n), diffusion coefficient (D), Faraday's constant (F), concentration of oxidized mediator (C_{ox}) or $[EES]_{ox}$, initial mediator concentration (C_o) or $[EES]_0$, and reactor volume (V).

coating electrodes with conductive poly (3,4-ethylenedioxythiophene)-coating carbon nanofiber, carbon cloth, and gold microfluid electrodes, resulting in 38%, 64%, and threefold improvements [49, 50]. Cell boundary layers (3) and (4) are more difficult for soluble EESs to contend with since the chemical environment is drastically changed (Figure 3.2). As discussed in the later sections for *Shewanella* or *Geobacter*, complex natural systems have evolved equally complex transport systems to overcome cellular barrier limitations.

The shape of the CV curves of the EES changes in the presence of the microorganism (represented by x), giving rise to a profile with a typical sigmoidal shape, which is characteristic of a catalytic process [51]. Therefore, the reactions for this process are determined by:

$$\mathrm{EES_{Red}} \underset{k_b}{\overset{k_f}{\rightleftharpoons}} \mathrm{EES_{Ox}} + ne^- \tag{3.2}$$

$$S + \mathrm{EES_{Ox}} \xrightarrow{x,k'} P + \mathrm{EES_{Red}} \tag{3.3}$$

where $\mathrm{EES_{Red}}$ is the reduced form of EES, $\mathrm{EEE_{Ox}}$ is the oxidized form of the EES, S is the substrate, P is the product, and k' is the kinetic constant rate of the electrochemical process in the biofilm by x. Considering that EES oxidation occurs at the electrode, and the reduction is a process carried out by the x, then the equation representing the kinetic current i, according to Torres et al. [52], is shown here modified by the definition of the limiting current i_{lim} by Rousseau et al. [53]:

$$i = i_{\mathrm{lim}} \left\{ \frac{1}{1 + \exp\left[\frac{nF(E-E^0)}{RT}\right]} \right\} = \frac{nFA_e\sqrt{8\left(\frac{k' C_{\mathrm{micro}} \Gamma_{\mathrm{micro}} A_e}{DC_{\mathrm{mediator}}}\right)}}{1 + \exp\left[\frac{nF(E-E^0)}{RT}\right]} \tag{3.4}$$

where R corresponds to the ideal gas constant, T is the absolute temperature, E is the potential applied in CV, E^0 is the standard thermodynamic potential of the EES, C_{micro} and Γ_{micro} are the concentration of x and the biofilm thickness, respectively, and C_{mediator} represents the concentration of the EES.

Thus, through the kinetic process given by k', Eq. (3.3) dominates over Eq. (3.2); therefore, the CV curve corresponds to that of Figure 3.3a (line in light blue color), while in the absence of x or S, the CV curve coincides with the profile of a reversible process (Figure 3.3a, dark blue color line). As in the thin-film profile, the response of microbial electrocatalysis is affected by the value of scan rate (v). This is because, at high values of v, diffusive process given by EES dominates the process. This occurs when k' is maintained at relatively low values in relation to v, causing the CV curves to lose their sigmoidal shape and return to the typical shape of a reversible process (Figure 3.3b). However, if the value of k' increases in such a way that the relationship $k'/v = \lambda$ is greater than 0, the catalytic profile is recovered because the regeneration rate of the EES dominates the process. According to Rountree et al. [55] for a molecular electrochemistry case, which can be analog for microbial electrochemical cases, the mechanism is maintained in the pure kinetic region. In this region, i does not depend on v, as indicated by Eq. (3.4) [55].

According to Pérez-García et al. [54], CV simulations using the definition given by Rousseau et al. [53] illustrate the change in the shape of the CV profile as k' function (Figure 3.3c). This CV profile coincides with electrochemical profiles described by Rountree et al. [55] and Savéant and Vianello [51], for a catalytic process in molecular electrocatalysis, where S does not determine the electron transfer rate when k' is high, since k' is an S function. This situation is named steady-state condition, and according to the authors, in order to gain access to this region, the S depletion on

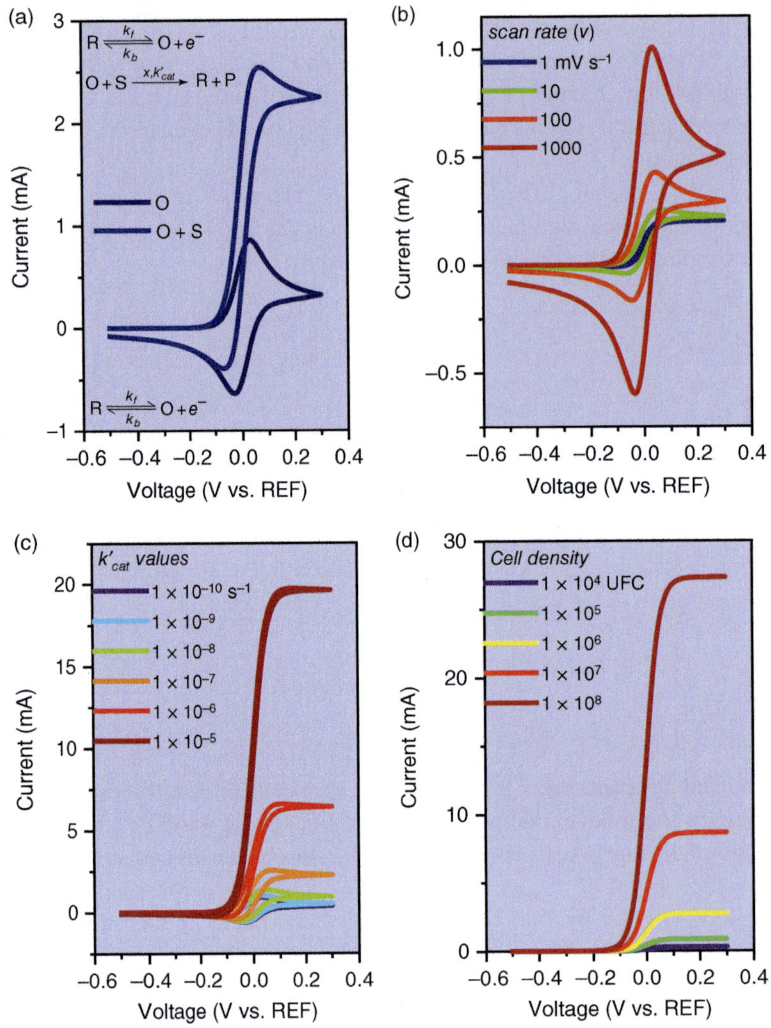

Figure 3.3 (a) Electrochemical response of soluble EES in the presence and absence of *x*. (b) Effect of *v* in the CV curve of an EES in the presence of *S* and *x*. (c) *k′* effect on the electrochemical profile of the EES. (d) C_{micro} effect on the CV curves of the EES. *Source:* Reprinted and adapted from Pérez-García et al. [54] Copyright 2021 Elsevier Inc.

the electrode must be avoided at all costs [55]. In the same way, it is possible to access a steady-state condition if there are high concentrations of *S* in relation to the EES concentration, in such a way that *S* remains steady-state [51].

Note that, according to Figure 3.3d, if the steady-state condition for *S* is maintained, then i_{lim} in CV curves are proportional to C_{micro}. So, for known values of C_{micro}, experiments at different values of *v* could lead to the experimental values of *k′* by analyzing i_{lim}. In real experiments, the Rousseau model has complications due to the high values of *x* demanding a high consumption of *S* [53]. However, a continuous feeding system could prevent such a problem. In addition, maintaining high values of *x* without interfering with the diffusion of the EES could help improve the catalytic process.

3.2.2 Effects of Substrate Depletion, Redox Mediators, and Catalyst Wear on EES-Mediated EET

Redox-mediated EET process often presents complications that make it difficult to obtain CV profiles that match the theoretical profiles, complicating the comparison between experimental measurements and adjustments. These complications are mainly due to additional contributions derived from adsorption processes, substrate depletion, cell density, and microbial affinity for the substrate. In addition, cell growth and the rate of microbial reduction or oxidation of EES can contribute to these complications as well. These contributions should be considered for a detailed and correct interpretation of the EET processes obtained in CV measurements. That is why, in the study of biological reactions, a steady-state approximation is generally applied due to the rapidity of these reactions and the maintenance of excess of S. However, it is important to recognize that this approach is not always valid in all systems [56]. S depletion can lead to a false interpretation of the reaction order given by S and x, being relevant in processes where high affinity for S leads to its fast, significant consumption [57]. When there is no depletion of S (steady-state condition), the CV response follows the classical sigmoidal behavior (Figure 3.4a). However, the response may deviate in this way due to secondary phenomena. For example, in Figure 3.4a, it is observed that the influence of S depletion increases as its concentration decreases [55]. The plateau stream shows an initial increase proportional to the concentration of S before reaching a limit, where the kinetic control changes from the reaction of EES reduction by x to the reaction of anode oxidation, thus following a behavior similar to the *Michaelis–Menten* equation for reaction rates [57].

This behavior has been previously observed in experimental studies. For example, Marsili et al. [58] studied a biofilm of *G. sulfurreducens* and observed that, when decreasing acetate substrate concentrations, the limiting anodic current increased linearly when acetate was <200 μM. However, at concentrations above 3 mM, the limiting anodic current did not change, suggesting a change in the reaction mechanism dependent on the acetate concentration (Figure 3.4c). In addition, Yang et al. [59] reported similar observations in biofilms of *S. oneidensis* by varying the concentration of the co-substrate or EES [pyocyanin (PYO)] under conditions of excess S. In this case, the output current increased proportionally with the increase in the concentration of the EES, but once the concentration exceeded 300 nM, the increase was significantly suppressed (Figure 3.4d) [59]. In contrast, Hamelers et al. [60] made observations about the dependence of the maximum current as a function of the concentration of S, and they were able to determine the apparent K_S (affinity constant for the substrate) by fitting curves. The values obtained for the apparent K_S were 0.37 mM at a potential of −0.4 V vs. Ag|AgCl, and 2.2 mM at −0.3 V vs. Ag|AgCl [60].

Additionally, it has been suggested that the microbial electrocatalysis process may be limited by the concentration of the catalyst x. This may be due to the cell life cycle, where electrocatalysis could be limited by the lag phase of x, and by cell death at the end of their life cycle. To address the breakdown of x, Marsili et al. [58] conducted experiments with different values of x on the electrode immobilizing the cells of *G. sulfurreducens* with pectin, which could fulfill an additional function as an EES. They observed that the plateau current increased with the increase of x in coatings, but approached saturation values by adding amounts greater than 1×10^{-9} cells cm^{-2} [61]. Likewise, Zheng et al. [62] studied biofilms of *P. aeruginosa* PAO1 with genetic modification, which resulted in a different catalytic behavior compared to unmodified biofilms. The current increased significantly in the genetic variant, although it failed to achieve sigmoidal behavior, suggesting an improvement in substrate mass transport, which would allow greater substrate assimilation and thus higher x at the electrode, leading to S depletion and generating the observed profile [62]. In contrast, Pérez-García et al. [63] demonstrated that mass transfer issues were not

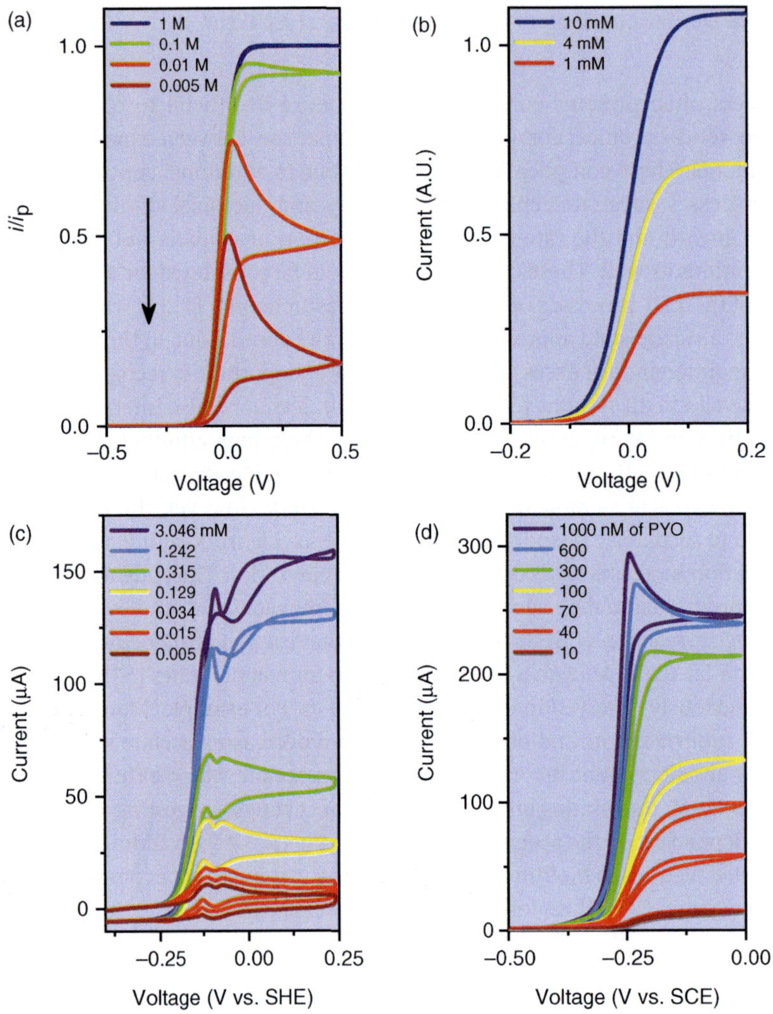

Figure 3.4 (a) Theoretical CV curves where the effect of *S* depletion on the CV response of the catalytic process is observed and (b) theoretical CV curves showing the effect of the catalyst concentration under steady-state conditions. *Source:* Adapted with permission from Rountree *et al.* [55] Copyright 2014 American Chemical Society. (c) Experimental CV curves of *Geobacter sulfurreducens* biofilms under varying acetate concentrations. *Source:* Adapted with permission from Marsili et al. [58] Copyright 2008 American Society for Microbiology. (d) Experimental CV curves of *Shewanella oneidensis* biofilm using pyocyanin (PYO) as EES, under varying PYO concentrations. *Source:* Reprinted from Yang *et al.* [59] Copyright 2017 Elsevier B.V.

related to *S* depletion, as glucose levels remained adequate. However, they suggested that PYO adsorption on the electrode might play a role in the observed limitations, although the specific mechanism remains unclear. Additionally, the shape of the CV profile revealed a limiting process possibly associated with biofilm deactivation. These observations align with the findings of Saunders et al. [64] and Meirelles and Newman [65], who highlighted those interactions between phenazines and components of the cellular microenvironment, such as extracellular DNA, that can influence redox functionality both positively and negatively. Meirelles and Newman [65] further observed that while high concentrations of PYO can cause cellular self-poisoning, under conditions of oxidant limitation, this EES can favor phenazine retention and biofilm survival.

3.2.3 Extracellular Electron Transfer Through a Solid Conductive Matrix

In contrast with redox-mediated EET, electroactive bacteria that form biofilms do not require the presence of mediators to transfer electrons. Instead, the biofilm itself acts as a conductive matrix that establishes direct contact with the electrode. The components of the biofilm matrix in anodic EET (the most researched biofilm EET), including nanowires that have been primarily studied in the model electroactive bacterium *G. sulfurreducens*, are discussed in more detail in Section 3.6.

As biofilm EET involves direct attachment between the redox catalysts (i.e. bacterial cells) and a conductive matrix, this process can be electrochemically modeled similarly to other catalysts that are attached to the electrodes that are not dependent on mass transfer (Figure 3.5). However, due to the biological activity of the biofilm, factors that influence cell metabolism must also be incorporated to model the current. This has led to the development of several models that incorporate both electrochemical and biological factors contributing to the observed current during biofilm EET. Marcus et al. [66] established the Nernst–Monod equation for anodic EET, in which the electrochemical potential for energy gain during EET, predicted through the Nernst equation, is incorporated into the biological growth frequently modeled using the Monod equation:

$$i = i_{max}\left(\frac{1}{1+\exp\left[-\frac{F}{RT}(E-E_{KA})\right]}\right)\left(\frac{S}{K_S+S}\right) \tag{3.5}$$

In this equation, i_{max} is the maximum possible current density (expressed in A m^{-2}), K_S is the empirical Monod half-saturation constant, S is the substrate (electron donor) concentration, and E_{KA} is the electron acceptor potential at which $i = \frac{1}{2}\, i_{max}$.

Under similar premises, Hamelers et al. [60] developed the Butler–Volmer–Monod equation, in which electron transfer kinetics predicted by the Butler–Volmer equation are utilized alongside the Monod equation instead.

$$i = i_{max}\left(\frac{1-\exp\left[-\frac{F}{RT}(E-E_{SP})\right]}{K_1\cdot\exp\left[-(1-\alpha)\frac{F}{RT}(E-E_{SP})\right]+K_2\cdot\exp\left[-\frac{F}{RT}(E-E_{SP})\right]+\frac{K_M}{S}+1}\right) \tag{3.6}$$

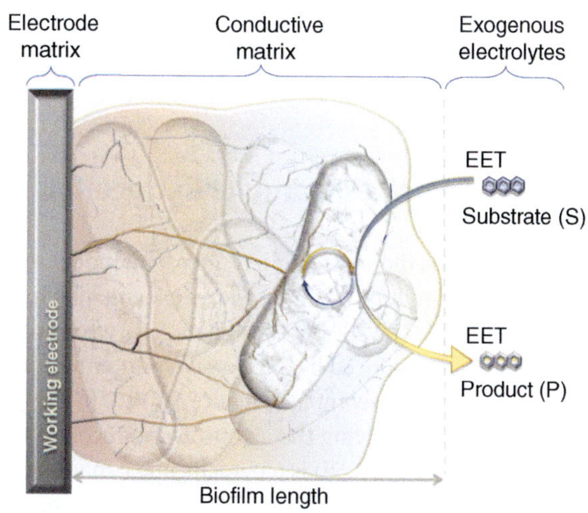

Figure 3.5 EET representation for microorganisms attached to the electrode through a biofilm.

In this case, E_{SP} is the theoretical anode potential at equilibrium conditions with the electron donor [67], α is the transfer coefficient from the Butler–Volmer equation, K_M is the apparent substrate affinity constant (derived from the enzymatic Michaelis–Menten equation), and K_1 and K_2 are kinetic parameters encompassing reactions between substrate and products (for a more detailed formula of these parameters and their significance, see [60]). Compared with the Nernst–Monod equation, the Monod parameters in Eq. (3.6) are dependent on the anode potential, providing a more accurate fitting under low overpotentials and when the biochemical rates are considerably faster than the electrochemical rates [60].

Other models that have been derived from these equations include the Nernst-ping-pong model, which accounts the effect of substrate in shifting half-wave potentials and possible inhibition by substrate [68], the multiple EET mechanism model (which is suitable for mixed communities and microorganisms with multiple EET pathways such as *S. oneidensis*), [69] and the integrative model developed by Korth et al., which pairs the Butler–Volmer equation with biochemical parameters, including the concentration of intracellular electron carriers such as NADH and cytochromes to provide a more specific contribution of biological reactions than the empiric Monod equation [70]. For the majority of the models, under non-limiting substrate conditions, the equations follow the sigmoidal shape frequently observed in biofilm EET experiments, meaning that most potential losses observed are intracellular, in contrast with mediated EET that is highly dependent on diffusion mechanisms [18].

Studies after the development of these models have discovered and outlined the presence of multiple respiratory pathways in *G. sulfurreducens* [71–73], often used concomitantly in a biofilm. These pathways have characteristic midpoint potentials within the range of −0.23 to −0.10 V vs. standard hydrogen electrode (SHE), and additional, uncharacterized pathways are expected to be discovered in the future. The presence of multiple pathways within a *G. sulfurreducens* biofilm might explain deviations from the theoretical Nernst–Monod model in its electrochemical response. Recent studies have now implemented the use of up to three individual Nernst–Monod equations (at the respective mid-point potentials) to model the biofilm's net electrochemical response [74].

As it will be observed in Section 3.6, the biofilm matrix is generally modeled as a conductor to semiconductor material with a conductivity (κ), in which most models agree that it should not influence current responses unless it possesses low values ($<0.5\ \mathrm{mS\,cm^{-1}}$) [18, 70]. Nevertheless, further understanding of electrochemical and biological components involved in biofilm EET, especially at high current densities, is still necessary to develop more comprehensive models of this process enabling to achieve more predictable current outputs that can be integrated into whole cell models for METs.

3.3 *Shewanella* sp. and its Extracellular Electron Shuttles

Shewanella sp. is the most studied microbial model to understand EES-mediated EET. This is due to its ability to produce flavin-derived molecules such as flavin mononucleotide (FMN), flavin adenine dinucleotide (FAD), and riboflavin (RF), which perform the function of EES. These extracellular mediators are reduced by *Shewanella's* outer membrane proteins, which belong to the MtrCAB and MtrFDE protein complex (Figure 3.6) [77]. The protein MtrA (Figure 3.6a) is a cytochrome *c* responsible for receiving the electrons produced in cellular respiration from inside the cell. The MtrC protein (Figure 3.6b) is a cytochrome *c* made up of 10 heme groups (deca-heme) located on the outer cell membrane, responsible for performing EET with the EES. Finally, the protein MtrB

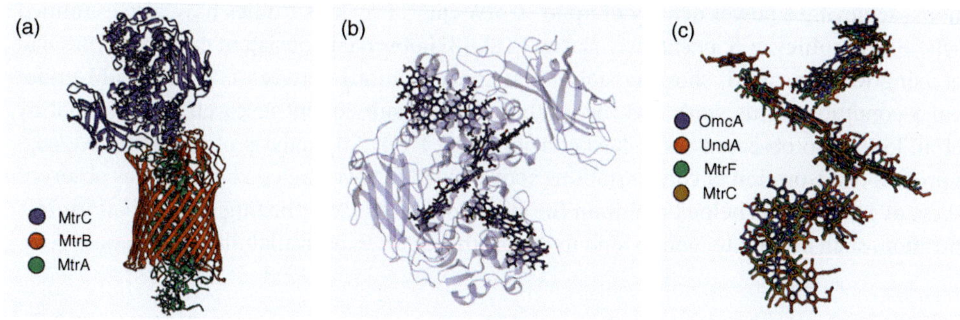

Figure 3.6 (a) Crystal structure of the MtrCAB complex. *Source:* Reprinted from Edwards et al. [75] Copyright 2020 Published by Elsevier Inc. (b) Arrangement of the heme group chain distributed within the MtrC crystal structure. *Source:* Reprinted from Clarke et al. [76] Copyright 2011 National Academy of Sciences. (c) Overlapping of the chain of heme groups of MtrC, MtrF, OmcA, and UndA, showing the step-cross heme distribution. Molecular structures were taken from PDB database, protein IDs: 6R2Q, 3PMQ, respectively, and for overlapping were used protein IDs: 4LMH, 3UCP, 3PMQ, and 4LM8, respectively. Figures were made using PDB 3D view.

is a porin crossing the outer membrane that organizes and stabilizes MtrA and MtrC, in addition to serving as an electron transport conduit between both cytochromes [78]. The MtrFDE complex has a similar structure. In this case, the deca-heme MtrF performs EET with EES.

During the process of EET, several cytochromes participate in transporting electrons to the extracellular environment. The tetra-heme CymA participates in the oxidation of quinones at the inner membrane. Periplasm protein mediators, SCT and FccA, oxidize CymA independently and diffuse to the MtrA/D to be oxidized. Subsequently, MtrA/D directs the electrons to the cyt c deca-*heme* MtrC/F through the outer membrane porin MtrB/E. The EET process is completed when the EES diffuses to the electrode and delivers the electrons. In addition to these protein complexes, *Shewanella* sp. may have other cytochromes such as OmcA and UndA, which are present in the vicinity of the MtrCAB and MtrFDE complexes. According to their molecular characteristics, these cytochromes are also capable of exchanging electrons with an EES [79].

The similarity in the crystal structures of MtrC, MtrF, OmcA, and UndA is remarkable. When analyzing their crystal structures, groups are arranged side by side in a sequence intended for the directional flow of electrons (Figure 3.6c). However, the arrangement is not simply a linear string of heme groups; rather, it presents a formation of "staggered cross" among them. This arrangement favors electron transfer among heme groups by having distances close enough for tunneling (<6 Å). The staggered arrangement of heme groups highlights the presence of potential binding sites for EES at opposite ends of the protein structure (Figure 3.6c). These ends, oriented toward the β-barrels (as in the MtrCAB and MtrFDE complexes), are designed to interact with extracellular redox molecules, including flavins or structurally similar compounds [75]. Electron transfer between the cytochromes and EES could occur through electrostatic interactions. These interactions result in a coupling between the EES and the heme group because of the difference in their redox potential. Such electronic coupling arises from the specific electrostatic environment of the heme group delivering electrons [78].

A few examples for the use of EES in *S. oneidensis* have been shown by Wang et al. [80], who developed a graphene electrode functionalized with RF, achieving a power density of 257 mW m^{-2} and current densities of 0.3 mA cm^{-2} by significantly enhancing charge transfer and cell adhesion. Zhang et al. [81] proposed an innovative concept of microbial flow fuel cells without biofilms, employing artificial EES in a fluid medium, which overcame mass transport and internal resistance limitations, achieving a power density of up to 10 mW cm^{-2}. Previous studies have also examined electrochemical profiles in *S. oneidensis*. Baron et al. [82] observed consistent catalytic sigmoidal profiles using RF as an EES, showing stability at concentrations between 0.25 and 1 μM under excess of *S* conditions. This work laid the groundwork for subsequent research, such as that by Yang et al. [59], who observed a similar behavior using PYO. An analysis of the CV curve suggests a process of adsorption at concentrations above 300 nM. Nevertheless, both studies observed that excess of *S* conditions helped maintain the sigmoidal profile, contrasting with variations in *S* concentration, such as acetate, which tend to distort these curves as availability decreases.

3.4 Phenazine-Mediated Extracellular Transfer in *Pseudomonas aeruginosa*

Recent studies have underscored the central role of PYO as an EES in EET in *P. aeruginosa*. Yong et al. [83] demonstrated that an integrated aerobic startup followed by anaerobic discharge in MFCs achieved a current density of 99.80 mA m^{-2}, representing a 91.6% increase compared to conventional anaerobic startup. This increase correlated with a maximum PYO concentration of 13.2 mg mL^{-1} in the early cycles, approximately 76% higher than conventional MFCs.

Zheng et al. [62] explored strategies to increase rhamnolipid production through overexpression of the *rhlA* gene, which enhanced PYO production and facilitated bacterial adhesion to the anode, improving EET. Although the study did not conclude that transfer issues were related to *S* depletion, the CV curves suggest this possibility when analyzed using the Nernst–Monod equation. Yong et al. [84] also highlighted cofactor manipulation through *nadE* overexpression, which increased PYO concentration by 50%, reduced charge transfer resistance, and improved electrochemical activity to 166.68 μW cm^{-2}. These findings emphasize the need for strategies to maximize PYO production, balancing its concentration to optimize conditions and mitigate adverse effects such as inhibition, and provide convincing evidence that EES concentrations limit current densities in *Pseudomonas* electroactive biofilms.

However, except for the study by Zheng et al. [62], CVs performed on the anode primarily revealed the oxidation and reduction processes of PYO, without evidence of significant catalysis or EET. Meirelles and Newman [65] emphasized that high concentrations of PYO do not guarantee improved EET performance and, instead, can cause cellular self-poisoning, affecting biofilm functionality. These results highlight the need to carefully optimize PYO concentrations to maximize its effectiveness while minimizing associated risks. For this reason, the biochemical process of EET of *P. aeruginosa* employing phenazines as EES has been intensively investigated, focusing on the process of microbial reduction of EES by transmembrane or intracellular metabolic enzymes. Research suggests that the reduction reaction of PYO could be catalyzed by flavoproteins and hemoproteins such as pyruvate dehydrogenase and α-ketoglutarate dehydrogenase. Glasser et al. [36] propose that these enzymes can reduce PYO and phenazine-1-carboxylic acid (PCA) during the EET process. According to Glasser et al. [85], the intracellular enzyme α-ketoglutarate dehydrogenase (LpdG) reduces PYO because of electron transport for ATP synthesis. Subsequently, the MexGHI–OpmD complex, which is a transmembrane porin, leads reduced PYO into the extracellular space (see Figure 3.7b) [85]. In contrast, in a study conducted by Chukwubuikem et al. [86], it is suggested that the oxidation of glucose to gluconate mediated by the periplasmic enzyme glucose dehydrogenase (Gcd) promotes the flow of extracellular electrons, taking advantage of the pyrroloquinoline quinone (PQQ) redox cycle. This molecule reduces the membrane transporter ubiquinone (UQ), which diffuses across the membrane to deliver electrons to cytochrome *bc1*,

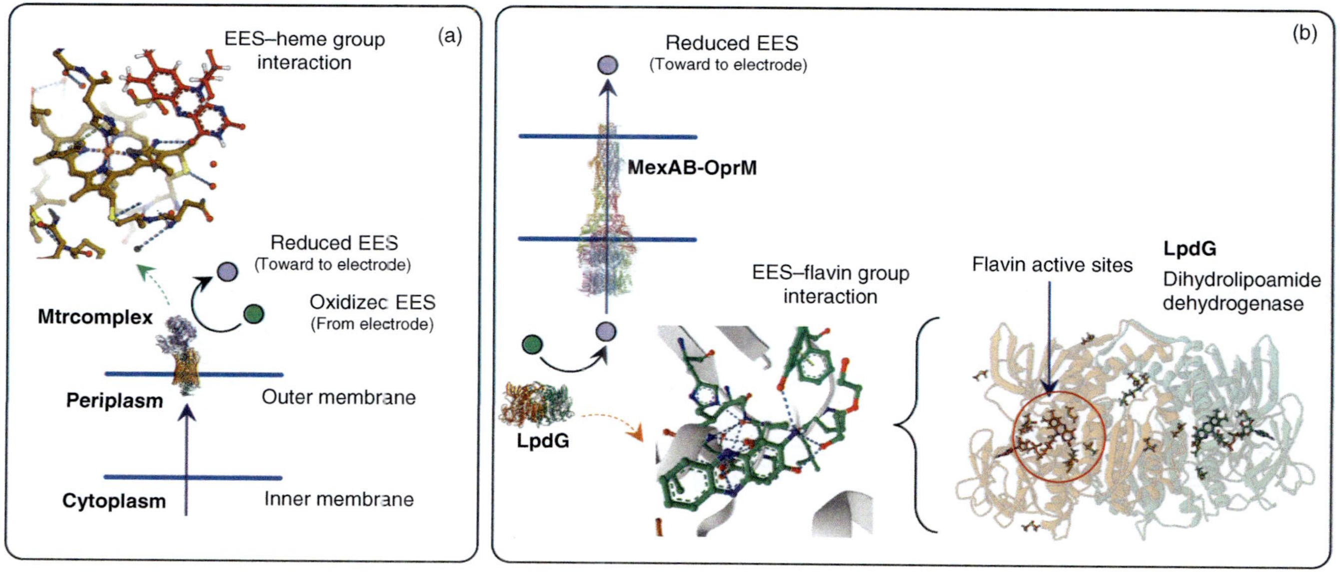

Figure 3.7 Protein – mediator interaction, representative comparison between *S. oneidensis* (a) – Molecular structure from Edwards et al. [75] and *P. aeruginosa* species (b) – Molecular structure from Glasser et al. [36]. Structures were obtained from the RSCB Protein Data Bank (PDB), protein IDs 6R2Q (MtrCAB complex), 6IOL (MexAB–OprM), and 5U8U (LpdG), and figures were made using 3D view from RSCB.

and thus integrates them into the electron transport chain. During this process, phenazines can function as electron acceptors in the PQQ cycle, and then promote oxygen reduction or be used for EET. In this case, once phenazines are oxidized in the extracellular environment, they return to the periplasm to be reduced and continue the cycle [86].

Therefore, *P. aeruginosa* employs intracellular enzymes to reduce phenazine-derived EES, including PYO. Notably, the reduction of PYO by α-ketoglutarate dehydrogenase is mediated by FAD groups in its active site (Figure 3.7b). Meanwhile, *S. oneidensis* utilizes outer membrane protein complexes to interact with flavin-based EES, where electron transfer is facilitated by heme groups within the cytochrome *c* proteins. The versatility of *P. aeruginosa* lies in its ability to adapt to various metabolic demands by employing multiple EET pathways. These include glucose oxidation via the PQQ cycle, where phenazines serve as versatile electron acceptors, enabling redox cycling in diverse environmental niches. Meanwhile, *S. oneidensis* achieves greater efficiency by minimizing energy losses through its tightly coupled cytochrome network mediated heme-EES interactions (Figure 3.7a). While *S. oneidensis* excels with its highly organized and efficient EET system, *P. aeruginosa* compensates with the chemical adaptability of its phenazine-based EES, enabling it to thrive across diverse environmental conditions (Figure 3.7b).

P. aeruginosa is known for its ability to form biofilms, in which it faces diverse, changing, and challenging conditions [87]. To counteract these conditions, biofilms define the immediate extracellular environment of microorganisms, and their composition and properties can influence the mobility and reactivity of EES. In this extracellular environment, aqueous diffusion competes with surface interactions, such as biofilm components, minerals, organic matter, or host tissue, thus determining the transport of EES. The characteristics of these complex environments, such as the redox state, pH, ionic strength, and temperature, will affect the behavior of EES, impacting their usefulness for the cell that produces them. In the specific case of *P. aeruginosa*, PYO is produced, which binds to extracellular DNA in the biofilm. This binding accelerates PYO-mediated electron transfer through the biofilm, creating a favorable PYO concentration gradient toward the biofilm depth (Figure 3.8). This mechanism retains PYO in the biofilm, preventing its loss to the outside. Phenazine electron transfer mediated to extracellular DNA can be an effective strategy to alleviate reductive stress in the anoxic depth of the biofilm, which allows to counteract the low availability of oxygen [64, 88].

In *P. aeruginosa* biofilms, pH can vary significantly as well. pH gradients from the depth to the edge of the biofilm could induce electrostatic changes that affect the penetration of small molecules (Figure 3.8) [89, 90]. The adsorption capacity of PYO and other phenazines on surfaces could induce beneficial electrostatic changes in mesopores, micropores, and the biofilm matrix. These changes could regulate the dynamics of mass transport, influencing pH, oxygen, ion diffusion, and substrates [91]. The improvement in the EET of *P. aeruginosa* when PCA is adsorbed in the biofilm environment, even at low concentrations ($\leq$1 μM), suggests that adsorption could play an important role, although this property has been little explored to date [92]. In this regard, electrochemical methods can play an important role in the exploration of EET.

3.5 Aiding the Movement of Electrons with the Addition of Extracellular Electron Shuttles (EESs)

3.5.1 Characterization of EES Candidates

Although certain microbial systems have evolved natural EET mechanisms, bioelectrochemical connections through extracellular electron shuttle(s) (EESs) can also be made for microbes

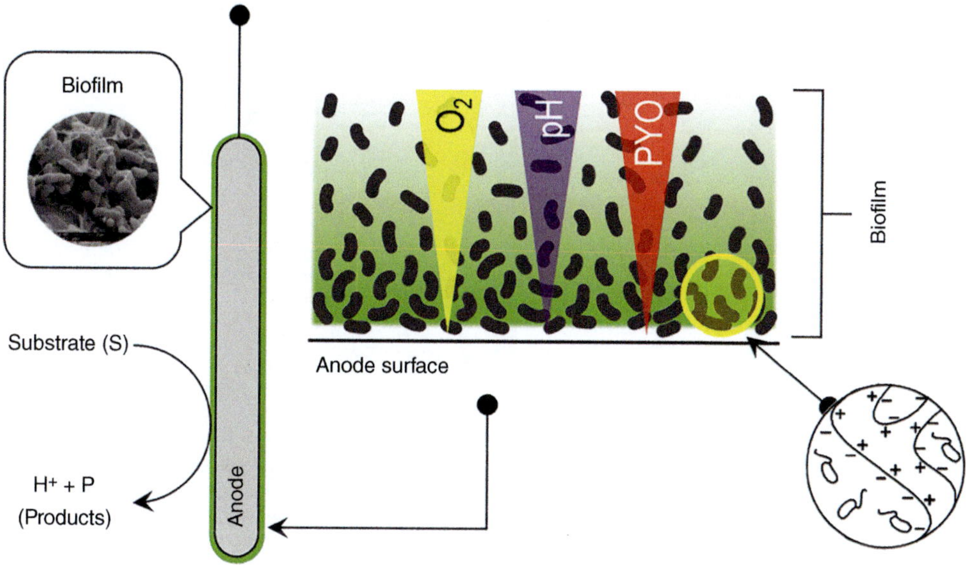

Figure 3.8 Representation of the interfaces present in the biofilm, concentration gradients, pH, and O_2. Close-up shows a representation of porosity and charge distribution in the biofilm.

that possess limited EET strategies. This section aims to better describe what to consider when using EES for microbial systems. Research will be highlighted to help demonstrate EES design and characterization. EESs reviewed in this section were found to generate between 0.001 and 0.600 mA cm^{-2}. However, high current densities may not be the only goal in the study of EESs. Jeuken et al. [93] emphasized that, in addition to respiration, EESs can be employed for $NADH/NAD^+$ redox balancing, nutrient uptake (like iron bioavailability), or to act as a reservoir for energy balance.

For hybrid/artificial or supplemented EES system, many approaches (see Figure 3.9) have been employed to introduce electron transport. Single EES molecules are often used, but it has been found that introducing cofactors or a combination of molecules can also enhance the EES's processes (as later discussed) [94–96]. An EES can be organic [97–99], inorganic [100], and/or charged to allow electrons to travel through diverse environments [101–103]. The shuttles can be simplistic, or complex large organic molecules, or even macromolecules (i.e. cytochromes) [100]. Some systems have EESs that are freely mobile, while others are tethered [50]. Other modifications of microbial systems involve cell surface modifications [104] and genetic and metabolic engineering [105]. To further complicate the matter, the shuttling process can occur in a single step or via multiple steps in the mediation pathway between a working electrode and the intracellular site.

3.5.2 What Makes a Good EES Candidate?

The selection of an EES for a cellular target should begin with a full understanding of its electrochemical potential [44], spectroscopic characteristics [106, 107], and structural contributions to its chemical behavior, along with some predictions on how it will interact with the microbial system it encounters. Therefore, it is best to gather all the necessary data to characterize the chosen EES. The characterization can be achieved through previously published experiments, or in the case of novel EESs, a fair amount of testing may be necessary to characterize them properly. A good EES candidate requires trade-offs between osmotic and toxic effects in order to maintain the cell's stability. The following points of characterization should be clarified before moving forward with any EES candidate.

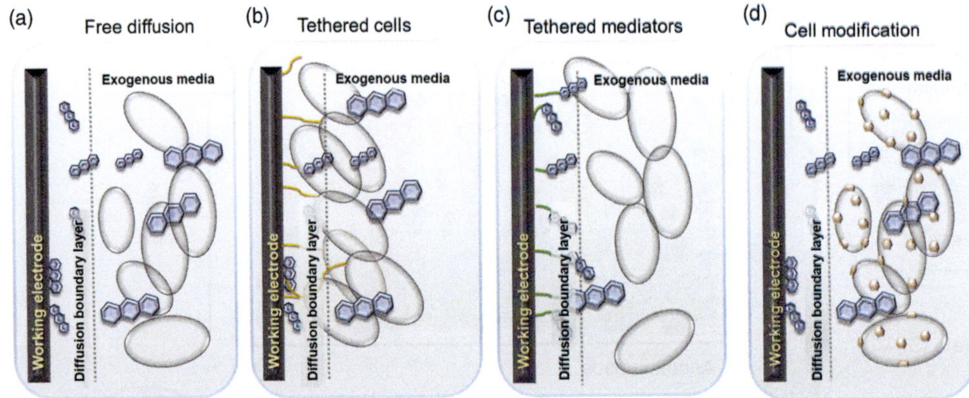

Figure 3.9 Methods to mediate electron transport in EESs. (a) Free diffusion governed by Fick's law in endogenous media and often limited by cellular barriers (i.e. cell membrane). (b) Tethering cells to overcome some of the issues with cell concentration near the diffusive boundary layers. (c) Mediators can also be tethered, reducing the diffusion boundary layer but may not react well with planktonic cells. (d) Cell modification where different manipulations of the cell are attempted to increase the turnover process and overcome barriers at the cell.

3.5.2.1 An EES Must Reduce/Oxidize at the Correct Cellular Target Potential Within Its Given Chemical Environment

The cellular target's potential ($E^{o\prime}$) and EES's $E^{o\prime}$ will need to have energetic correlation to work. Additionally, the EESs should rapidly cycle back to the electrode to regenerate the oxidized or reduced form. Often, in underdeveloped microbial systems, $E^{o\prime}$ is not well known. In those cases, a library of EES analogues (with small functional changes to tune them) may be necessary to screen for new cellular targets. Kracke et al. [96] provide an extensive list of eight well-known electrogenic microbes and the redox meeting points of previously characterized organic and inorganic compounds. It pairs the $E^{o\prime}$ of microbial redox points, the organic and inorganic electron donor/acceptor $E^{o\prime}$, as well as the inorganic and organic EES's $E^{o\prime}$ potentials for referencing. For novel EESs, understanding and characterizing the molecule may also be necessary. Most EES are organic and aromatic, and thus they can be altered with functional group additions or deletions. Functionalization alters the distribution of electron density, tuning its redox potential [103]. Functional groups that withdraw electron density from an aromatic ring are called *captors*. Captors will inductively pull electron density from a molecule's pi-orbital network. This will shift electrochemical potentials more negatively. Alternatively, functionalizations that act as electron donors are called *dative* groups and will increase in electron density into a molecule's pi-network of their aromatic ring. This promotes a positive shift in the electrochemical potential. When the structure possesses both captors and dative groups, it renders a *captodative* effect and can shift electrochemical potential differently than anticipated [108–110]. Further discussion on tuning redox potential is provided in the quinone section (3.5.3.2). Density functional theory (DFT) has commonly been employed to help predict where $E^{o\prime}$ will theoretically occur [101–103, 110, 111], in addition to experimental CV [44].

3.5.2.2 An EES Should be Both Electrochemically Reversible and Stable

Cyclic electron shuttling in the presence of a working electrode offers two key advantages. First, concentration levels of EESs can be lower, with less need for EES replacement. Second, one can monitor the concentration of the EES's charged state in real time. The reversibility must be considered in both aqueous and lipidic environments because microbial media is typically aqueous, while cellular redox targets are usually found in lipidic environments. Not only will the solvent affect the potential, but pH and electrolyte changes will shift EES potentials either positive or negative. A CV measured in the correct media will determine the potential at which the electron will be delivered. That potential also denotes the success of the redox exchange at the target site. Molecules with more negative potential will pass electrons to acceptors that have a more positive potential. Although it is true that the larger the difference, the faster this process will occur, there are limits to the overpotential as the driving force because the Marcus inverted region needs to be avoided.

3.5.2.3 An EES Must be Soluble and Diffusible with Limited Kinetic Loss to "Shuttle"

Most EESs are expected to diffuse through the exogenous media, lipid bilayers, and cytosolic environments. This requires consideration of the shuttle's size, stereochemistry, carbon count, functionalization, and hydrophilicity. Smaller organic molecules will diffuse more readily. Larger bulkier molecules may have issues with solubility and steric hindrance at redox exchange sites. Although carbon-laden molecules can diffuse through lipid bilayers more easily, they may have a propensity to permanently lodge within a lipid bilayer, and the cycling ability of the shuttling is lost. Alternatively, molecules that are too hydrophilic will be repulsed by lipid bilayers and the shuttling efficiency can decrease. Aqueous to lipophilic characterization through partition coefficient models aids in understanding the ability of an EES candidate to be soluble in different solvent environments. Ikonen et al. demonstrated the ease at which specialized electrochemical pipettes can be used to compare *P* values (ratio of partition equilibria in different media at the electrochemical interface) of different target molecules [112, 113]. New advances on DFT, using COSMO-RS (acronym for COnductor like Screening MOdel for Real Solvents), can predict partition coefficient and account for some accuracy of temperature change of larger particles (C-5 to C-10) without the need for in-lab experiments [114].

3.5.2.4 Ensure that an EES Addition is Nontoxic and Does Not Influence Other Cellular Processes

Many electron shuttles (i.e. quinones) are excellent electron shuttles but are well documented for their role in propagating reactive oxygen species (ROS), leading to several deleterious outcomes: lipid peroxidation, the destruction of proteins, and cell fatalities. As such, most quinone-based EES systems are run anaerobically to avoid negative outcomes [115–119]. Other EESs are known intercalators (that can disrupt signaling and nucleic acid processes). The ability to dispense redox charge exchange can have multiple outcomes when it comes to concentration, and there are levels at which the redox chemistry can coincide with unintended processes in the cell, thus rendering the EES toxic [120]. Photosynthetic and aerobic EES-MET systems often need to proceed with more caution due to ROS-related issues [105, 119, 121–124].

3.5.2.5 EES Must Function in Dynamic Systems

One must keep in mind that concentration ratios of oxidized to reduced shuttles are dynamic in the media and cellular compartments. We must also consider that there are competing detractions or amplifications on efficiency occurring (i.e. adsorption or electron hopping, respectively). Glassier et al. [36] provide the details to consider when factoring dynamic microbial systems.

3.5.3 Examples of Exogenous Electron Shuttles Used in METs

3.5.3.1 Humic Acid [$E^{o\prime} \sim -200$ to $+300$ mV vs. Standard Hydrogen Electrode (SHE)]

Humic acid (HA), named after humus (Latin for soil), is a byproduct of decaying biomass, and has been studied since the sixteenth century for its unique ability to affect biological organisms in both natural soils and bodies of water [125]. It was one of the first documented EESs known to

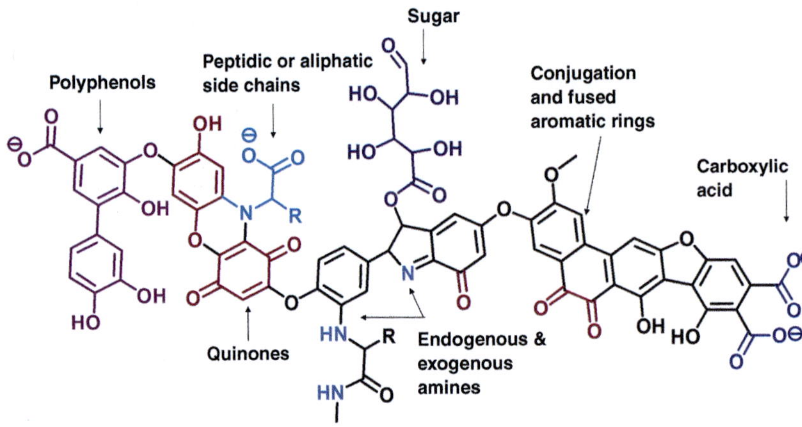

Figure 3.10 Humic acid with different redox components identified by color: polyphenols (purple), quinones (burgundy), peptic/aliphatic side chains (pink), sugar (dark violet), and carboxylic acids (blue) with an $E^{o\prime}$ ranging between −200 and + 300 mV vs. SHE.

aid in electron trafficking between microbes both found in soils or natural bodies of water [126]. When considering the complexity of HA's structure (Figure 3.10), it should be noted that it could include polyphenols, fused aromatic rings, conjugation, endogenous/exogenous amines, quinones, peptidic side chains, sugars, aliphatic side chains, and carboxylic acids with multiple redox sites. Consequently, various redox couples are present, ranging between −200 and + 300 mV vs. SHE. In the 1990s, HA was found to shuttle redox exchanges between microbial communities where iron reducing, fermenting, and sulfate reducing bacteria donate electrons [127], and alternative organisms or even an aerobic environment can then oxidize it [101]. In fermentative processes, HA has been shown to increase the oxidized to reduced product's ratio with increased ATP yield. Like any redox mediator, HA's ability to shuttle charge is affected by temperature, aerobic vs. anaerobic conditions, ionic strength, and pH. As the pH lowers, HA will become more hydrophobic and even create micelle-like environments. Although HA has been shown to reduce ROS, it is dynamic and decomposes further into *fulvic acid* and *humin*, and many of its functional groups will oxidize to smaller quinones that participate in propagating ROS [116].

3.5.3.2 Quinones ($E^{o\prime}$ ~ −300 to +200 mV vs. SHE)

Benzoquinones (BQ) and naphthoquinones (NQ) are found in natural electron transport systems where they are enzymatically prenylated (15 or 20 carbon chain additions; or combinations thereof) in order to anchor them within a lipid bilayer (Figure 3.11). Three main types of BQs are typically found: plastoquinones, UQs, and rhodoquinones. Plastoquinones are only evidenced within cyanobacteria and plants [128], while UQs [129] are the most widely distributed quinone found in both eukaryotic and prokaryotic organisms. Rhodoquinones are less frequent (found in select proteobacteria and a small number of eukaryotic organisms). NQs have fused bicyclic rings and are found in Gram-positive bacteria and archaea.

Natural electron transport chains utilize a range of quinones to arbitrarily shuttle electron flux between more precise redox protein machinery within lipidic environments (i.e. electron transport chains). These same redox sites can be excellent EES targets when aiming to interact with a specific metabolic system [105]. EES-based quinones should be unprenylated soluble structures, functionalized to tune the energy level up or down [130]. A wide variety of monocyclic, bicyclic, and tricyclic

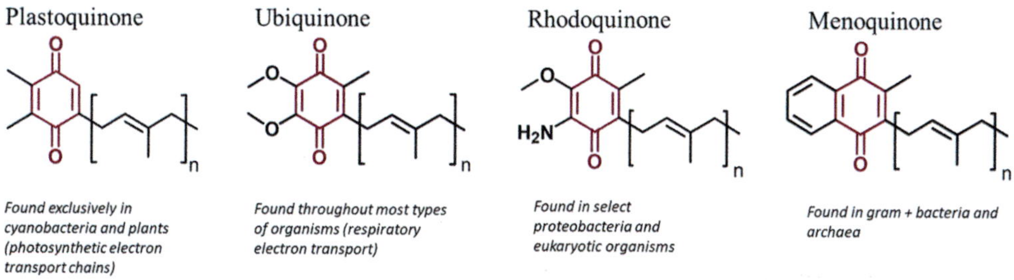

Figure 3.11 Plastoquinone, ubiquinone, rhodoquinone, and menaquinone are found in natural EES within electron transport chains.

BQs are utilized as EESs in MET systems. However, as previously mentioned, solubility can become an issue as carbon count gets larger.

Although quinones can double up their redox exchanges, they are sensitive to the mechanistic path they follow when exchanging electrons according to their surrounding pH, protic availability [131], and functional groups [128, 132–134]. This not only affects EES design but also governs how natural quinone shuttles are housed in EES-targeted electron transport chains. A full understanding of both cellular processes and EES structures provides insight into clever MET design. For example, 2,6-dichloro-[1,4]-benzoquinone (DCBQ), a well-known inhibitor of PSII in light activated photosynthesis (see Figure 3.12), can also function as a quinone shuttle. Longatte et al. [135] reasoned that they could use an anode and DCBQ as an EES in *Chlamydomonas reinhardtii* to cycle to the PSII-plastoquinone pocket (where DCBQ typically inhibits). They extracted a portion of the electron flux that was typically dissipated by natural photosynthesis and generated 0.06 mA cm^{-2} of current density. Natural quinone shuttles are compartmentalized, typically limited to cycling between specialized redox proteins, they can be regulated because they are sensitive to their nine redox states, and the interplay between pH, pK_a, and reduction potentials is important (Figure 3.12) [131]. Gaudin et al. [136] utilized pH, pK_a, and Pourbaix profiling to compare 56 different quinones to predict redox potentials. *Escherichia coli*, for example, can use demethylmenaquinone, menaquinone, and UQ disperse from the same location, except the EET is channeled via three distinctly different redox triggers [137].

If your EES or your redox target is a quinone in your microbial electron transport design, understanding how these nine redox states play a role is important (Figure 3.13). This is because it is essential to quantify the amount of charge that passes per EES molecule, if that number is ambiguous (because of semiquinone involvement), the explanation becomes complex. CV can help with defining the quinone state in your EES system [138]. Cited here are suggested literature to better understand quinone's innate characteristics [133, 138–143].

Huynh et al. [111] compared experimental CV measurements to DFT models [103], to demonstrate the accuracy of reductive potential prediction through understanding the structural contributions. The study compared 135 quinone structures to define the path of reductive potential in the presence of a protic solvent. In doing so, they found a linear correlation between the reduction potential of the first electron of the quinone with that of the Hammet potentials and pK_as. Functional groups that participated in hydrogen bonding shifted to more positive electrochemical potentials when compared to 2,4-*p*-benzoquinone, while molecules with more steric bulk (or higher carbon count) shifted more negatively. Halogenated quinones shifted negatively after $2e^-$ reductions occurred in protic solvent.

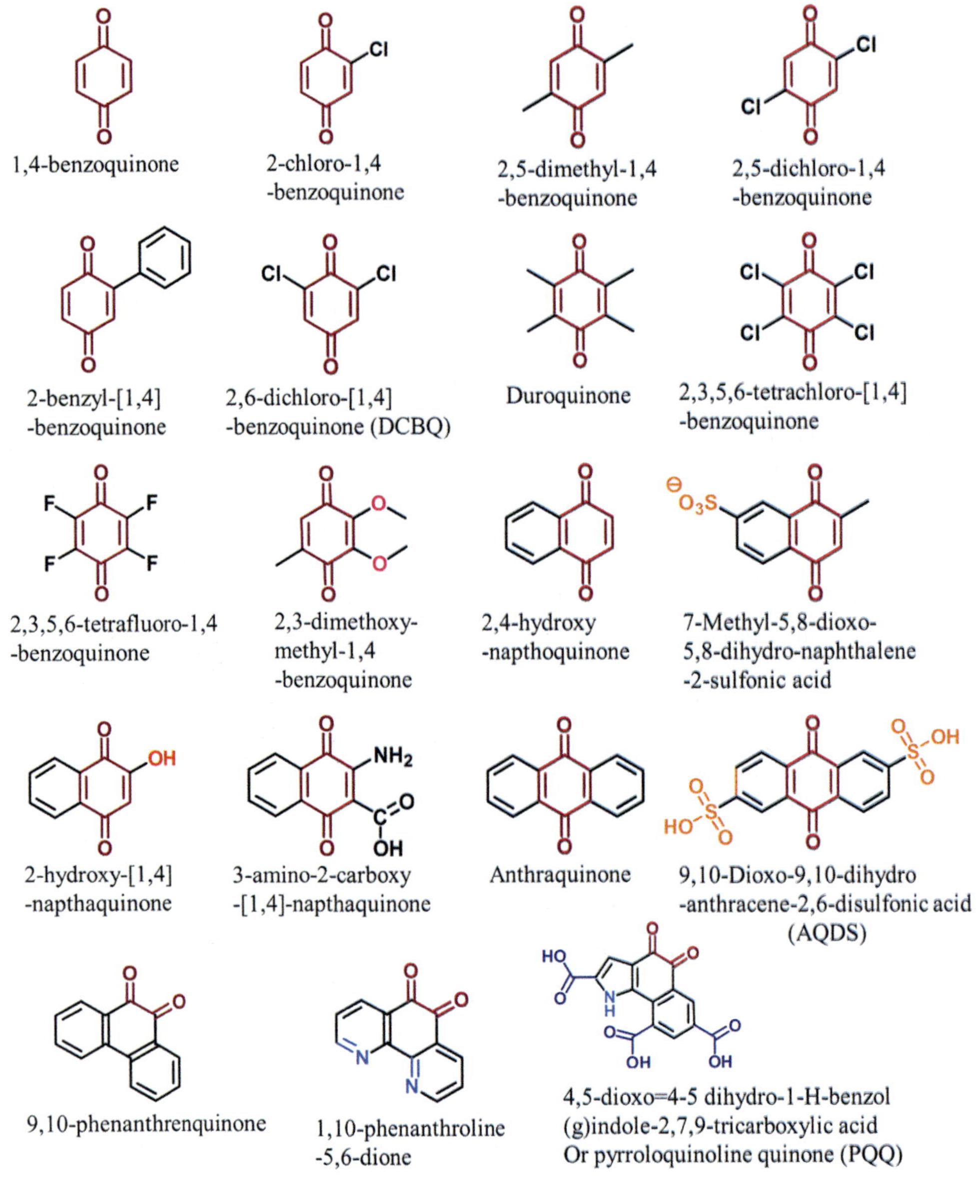

Figure 3.12 Common quinone EESs used in microbial shuttling.

Using an *in vitro* DET method, Takeda et al. [144] demonstrated that only one electron was transferred using a PQQ, thus producing semiquinones. Any prolonged period in the semiquinone state further complicates the process. For a microbial cell, this is important when considering the Faradaic efficiency, where the number of electrons passed is compared to the amount of yielded product [38, 43, 145, 146]. It is also important to note that quinones (especially those in a radical, high energy state) in combination with light and oxygen are infamous propagators of ROS, which will immediately break down lipid bilayers and important protein machinery [117, 123, 124, 147].

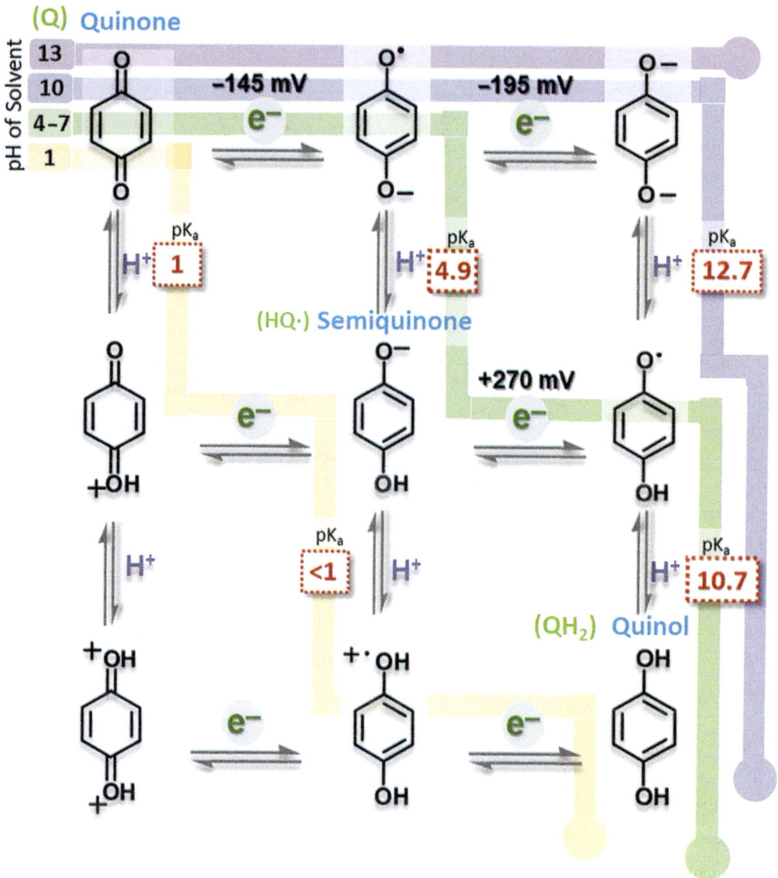

Figure 3.13 The nine redox states of quinone. Most microbes will be in an aqueous environment indicating a protic-assisted 2e$^-$, 2H$^+$ reduction; however, the pH quinone's pK_a may change the path can be dictated. At pH of 1 (gold), the quinone protonate (2H$^+$)→(2e$^-$) then reduces to a semiquinone. At pH of 4–7 (green), the reduction will proceed in a stepwise fashion 1e$^-$/1H$^+$ at a time. At pH 10 (blue), 2e$^-$ will reduce followed by 2H$^+$ additions, while at pH of 13 (purple), the quinone will oxidize completely if provided enough energy. This chart is the commonly known quinone redox chart, pK_a values are from Gunner et al. [96, 131].

3.5.3.3 Phenazines, Flavins, and Dyes (E°′ Range Approximately Between −100 and +500 mV)

Phenazine and flavin structures have integrated endogenous amines within the confines of fused tricyclic aromatic cores (Figure 3.14). Their color and the midpoint potential will range according to the functional groups present. Phenazines are found as natural secondary metabolites that can be biosynthesized by several bacteria to perform a variety of tasks. If considering phenazine as EES, their concentration levels and solubilities may be limited. DFT studies have shown that phenazines that are sulfonated or charged are more likely to solubilize than those that are not [102]. Mavrodi et al. [148] helped to characterize the biosynthesis pathway within *Pseudomonas*, *Burkholderia*, *Pectobacterium*, *Brevibacterium*, and *Streptomyces*, and consequently could conclude that most phenazine-producing bacterium was found to be either soil-dwelling or plant-associated species.

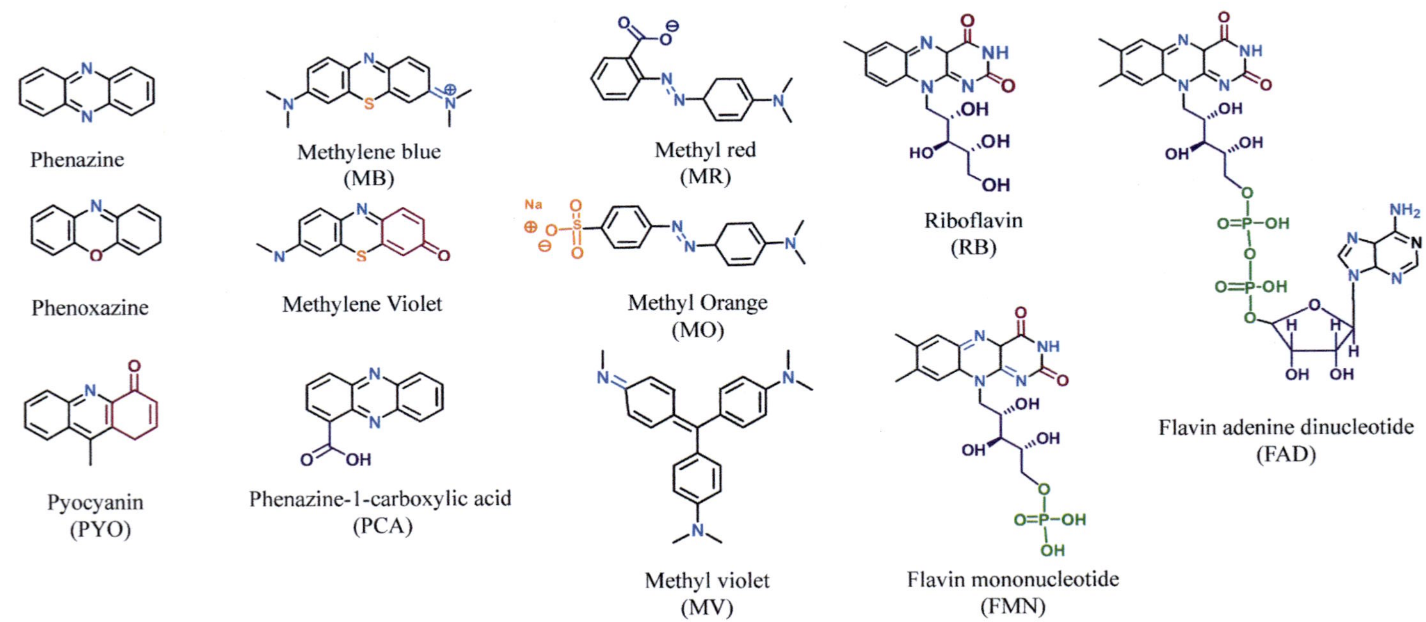

Figure 3.14 Common phenazines, flavins, and dye EES analogs.

Two endogenous phenazines, in particular, PCA and PYO, have been extensively studied because of their known biosynthesis and success in carrying charge for live cells. Clifford et al. [149] demonstrated that the use of phenazines in light activated systems can improve current density as much as fourfold in cyanobacteria. Their team then introduced the genetic modifications for endogenous PYO production but found that O_2 side reductions resulted, which competed with photocurrent generation. The results, however, showed promise in considering the design of different MET systems in the future. RF, also known as vitamin B_2, is a pteridine-based compound (typically yellow in color). It has a tricyclic isoalloxazine ring with four endogenous nitrogens and a sugar. Other well-studied flavins include *FMN* (flavin mononucleotide) and *FADH*$_2$ (flavin adenine dinucleotide). Flavins are reported in electron shuttling in diverse environments. Research indicates that often flavin activity can be complex as they can bind as a cofactor to incite different levels of one or two electron transfer processes. When not bound as a cofactor, flavins can also act as a simple EES [95, 150].

Many of the EESs that were first identified were known and named by their colors, as well as their noticeable color changes. When EESs cycle between the oxidized and reduced states, color changes are not uncommon. This is due to the distinct differences of the oxidized and reduced structures: resonance energy, conjugation, and HOMO → LUMO gap. If the light energy interacting with the reduced and oxidized state is absorbed and transmitted in the visible spectrum, we can visually detect these energy changes. In the case of many quinones, EESs change from colorless to colored. Glasser et al. [36] provide an informative review focusing on the chemistry and colors of EESs.

Nearly two decades ago, the use of neutral red (NR) as an EES was studied by Park and Zeikus [151]. They demonstrated that NR could extract electrons from *E. coli*'s NADH reserves and deliver them to the anode at a yield that was five times higher than that of thionine. NR is used in microbial electrochemistry studies in both anodic and cathodic configurations [152]. Simoska et al. [153] reported its use as an EES in *E. coli*. They demonstrated that not only does NR generate more current density, ~ 0.008 mA cm^{-2}, than phenazine (0.006 mA cm^{-2}), but also it did so at half the concentration of phenazine; 0.1 vs. 0.2 mM, respectively. Furthermore, NR enhanced growth rates, while phenazine increased toxicity. Zheng et al. tested three redox mediators (NR, RF, and methylene blue) for their ability to transfer electrons to *Actinobacillus succinogenes* for conversion of glycerol to succinate [154]. NR was found to be vastly superior to the other mediators in this case, proving current densities of up to ~0.1 mA cm^{-2}. Optimal product distribution was observed with 0.1 mM NR. Similarly, Ma et al. utilized NR as an EES for cathodic-assisted electro-fermentation to produce caproate from acetate. In contrast, methylene blue used by *E. coli* to enhance transfer to an anode resulted in much lower current densities (~0.00015 mA cm^{-2}), consistent with previous studies that outline the benefits of NR [155].

3.5.3.4 Ferrocene Analogs ($E^{o\prime}$~ −200 to +500 mV vs. SHE)

Ferrocene is an organometallic compound used in electrochemical systems. It is found to be stable in acidic, alkaline, or high temperature conditions and it has a reliably reproducible voltammetry [156]. When looking at its structure, the cyclopentadienyl rings are in a staggered configuration, lending to lone pairs that resonate. Due to the availability of the electrons in the unique sandwich configuration, ferrocene resists shifts of $E^{o\prime}$ regardless of the solvent system. Since ferrocene was discovered, it was adapted for use in cells and pharmaceutical research since it could be formulated to have low toxicity levels in humans [157, 158]. As its applications rose, other metal complexes, synthesized to be similarly sandwiched, were studied for their usefulness in sensors and redox shuttling [158]. In fact, ferrocene analogs are utilized for several human medicines for use as antibiotics – which should be a warning in microbial electrochemistry as it can also speed up oxidative reactions, like ferroptosis (iron-related cell death) [159].

Regardless, the practical applications of ferrocene's use as an internal reference for characterizing midpoint potentials are useful, and consequently cannot be discounted in the field of electrochemistry. It should be noted, however, that without functionalization, ferrocene alone is not soluble in aqueous solvents. In the case of aqueous solvents, the use of ferrocenylmethyl trimethylammonium (FcTMA+) or ferrocenecarboxylic acid (FcCOOH) is preferred [44, 160]. Ferrocenes stability remains quite useful in biosensor systems, motivating its use for *in vitro* bioelectrochemical applications [43]. Additionally, with regard to EES-based systems, they are found to be incorporated as conjugates to enhance electrodes or cellular membranes to decrease kinetic limitations [161–163]. In 2024, Zarenezhad et al. [164] constructed conjugated oligoelectrolyte EES-based system whereby they enhanced the cellular membrane with either [4-[1-(3-methylimidazolium chloride)]butyl] ferrocene (MIF+) or Bis [4-[1-(3-methylimidazolium chloride)]butyl] ferrocene (BMIF+). They found MIF+ conjugated cells were able to generate three to four times the power when compared to the control (73 $mW\,m^{-2}$) and 6.1 $mA\,cm^{-2}$ of current density. BMIF+ cells reached 115 $mW\,m^{-2}$, roughly three to four times the power of the control, and 11.9 $mA\,cm^{-2}$ of current density.

3.5.3.5 EES-MET Systems, EES with Genetic Modifications, and EES Effects on Biofilms

Simoska et al. [165] modified *E. coli* to express seven of the necessary gene loci and accessory genes for synthesis of PCA and PYO. This allowed for the endogenous expression of the PCA and PYO and resulted in 0.526 $mA\,cm^{-2}$, threefold higher than nonmodified cells [165]. Kim and Song et al. [166] compare the EES effect of supplementing a mixed culture containing both 2,4-hydroxy-naphthoquinone and NR. It was found that coulombic efficiency increased over a 3-month period by approximately 87% with EES additions to valorize acetate from CO_2, when compared to an unmediated system. However, the biofilm showed a high dead-to-live cell count and, therefore, it was determined that planktonic cells were more benefited. Zhu et al. [167] explain, however, that mobile EESs at the wrong concentrations can sometimes be inhibitory to biofilms. Their team progressively increased concentration levels of 9,10-anthraquinone-2-sulfonic acid (AQS) in the presence of *S. oneidensis* and found the EESs able to enhance biofilm growth but with inhibitory effects at higher concentrations. They modeled EES concentration levels using a Michaelis–Menten function to demonstrate how to find ideal EES concentrations to enhance biofilm formation. They measured a 9- to 17-fold increase in biomass with their methodology, a 21- to 80-fold increase in current density with the addition of AQS, and current density levels reached as high as 0.67 $mA\,cm^{-2}$ [167].

3.5.4 Closing Perspective on Exogenous EES

Discovery of EES in natural systems is emerging with diverse microbes for many types of purposes (i.e. respiration, nutrition uptake, energy balance, etc.) that may be useful to exploit. Making a suitable EES selection to connect cells to electrodes requires comprehensive knowledge and a complicated screening process. This, of course, begins with ensuring that the EES can properly cycle between the targeted redox process and the exogenous environment at the electrode. A thorough understanding of the EES's structure, its functional groups, and how pK_as, and pH affects redox exchanges is also necessary. Many molecules can serve as an EES; however, the effectiveness varies on the metabolic process targeted and the EES used. An EES should not be toxic. Toxicity concerns rise with EES concentrations and can become a limiting factor. A good EES-based MET will eliminate limitations and maximize efficiencies. Although research has shown that natural systems can use EES effectively, EES transport relies predominately on diffusive properties. Over time, researchers have focused on the cellular membrane and the structure of the molecules because they have been shown to be the most effective in transporting electrons at desired potentials. The utility of computer-based screening of the electrochemical characteristics of an EES (DFT) allows us to determine partition coefficients and the ability to predict diffusion through different chemical environments. These computer-based strategies aid in understanding and tuning large EES libraries for effective cell targeting. EES transport may be enhanced by cofactors (as in the flavin example) and alternative processes, like electron hopping. Artificial cellular membrane conjugates can be added to enhance EES transfer efficiency. Outer membranes have been modified with EESs to allow for faster diffusion. Endogenous EES loci have been introduced into new, versatile microbial chassis to produce new products of interest. New modeling of effective EES concentration levels aids us in understanding how to utilize biofilms effectively for higher current density production. These new advances in engineering approaches make EES-MET systems comparable to solid conduction systems. The search for optimal EESs continues and our ability to understand and design new molecules can lead to future breakthroughs in this field, allowing new metabolic interactions between cells and electrodes.

3.6 *Geobacter sulfurreducens* and Nanowires

From its isolation and characterization in 1994, *G. sulfurreducens* has been one of the most studied microorganisms known to perform EET by transferring respiration electrons to solid, conductive particles such as metals [14]. This capability has been harnessed to grow *G. sulfurreducens* in the anode of METs, where the electrons transferred by *G. sulfurreducens* drive the generation of electrical current [168]. It was quickly discovered that *G. sulfurreducens* produce high current densities (>0.15 mA cm^{-2}) compared to other electroactive bacteria [61, 169–171] and, likewise, *Geobacter* species are frequently enriched in high current-generating METs inoculated with sources such as soil and domestic wastewater when acetate (a nonfermentable substrate) is utilized as the electron donor [172–174]. *G. sulfurreducens* has thus become a model microorganism to study high-performance anodic EET.

One of the most defining characteristics of *G. sulfurreducens* is its ability to form thick, visible biofilms (>50 μm) when respiring on the anode of METs [175]. Besides direct contact with the anode, bacteria form biofilms to perform multilayered and long-distance EET, since *G. sulfurreducens* do not produce soluble mediators [176]. By analyzing EET kinetics involved in electron transfer to anode surfaces, Torres et al. concluded that, in order to generate the current density observed in typical METs, the only viable pathway is through several layers of bacteria within the biofilm, allowing them to transfer electrons over several micrometers [18]. As such, this biofilm has been defined as a solid conductive matrix or extracellular conductive matrix [18]. Manipulating matrix conductivity has directly resulted in anodic biofilms that yield higher current production [177]. Identifying bacterial components and their function within that matrix has become the main goal to improve electrochemical yields and maximize current production. Herein, we report the most recent microbial components and configurations within the extracellular conductive matrix of *G. sulfurreducens* as well as their role in establishing a high-current-density anodic biofilm.

3.6.1 Components of Extracellular Conductive Matrix of *G. sulfurreducens*

3.6.1.1 Protein Filaments (pili)

When thick biofilms (>20 μm) of *G. sulfurreducens* were observed to produce high current densities, one of the first objectives was to elucidate bacterial distribution within the biofilm as well as the mechanisms that allow cells in the external layers to transfer electrodes to the anode. After gene and protein expression analyses supported by microscopic observations, it was determined that *G. sulfurreducens* produces protein filaments, termed nanowires, when growing in insoluble metals and anodes instead of soluble electron acceptors such as fumarate [21, 175, 178]. Initially, it was proposed that nanowires were electrically conductive type IV pili composed of protein pilin (PilA) assemblies, which achieve electron transfer through electron hopping between adjacent (3.5–8.5 Å) aromatic residues [170, 178, 179]. Disturbing PilA structure through mutation of key-structural amino acids seemed to decrease current output by *G. sulfurreducens*, achieving a maximum current of <50% of the wild type [170, 179]. However, recent studies using cryo-electron microscopy (cryo-EM) and atomic force microscopy (AFM), showed that pilin assemblies were no different than nonconductive pili from other species, and the initially proposed conduction model through aromatic residues was found to be incompatible with the observed structures [39, 180]. Currently, there is a controversy on whether *G. sulfurreducens* type IV pili are conductive or not and/or whether they are used to conduct electrons to an anode or metal oxide surface [35, 181, 182]. Studies argue that previous observations of alleged pili were in fact multi-heme cytochrome nanowires, mainly OmcS, OmcZ, and OmcE, and even some filament observations were attributed to extracellular DNA (eDNA) [39, 180].

While *G. sulfurreducens* pili might not be electrically conductive, they still appear to contribute structurally to the formation of thick biofilms. As shown by Ye et al., when pili assembly is disrupted (through deleting the adjacent gene *pilB*), biofilm thickness decreased from 35 to 25 μm even though current per cell was maintained, resulting in an overall lower current (1.44 ± 0.02 vs. 1.63 ± 0.11 mA by the wild type) [187]. Additionally, it has been proposed that pili are instead related to cytochrome nanowire secretion and/or formation of filaments [180]. Other less studied (classified as "minor") pilin proteins, such as PilV, appear to similarly contribute to proper adhesion during biofilm formation as well as cytochrome expression, eventually impacting the maximum current generated (up to 63.2% decreased current production by the *pilV* mutant vs. the wild-type strain) [188]. Regardless of the lack of consensus about the actual conductivity of expressed pili, the role of these appendages in the establishment of high current biofilms must continue to be studied to establish its relationship with other conductive moieties.

3.6.1.2 Outer Membrane Cytochromes

G. sulfurreducens expresses multiheme *c*-type cytochromes to perform EET. These proteins typically have multiple heme (iron-containing) groups to receive electrons from inner and periplasmic *c*-type cytochromes, which likewise function as metabolic electron carriers across the inner cell membrane. Depending on the electron acceptor utilized, different outer membrane and extracellular cytochromes are expressed to allow EET to iron (III) and manganese (IV) oxides, including OmcB, OmcC, OmcE, OmcS, and OmcZ [40]. The last two are commonly expressed when *G. sulfurreducens* respires on anodes, becoming crucial components when cells are in direct contact with the electrode. However, there are differences between their structure and importance in long-distance EET within the extracellular conductive matrix.

One of the most studied extracellular cytochromes during anode respiration by *G. sulfurreducens* is OmcS, as this cytochrome is highly expressed when bacteria respire on insoluble materials

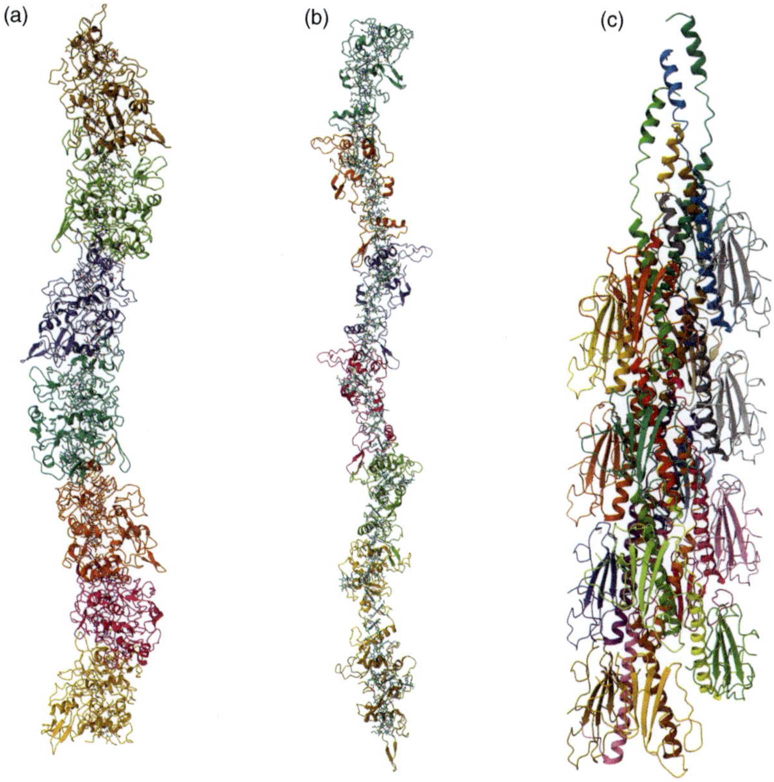

Figure 3.15 Structures of filaments that have been proposed to allow *Geobacter sulfurreducens* to perform long-distance extracellular electron transfer and are involved in establishing a high current, conductive biofilm matrix during anode respiration. Only filaments successfully observed using high-resolution cryo-EM are shown. The monomer component of each filament is, from left to right, (a) OmcS [183], (b) OmcZ [184], and (c) PilA [180]. Structures were obtained from the RSCB Protein Data Bank (PDB) [185], protein IDs: 6EF8, 8D9M, and 6VK9, respectively. *Source:* Images were created using the Mol* Viewer [186].

such as iron oxides. As a monomer, OmcS has six heme groups and possesses a measured *in vivo* reduction potential of − 0.13 V vs. SHE, which is more positive than periplasmic cytochromes to allow electron transfer [189]. Originally, OmcS monomers were proposed to be attached to pili to facilitate EET [190]. However, they are now confirmed to assemble directly into conductive nanowires with a diameter of ~3–4 nm (Figure 3.15) [181, 183]. Upon assembly, the heme cofactors appear to be stacked either in parallel (at a distance of 3.4–4.1 Å between heme centers) or T-stacked (5.4–6.1 Å between hemes) (Figure 3.16b) [183, 187]. Through this arrangement, the heme groups form a conductive core surrounded by the insulating protein, effectively acting like a two-end cable. The heme stack distribution also follows a pattern that coincides with other *c*-type cytochromes, such as the tetra-heme OmcE, which also forms extracellular filaments but is not apparently required to perform anodic EET [191, 192]. The potential range along the filament is around 360 mV, spanning a potential window from −0.31 to +0.05 V vs. SHE [189].

Comparatively, it appears that OmcZ has a higher importance for long-range electron transfer. OmcZ has been frequently associated with anode-respiring *G. sulfurreducens* [175], being detected in high concentrations at the interface between the biofilm and the anode [193], as well as capable of forming nanowire filaments [22, 187]. OmcZ possesses two forms: a high molecular weight (50 kDa) precursor form denominated as $OmcZ_{50}$, and a lighter form, $OmcZ_{30}$ (30 kDa), which is

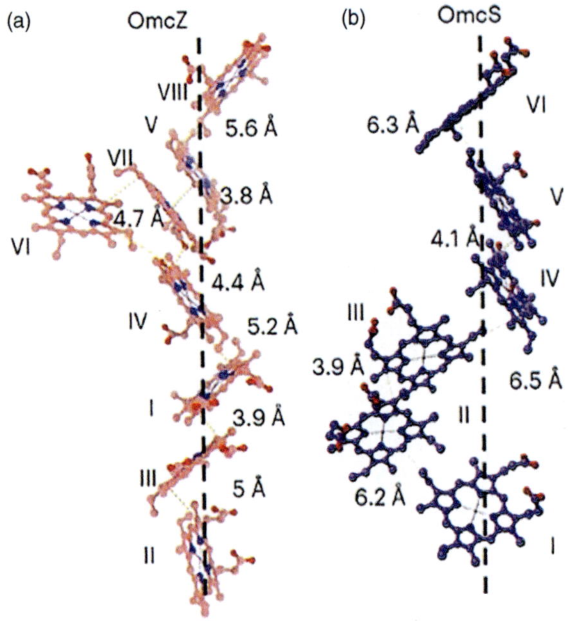

Figure 3.16 Heme center arrangement comparison between nanowires formed by (a) OmcZ and (b) OmcS. *Source:* Reprinted with permission from Gu et al. [22] Copyright 2023 Springer Nature.

produced when $OmcZ_{50}$ is cleaved by a serine protease [22]. The midpoint potential of $OmcZ_{30}$ is comparable to OmcS at −0.24 V vs. SHE *in vitro* [193]. Ultimately, $OmcZ_{30}$ is able to self-polymerize when secreted to form OmcZ nanowires. These nanowires have a diameter of 2.3 nm, which is comparatively smaller than OmcS nanowires (Figure 3.15), as well as being more compact, with the stacked OmcZ units allowing shorter distances between heme centers (~3.8 Å between hemes in parallel, and 5–5.6 Å when T-stacked) [22]. It is predicted that heme stacking, significantly different from OmcS and OmcE, exposes one heme out of the insulating protein cover (Figure 3.16a), which has been hypothesized to allow a branched connectivity [22]. OmcZ nanowires have been tested to possess a conductivity of >30 $S\,cm^{-1}$ and have been proven to possess both a structural and conductive role in the development of conductive biofilms, as OmcZ mutants severely decreased biofilm thickness (~18 μm) as well as current output (0.12 ± 0.01 mA) [187].

3.6.1.3 Other Conductive Matrix Components in Anodic *G. sulfurreducens* Biofilms

Besides cytochrome nanowires, there are other bacterial components whose importance in establishing an effective, conductive biofilm remains to be completely elucidated. The biofilm matrix is also composed by a polysaccharides network that participates in initial cell attachment, scaffolding, and even anchoring singular cytochrome nanowires (such as OmcZ) within the biofilm [194]. One of them is the lipopolysaccharides, which are attached to the outer membrane and whose modulation allows *G. sulfurreducens* to change its hydrophobicity to distribute cell growth within a biofilm [195]. Therefore, identifying other polysaccharides involved as well as the conditions that maximize their expression should allow a more effective biofilm establishment on anode surfaces to potentially support high current density-producing matrices.

3.6.2 Importance of Nanowire Conductivity in Establishing High-Current Biofilm Matrix

To assess the electrochemical performance of METs, it is assumed that anodes (e.g. carbon-based) pose as an "infinite" source of electron acceptor for the electroactive bacteria to transfer electrons.

As such, the energy gained for metabolic processes depends, in theory, on the difference between the anode potential (E_{AN}) and the oxidative potential of the electron donor, in the case of *G. sulfurreducens*, either acetate (−0.29 V vs. SHE at standard biological conditions) [196], formate (−0.43 V vs. SHE) [197], or hydrogen gas (−0.41 V vs. SHE) [196]. As described in Section 3.2.3, the current generated by metabolic respiration (i) has been modeled around equations that relate to the cell's metabolic activity, including the Nernst–Monod equation and the Butler–Volmer–Monod equations [60, 66]. However, losses between the outer-membrane cytochromes (the last electron storing components before EET) and the anode potential limit the effective metabolic potential to $E = E_{OM}$, and the difference between E_{OM} and E_{AN} results in energy dissipated as heat [18].

Due to its heterogeneous components within the biofilm, the extracellular conductive matrix is considered to possess semiconductive properties [conductivity (κ) between 10^{-6} and 10^{6} mS cm^{-1}]. As such, electron transfer rates [as current density (i)] through the biofilm have been initially modeled using Ohm's law to account for potential differences between the cellular cytochromes and the anode surface [18]:

$$i = -\frac{\kappa_{bio}(E_{OM} - E_{interface})}{\Delta z} \tag{3.7}$$

where κ_{bio} is the conductivity of the biofilm matrix, E_{OM} is the potential at the cell's outer membrane (before EET), $E_{interface}$ is the potential at the interface between biofilm and anode (which can be assumed equal to the anode potential, E_{AN}), and Δz is the distance between bacteria and the anode surface. Therefore, to agree with the current density observed in *G. sulfurreducens* biofilms, κ_{bio} should be at least 0.5 mS cm^{-1} to make potential losses negligible as seen in electrochemical studies [18, 66]. Table 3.1 includes reported κ values from diverse studies for the total biofilm matrix as well as the alleged conductive nanowires.

As expected, κ values of the biofilm confirm its conductive properties akin to a semiconductive matrix. Although pili offers comparable conductivity to OmcS and OmcZ nanowires, most of the studies have assigned the filaments to be composed of either of the latter proteins due to the similar size and structure observed [41]. Comparatively, cytochrome nanowires are generally several orders of magnitude more conductive than the total biofilm. It appears that OmcZ filaments possess a dramatically higher κ than OmcS and previously reported PilA. However, κ is calculated by normalizing through transversal area [198]. As OmcZ diameter is at least 1 nm smaller than OmcS and considering that only the heme groups are conductive, the insulating protein structure may obscure the true κ observed in these filaments [184]. Nevertheless, the high values reported so far suggest the possibility of further engineering *G. sulfurreducens* biofilms to increase its metabolic rates and hence its current density, as well as the potential isolation of filaments to manufacture conductive materials, which is currently an attractive option for the development of bioelectronics [10, 42]. However, there are some considerations that need to be accounted for regarding κ measurements and implications as discussed below.

3.6.3 Current Efforts to Understand Nanowire Utilization

While knowledge and structure of components that *G. sulfurreducens* use to form a conductive biofilm matrix when respiring on anodes is steadily being elucidated, there are many knowledge gaps that need to be filled to maximize this microorganism potential for use in METs. Some of the most important are expressed below:

a) Establishing a consensus on nanowire identification and systematic properties within the biofilm matrix.

Table 3.1 Examples of published conductivity (κ) values of nanowire filaments and conductive matrices (biofilms) of *Geobacter sulfurreducens*.

Constituting protein[a)]	Author(s)	Strain	Diameter (nm)	Measuring device	Maximum reported κ(mS cm^{-1})
OmcZ	Yalcin et al. [181]	Δ*omcS*	2.5 at pH 7	CP-AFM on patterned gold electrodes	3×10^4 at pH 7 4×10^5 at pH 2
OmcS	Wang et al. [183]	Wild type	~4	Two electrodes and a probe station connected to a semiconductor parameter analyzer	~30
	Yalcin et al. [181]	W51W57	3.5 at pH 7	CP-AFM on patterned gold electrodes	~20 at pH 7 300 at pH 2
PilA (unconfir-med)	Malvankar et al.[b)][198]	KN400	Not measured	Gap between gold anodes (dried pili)	$\sim 6 \times 10^{-3}$
		CL-1	Not measured	Gap between gold anodes (dried pili)	$\sim 6 \times 10^{-3}$ at 300 K ~1 at 270 K
	Lampa-Pastirk et al.[b)][199]	Wild type	2.5–3.5	CP-AFM on patterned gold electrodes	$1.4–4.3 \times 10^3$
	Adhikari et al.[b)][200]	Wild type	3	Gold nanoelectrodes	51.05 at pH 7 187.75 at pH 2 0.04 at pH 10.5
	Tan et al.[b)][201]	W51W57	1.5	Nanoelectrodes on imprinted lithography	9.77×10^5 at pH 2
	Gu et al. [180]	Δ*omcS*	6	Interdigitated array electrodes	~1
Biofilm	Malvankar et al. [198]	KN400	N/A	Gap between gold anodes	~5
		CL-1	N/A	Gap between gold anodes	~5 at 300 K ~0.5 at 280 K
	Ye et al. [187]	Wild type	N/A	Interdigitated diode arrays	1.06×10^{-3}

a) Constituting protein as reported by the original article. The label "unconfirmed" indicates that the constituting protein originally reported has been disputed later by an author.
b) In some studies, PilA was not confirmed to be the protein being analyzed.

Due to the heterogeneity and diversity of macromolecules within the conductive biofilm matrix, elucidating the functions of each component has been a complex and sometimes contradictory task, especially regarding nanowires. It is not clear if the nanowires characterized in early studies were misidentified as pili, as high-resolution visualization has provided a more accurate structure of both PilA and cytochrome filaments [39]. Currently, cytochrome nanowires appear to be directly responsible for forming a high current matrix, particularly OmcZ filaments, which are highly expressed during anode respiration and may provide both a conductive and structural role within the biofilm [187]. However, ongoing limitations of diverse characterization techniques such as preparation conditions for microscopic visualization, conductivity measurements, and the generation of mutant strains, still generate sources of contention to identify the structure and function of biofilm components [35, 42, 202]. It is acknowledged that, due to the biological nature of nanowires, conditions such as low pH and temperature (which are

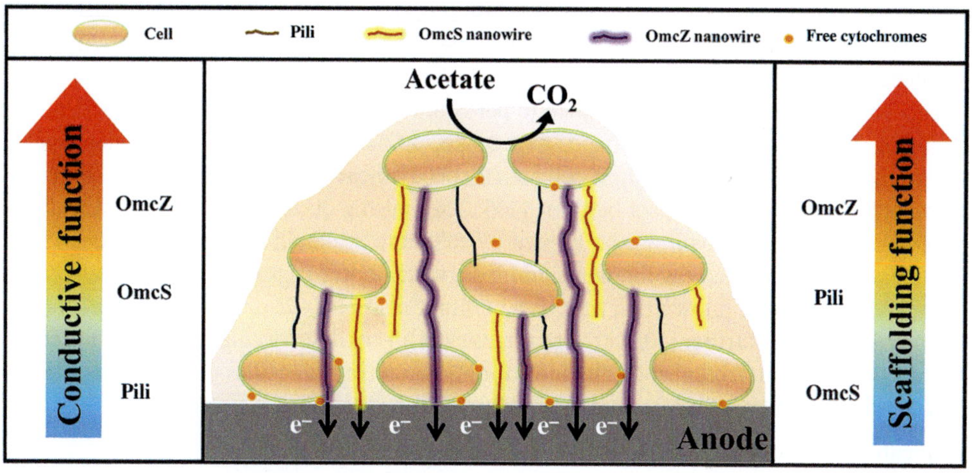

Figure 3.17 Proposed model of the extracellular biofilm matrix of anode-respiring *G. sulfurreducens* with pili, OmcS, and OmcZ filaments/nanowires serving either as a structural (in the case of pili and OmcZ) or conductive (OmcS and OmcZ nanowires) scaffolds for high current output. *Source:* Reproduced with permission from Ye et al. [187] Copyright 2022 The Authors https://creativecommons.org/licenses/by/4.0/.

sometimes needed for sample preparation) significantly alter filament structure and properties, as observed in conductivity (κ) values [181, 198]. Limitations in calculating κ, for example, the variability in probe distance during conductive probe AFM (CP-AFM), the measurement of air-dried samples, and the inherent normalization of conductance by transversal area (as previously discussed regarding OmcS vs. OmcZ nanowire conductivity) may also yield significantly different properties than what is observed *in vivo* [203]. Additionally, generation of gene-specific mutants may yield limited results and possibly false associations due to the multiple connections between components, as demonstrated when deletions of *pilA* affected nanowire formation due to its cytochrome nanowire-exporting properties [41, 180]. It is thus necessary to study further the regulatory mechanisms as well as the role of each component within the cell that leads to the development of conductive biofilm matrices.

Integrating the role of pili along with OmcS and OmcZ nanowires, a conciliatory model has been proposed by Ye et al., in which pili contribute to the structural integrity of the biofilm, OmcS filaments provide short-distance electron transfer, while OmcZ provides both structural and long-distance electron transfer (Figure 3.17) [187]. Yet, a similar model with the opposite placement of OmcZ and OmcS has been proposed by Yalcin et al. [41], based on previous studies that show OmcZ localized near the anode [193], while OmcS deletion inhibited production in early biofilm development. Thus, the specific spatial and electrochemical distribution of EET proteins is yet to be elucidated.

There is still limited information on connections between components, for example, nanowires connecting cells and biofilm-bound cytochromes, or even between different nanowire filaments. It is predicted from the currently proposed OmcZ nanowire structure that branched electron transmission can potentially occur through these filaments [184]. This may explain the increased current density observed in anode-enriched strains such as KN400 (7.6 vs. $1.4\,A\,m^{-2}$ by the wild type) [204] which allegedly expresses more OmcZ filaments than OmcS [41, 181]. However, this connecting model is yet to be proven. OmcZ structure may also facilitate proton-coupled electron transfer in nanowires, as suggested by Guberman-Pfeffer [42], but there is limited information about this capability in external cytochromes. It is also proposed that OmcS nanowires participate in interspecies electron transfer [41], adding another level of complexity to a precise representation of components within the biofilm.

Although every component may be adapted to perform different functions, the biofilm itself is far from being a metabolically homogeneous matrix. Recent nanoscale secondary ion mass spectrometry studies have shown that highest cell activity occurs near the anode surface and current production is higher at more compact biofilms [205, 206], raising questions on whether the cells in the outer layers (which should not be limited by substrate availability nor pH gradients) are able to express higher conductivity filaments and if so, which other factors may limit its electron transfer capability. It is well known that there is a biofilm thickness (~20 μm) at which the anodic current does not increase further [205, 207], so the roles of all conductive components and limitations within the whole matrix must be elucidated if higher current densities on *Geobacter* METs are desired and electron transfer is indeed limited by the distance from the electrode.

b) Identification and expression of nanowires beyond *Geobacter* species.
   Besides the high current density produced by *G. sulfurreducens* biofilms, one characteristic that has allowed this bacterium to be studied in further detail is that its genome has already been sequenced [208]. A genetic system has already been developed to create specific mutations aiming to study determined protein components [209]. By comparison, EET mechanisms in other microorganisms are considerably less understood, which is important to study further given the diversity of high-current producing electroactive bacteria that can be utilized in METs and that may transfer electrons through a conductive matrix with a similar configuration [210]. For example, *Thermincola ferriacetica* is capable of generating similar current densities as *G. sulfurreducens* [26, 211]. As this bacterium is classified as a thermophile (i.e. grows at high temperatures), it can be potentially utilized to design METs that process high-temperature influents such as textile industry wastewater [212]. Because *T. ferriacetica* is a Gram-positive bacterium, its electron transfer mechanisms between the inner membrane and the external components of its cell wall are expected to be notably different than *G. sulfurreducens*, requiring further understanding of the involved mechanisms to properly harness its capabilities in METs. Additionally, OmcZ appears to have homolog gene sequences spread across bacterial and archaeal species [22], suggesting diverse bacterial biofilms can utilize and optimize EET performance in METs aimed to perform specific functions (e.g. anaerobic methane oxidation) [22].
   Another option to harness high current density through nanowires is to express *G. sulfurreducens* genes in other microorganisms through gene-editing tools. By conferring this capability to other bacteria, a wider array of properties can be supplemented like the utilization of fermentable substrates, as *G. sulfurreducens* is limited to grow by using simple substrates such as acetate and hydrogen as electron donors [213]. Recently, OmcS from *G. sulfurreducens* was successfully expressed in *S. oneidensis* to increase its power output by threefold when compared to the wild type, potentially harnessing the capabilities of *S. oneidensis* (such as rapid growth, the use of an alternate substrate such as lactate, and higher oxygen tolerance) toward production of conductive filaments for bioelectronics [214]. Utilizing synthetic biology to confer electron-transfer capabilities to non-EET microorganisms that are easier to manipulate (e.g. *Escherichia coli* [215]) is likely to become a more common strategy to elucidate functions within the conductive biofilm matrix, as well as to enhance performance and potentially create novel processes for METs within the next years.
c) Is EET limiting current densities in *G. sulfurreducens* biofilms?
   Most of the conductivity measurements performed so far would indicate that EET should not be a rate-limiting step in current production by *G. sulfurreducens*. Based on our own

observations [216] and images of cells with nanowires, we estimate a nanowire density of at least 20 nanowires μm^{-2}. Yet, nanowires close to the anode would require transporting electrons from the cells above, which is a possible explanation why the more conductive OmcZ is found closer to the anode surface [193]. At a typical current density of 10 pA μm^{-2} (1 mA cm^{-2}), each nanowire would need to transport a current of 500 fA, but only near the electrode surface. Even lower conductivities shown by OmcS would lead to a potential loss of only ~2 mV mm^{-1} at the highest demand near the surface, or just 18 mV over a 50-mm-thick biofilm. This small potential loss would not be significant enough to change the Nernstian response observed by *G. sulfurreducens*. When measurements of single nanowires over short distances show their ability to transfer current at the nA scale, it is difficult to imagine that this is a rate-limiting step in anode respiration by *G. sulfurreducens*.

It is important to note that the reported κ values in Table 3.1 and calculations above are based on an Ohm's law correlation. Indeed, most of the measurements performed so far show a near linear correlation of current vs. applied voltage, suggesting the nanowires act similarly to Ohmic conductors and follow Eq. (3.7). Under this assumption, any limitations associated with nanowire transport would be alleviated by applying a higher potential to account for any voltage loss across the nanowire network. However, higher potentials do not correlate with higher current densities in *G. sulfurreducens* when anode potentials are above −0.1 vs. SHE [217], which would imply that other transport mechanisms are responsible for the limiting current densities in these biofilms.

While measurements have shown linear responses to applied voltage, it is clear that cytochromes should not behave as Ohmic conductors. As observed in Table 3.1, the measured κ values change by orders of magnitude through conditions such as pH and temperature which may not be biologically relevant (for a detailed discussion on this topic, please refer to a recent perspective by Guberman-Pfeffer [42]). Early on, Strycharz et al. developed an EET model based on common cytochrome redox transfer, based on the diffusion of charges along cytochromes [218]. Under this model, the EET transport reaches a maximum rate of transport that is limited by the concentration of redox sites within the biofilm. This is consistent with some measurements performed on OmcZ (and alleged PilA) nanowires, where the current transported by the filaments seems to saturate at higher applied voltages [181]. Under this model, EET would yield a sigmoidal response similar to that observed in biofilms, typically attributed to the cell's metabolism. Thus, it is possible that the *G. sulfurreducens'* electrochemical response is associated with EET, and this, in turn, is limiting current generation. It is also suggested that proton-coupled electron transfer, a property already observed in periplasmic cytochromes [213, 214], may explain the discrepancies observed between the potential and observed current density in *Geobacter* biofilms [219] as it could potentially aid in transferring protons for the latter to sustain growth. However, this property has only been recently proposed for cytochrome nanowires, so further experiments that confirm electron and possible proton transfer are required.

Finally, several studies have shown that overexpression of PilA and OmcZ leads to higher current generation [177, 220]. Yet, these studies do not show current densities that are considerably higher than those obtained in other studies using the wild type strain [171, 221]. Therefore, while a possible limitation to electron flow due to EET is possible, it seems unlikely based on current evidence and other factors in METs such as reactor configuration [222], medium components [223], operational electrochemical conditions (e.g. anode potential) [221], and reactor materials [224] may currently have a stronger influence in determining current outputs in anodic *Geobacter* METs.

3.7 Final Perspective on EET Approaches in Microbial Electrochemistry

Interactions between microorganisms and electrodes need an inherent EET mechanism to bridge the gap between intracellular electrons and the electrode's surface. So far, we have identified two general mechanisms developed by bacteria to transport electrons to an anode using either EES or a solid conductive surface. Our knowledge of these mechanisms is limited to just a few decades of research and a few model organisms discussed in this chapter. Yet, through this knowledge, many applications in the field of METs have been proposed and implemented successfully, stimulating the research to understand the true extent of EET in natural systems.

In contrast, researchers have found novel ways to extend EET to microorganisms not capable of performing it naturally. External additions of EES and careful design and utilization of novel shuttles have led to novel applications and current densities that rival those in natural systems. Synthetic biology approaches have transferred EES and nanowires genes to other microorganisms successfully, also leading to examples of high rate EET.

EES-based EET typically has lower current densities than a solid conductive matrix. EES transport is characterized by transport limitations, mostly through diffusion, that limits the total rate of electron transport. Low solubility and toxicity of common EES, both natural and exogenously added, lead to low concentrations of these shuttles in most systems studied ($<0.5\,mM$), which in turn limits transport by diffusion. However, approaches that lead to the overexpression of EES, the formation of biofilms at the electrode surface to decrease diffusion distances, or the careful design of exogenous EES, have led to significant increases in rates that rival those of *Geobacter* biofilms.

In contrast, a solid conduction mechanism for EET has been characterized for being faster yet more elusive to understand. Current densities of $>1\,mA\,cm^{-2}$ have been observed in *Geobacter*-based systems through careful design. Yet, it is not clear whether the EET process is limiting current densities, or whether other transport/growth processes limit current production. The recent discovery and characterization of cytochrome nanowires have changed the way we understand conduction-based EET, while it has opened up new questions about the expected behavior of these nanowires.

There are still many unknowns regarding the underpinnings of EET. Efforts within the microbial electrochemistry community have made great advancements in our understanding of EET, and future studies will help elucidate the missing details regarding these complex mechanisms. Through this understanding, we can envision future applications in which EET proteins are transferred efficiently to other microorganisms, better MET designs are developed to enhance EET processes, and EET proteins are used in fields of bioelectronics and biosensing. We consider this an emerging field of research with great possibilities to enhance future applications in the microbial electrochemistry community.

References

1 Rittmann, B.E. (2008). Opportunities for renewable bioenergy using microorganisms. *Biotechnol. Bioeng.* 100 (2): 203–212. https://doi.org/10.1002/bit.21875.

2 Strous, M. and Sharp, C. (2018). Designer microbiomes for environmental, energy and health biotechnology. *Curr. Opin. Microbiol.* 43: 117–123. https://doi.org/10.1016/j.mib.2017.12.007.

3 Fasim, A., More, V.S., and More, S.S. (2021). Large-scale production of enzymes for biotechnology uses. *Curr. Opin. Biotechnol.* 69: 68–76. https://doi.org/10.1016/j.copbio.2020.12.002.

4 Verma, J., Kumar, D., Singh, N. et al. (2021). Electricigens and microbial fuel cells for bioremediation and bioenergy production: a review. *Environ. Chem. Lett.* 19 (3): 2091–2126. https://doi.org/10.1007/s10311-021-01199-7.

5 Amoah, J., Kahar, P., Ogino, C., and Kondo, A. (2019). Bioenergy and biorefinery: feedstock, biotechnological conversion, and products. *Biotechnol. J.* 14 (6): 1800494. https://doi.org/10.1002/biot.201800494.

6 Bedendi, G., De Moura Torquato, L.D., Webb, S. et al. (2022). Enzymatic and microbial electrochemistry: approaches and methods. *ACS Meas. Sci. Au* 2 (6): 517–541. https://doi.org/10.1021/acsmeasuresciau.2c00042.

7 Li, X., Cheng, M., Jiao, X. et al. (2022). Advances in microbial electrochemistry-enhanced constructed wetlands. *World J. Microbiol. Biotechnol.* 38 (12): 239. https://doi.org/10.1007/s11274-022-03413-2.

8 Shrestha, N., Chilkoor, G., Vemuri, B. et al. (2018). Extremophiles for microbial-electrochemistry applications: a critical review. *Bioresour. Technol.* 255: 318–330. https://doi.org/10.1016/j.biortech.2018.01.151.

9 Liu, X., Shi, L., and Gu, J.D. (2018). Microbial electrocatalysis: redox mediators responsible for extracellular electron transfer. *Biotechnol. Adv.* 36 (7): 1815–1827. https://doi.org/10.1016/j.biotechadv.2018.07.001.

10 Lovley, D.R. and Holmes, D.E. (2020). Protein nanowires: the electrification of the microbial world and maybe our own. *J. Bacteriol.* 202 (20): e00331-20. https://doi.org/10.1128/JB.00331-20.

11 Hernandez, M.E. and Newman, D.K. (2001). Extracellular electron transfer. *Cell. Mol. Life Sci.* 58 (11): 1562–1571. https://doi.org/10.1007/PL00000796.

12 Potter, M.C. (1911). Electrical effects accompanying the decomposition of organic compounds. *Proc. R. Soc. Lond. B* 84: 260–276. https://doi.org/10.1098/rspb.1911.0073.

13 Galeotti, T., Cittadini, A., Dionisi, O. et al. (1971). Pathways of intracellular hydrogen transport in the Walker carcinosarcoma 256. I. The intramitochondrial electron transport and the translocation of reducing equivalents across the mitochondrial membrane. *Biochim. Biophys. Acta - Bioenerg.* 253 (2): 303–313. https://doi.org/10.1016/0005-2728(71)90034-X.

14 Caccavo, F. Jr., Lonergan, D.J., Lovley, D.R. et al. (1994). *Geobacter sulfurreducens* sp. nov., a hydrogen- and acetate-oxidizing dissimilatory metal-reducing microorganism. *Appl. Environ. Microbiol.* 60 (10): 3752–3759. https://doi.org/10.1128/aem.60.10.3752-3759.1994.

15 Myers, C.R. and Nealson, K.H. (1990). Respiration-linked proton translocation couples to anaerobic reduction of manganese(IV) and iron(III) in *Shewanella putrefaciens* MR-1. *J. Bacteriol.* 172 (11): 6232–6238. https://doi.org/10.1128/jb.172.11.6232-6238.1990.

16 Shi, L., Dong, H., Reguera, G. et al. (2016). Extracellular electron transfer mechanisms between microorganisms and minerals. *Nat. Rev. Microbiol.* 14 (10): 651–662. https://doi.org/10.1038/nrmicro.2016.93.

17 Ciemniecki, J.A. and Newman, D.K. (2020). The potential for redox-active metabolites to enhance or unlock anaerobic survival metabolisms in aerobes. *J. Bacteriol.* 202 (11): e00797–19. https://doi.org/10.1128/jb.00797-19.

18 Torres, C.I., Marcus, A.K., Lee, H.S. et al. (2010). A kinetic perspective on extracellular electron transfer by anode-respiring bacteria. *FEMS Microbiol. Rev.* 34 (1): 3–17. https://doi.org/10.1111/j.1574-6976.2009.00191.x.

19 Bond, D.R., Strycharz-Glaven, S.M., Tender, L.M., and Torres, C.I. (2012). On electron transport through geobacter biofilms. *ChemSusChem* 5 (6): 1099–1105. https://doi.org/10.1002/cssc.201100748.

20 Picioreanu, C., Head, I.M., Katuri, K.P. et al. (2007). A computational model for biofilm-based microbial fuel cells. *Water Res.* 41 (13): 2921–2940. https://doi.org/10.1016/j.watres.2007.04.009.

21 Reguera, G., McCarthy, K.D., Mehta, T. et al. (2005). Extracellular electron transfer via microbial nanowires. *Nature* 435: 1098–1101. https://doi.org/10.1038/nature03661.

22 Gu, Y., Guberman-Pfeffer, M.J., Srikanth, V. et al. (2023). Structure of *Geobacter* cytochrome OmcZ identifies mechanism of nanowire assembly and conductivity. *Nat. Microbiol.* 8 (2): 284–298. https://doi.org/10.1038/s41564-022-01315-5.

23 Baquero, D.P., Cvirkaite-Krupovic, V., Hu, S.S. et al. (2023). Extracellular cytochrome nanowires appear to be ubiquitous in prokaryotes. *Cell* 186 (13): 2853–2864.e8. https://doi.org/10.1016/j.cell.2023.05.012.

24 Badalamenti, J.P., Krajmalnik-Brown, R., and Torres, C.I. (2013). Generation of high current densities by pure cultures of anode-respiring *Geoalkalibacte*r spp. under alkaline and saline conditions in microbial electrochemical cells. *MBio* 4 (3): e00144–13. https://doi.org/10.1128/mBio.00144-13.

25 Carmona-Martínez, A.A., Pierra, M., Trably, E., and Bernet, N. (2013). High current density via direct electron transfer by the halophilic anode respiring bacterium *Geoalkalibacter subterraneus*. *Phys. Chem. Chem. Phys.* 15 (45): 19699–19707. https://doi.org/10.1039/c3cp54045f.

26 Parameswaran, P., Bry, T., Popat, S.C. et al. (2013). Kinetic, electrochemical, and microscopic characterization of the thermophilic, anode-respiring bacterium *Thermincola ferriacetica*. *Environ. Sci. Technol.* 47 (9): 4934–4940. https://doi.org/10.1021/es400321c.

27 Lusk, B.G., Khan, Q.F., Parameswaran, P. et al. (2015). Characterization of electrical current-generation capabilities from thermophilic bacterium *Thermoanaerobacter pseudoethanolicus* using xylose, glucose, cellobiose, or acetate with fixed anode potentials. *Environ. Sci. Technol.* 49 (24): 14725–14731. https://doi.org/10.1021/acs.est.5b04036.

28 Zhou, J., Smith, J.A., Li, M., and Holmes, D.E. (2023). Methane production by *Methanothrix thermoacetophila* via direct interspecies electron transfer with *Geobacter metallireducens*. *MBio* 14 (4): e0036023. https://doi.org/10.1128/mbio.00360-23.

29 Deng, P., Wang, L., Li, X. et al. (2023). *Geobacter grbiciae*—a new electron donor in the formation of co-cultures via direct interspecies electron transfer. *Microbiol. Res.* 14 (4): 1774–1787. https://doi.org/10.3390/microbiolres14040122.

30 Holmes, D.E., Zhou, J., Smith, J.A. et al. (2022). Different outer membrane c-type cytochromes are involved in direct interspecies electron transfer to *Geobacter* or *Methanosarcina* species. *mLife* 1 (3): 272–286. https://doi.org/10.1002/mlf2.12037.

31 Yee, M.O., Snoeyenbos-West, O.L., Thamdrup, B. et al. (2019). Extracellular electron uptake by two *Methanosarcina* species. *Front. Energy Res.* 7: 29. https://doi.org/10.3389/fenrg.2019.00029.

32 Zheng, S., Liu, F., Wang, B. et al. (2020). *Methanobacterium* capable of direct interspecies electron transfer. *Environ. Sci. Technol.* 54 (23): 15347–15354. https://doi.org/10.1021/acs.est.0c05525.

33 Popat, S.C. and Torres, C.I. (2016). Critical transport rates that limit the performance of microbial electrochemistry technologies. *Bioresour. Technol.* 215: 265–273. https://doi.org/10.1016/j.biortech.2016.04.136.

34 Torres, C.I. (2014). On the importance of identifying, characterizing, and predicting fundamental phenomena towards microbial electrochemistry applications. *Curr. Opin. Biotechnol.* 27: 107–114. https://doi.org/10.1016/j.copbio.2013.12.008.

35 Lovley, D.R. (2022). Untangling *Geobacter sulfurreducens* nanowires. *MBio* 13 (3): e00850-22. https://doi.org/10.1128/mbio.00850-22.

36 Glasser, N.R., Saunders, S.H., and Newman, D.K. (2017). The colorful world of extracellular electron shuttles. *Ann. Rev. Microbiol.* 71: 731–751. https://doi.org/10.1146/annurev-micro-090816-093913.

37 Gralnick, J.A. and Bond, D.R. (2023). Electron transfer beyond the outer membrane: putting electrons to rest. *Ann. Rev. Microbiol.* 77 (1): 517–539. https://doi.org/10.1146/annurev-micro-032221-023725.

38 Gaffney, E.M., Grattieri, M., Rhodes, Z., and Minteer, S.D. (2020). Editors' choice—review—exploration of computational approaches for understanding microbial electrochemical systems: opportunities and future directions. *J. Electrochem. Soc.* 167 (6): 065502. https://doi.org/10.1149/1945-7111/ab872e.

39 Wang, F., Craig, L., Liu, X. et al. (2023). Microbial nanowires: type IV pili or cytochrome filaments? *Trends Microbiol.* 31 (4): 384–392. https://doi.org/10.1016/j.tim.2022.11.004.

40 Ueki, T. (2021). Cytochromes in extracellular electron transfer in *Geobacter*. *Appl. Environ. Microbiol.* 87 (10): e03109-20. https://doi.org/10.1128/AEM.03109-20.

41 Yalcin, S.E. and Malvankar, N.S. (2020). The blind men and the filament: understanding structures and functions of microbial nanowires. *Curr. Opin. Chem. Biol.* 59: 193–201. https://doi.org/10.1016/j.cbpa.2020.08.004.

42 Guberman-Pfeffer, M.J. (2024). To be or not to be a cytochrome: electrical characterizations are inconsistent with *Geobacter* cytochrome 'nanowires'. *Front. Microbiol.* 15: 1397124. https://doi.org/10.3389/fmicb.2024.1397124.

43 Simoska, O., Lee, Y.S., and Minteer, S.D. (2023). Fundamentals and applications of enzymatic bioelectrocatalysis. In: *Comprehensive Inorganic Chemistry III*, 456–491. Elsevier. https://doi.org/10.1016/B978-0-12-823144-9.00057-1.

44 Bard, A. and Faulkner, L. (2012). *Electrochemical Methods: Fundamentals and Applications*, 2nde, 823. New York: John Wiley and Sons Inc.

45 Walker, R. and Holt, N. (1983). Determination of the Nenrnst diffusion layer thickness in the hydroson agitation tank. *Surf. Technol.* 22: 165–174. https://doi.org/10.1016/0376-4583(84)90053-0.

46 El-Latif, E.I.A., Alahmadi, M., Kebede, M.A. et al. (2025). Modeling the diffusion coefficient of charge carriers in batteries using the randles-sevcik equation. *Adv. Theory Simul.* 8: 2500346. https://doi.org/10.1002/adts.202500346.

47 Song, Y., Penmatsa, V., and Wang, C. (2014). Modeling and simulation of enzymatic biofuel cells with three-dimensional microelectrodes. *Energies* 7 (7): 4694–4709. https://doi.org/10.3390/en7074694.

48 Trachioti, M.G., Lazanas, A.C., and Prodromidis, M.I. (2023). Shedding light on the calculation of electrode electroactive area and heterogeneous electron transfer rate constants at graphite screen-printed electrodes. *Microchim. Acta* 190 (7): 251. https://doi.org/10.1007/s00604-023-05832-w.

49 Guzman, J.J.L., Pehlivaner Kara, M.O., Frey, M.W., and Angenent, L.T. (2017). Performance of electro-spun carbon nanofiber electrodes with conductive poly(3,4-ethylenedioxythiophene) coatings in bioelectrochemical systems. *J. Power Sources* 356: 331–337. https://doi.org/10.1016/j.jpowsour.2017.03.133.

50 Cai, T., Meng, L., Chen, G. et al. (2020). Application of advanced anodes in microbial fuel cells for power generation: a review. *Chemosphere* 248: 125985. https://doi.org/10.1016/j.chemosphere.2020.125985.

51 Savéant, J.M. and Vianello, E. (1963). Potential-sweep chronoamperometry theory of kinetic currents in the case of a first order chemical reaction preceding the electron-transfer process. *Electrochim. Acta* 8 (12): 905–923. https://doi.org/10.1016/0013-4686(62)87046-7.

52 Torres, D.I., Marcus, A.K., Parameswaran, P., and Rittmann, B.E. (2008). Kinetic experiments for evaluating the Nernst-Monod model for anode-respiring bacteria (ARB) in a biofilm anode. *Environ. Sci. Technol.* 42 (17): 6593–6597. https://doi.org/10.1021/es800970w.

53 Rousseau, R., Délia, M.-L., and Bergel, A. (2014). A theoretical model of transient cyclic voltammetry for electroactive biofilms. *Energy Environ. Sci.* 7 (3): 1079–1094. https://doi.org/10.1039/c3ee42329h.

54 Pérez-García, J.A., Bacame-valenzuela, F.J., Palomares-Hernández, A. et al. (2021). Microbial electrochemical technologies for wastewater treatment. In: *The Future of Effluent Treatment Plants*, 287–305. Elsevier. https://doi.org/10.1016/B978-0-12-822956-9.00017-9.

55 Rountree, E.S., McCarthy, B.D., Eisenhart, T.T., and Dempsey, J.L. (2014). Evaluation of homogeneous electrocatalysts by cyclic voltammetry. *Inorg. Chem.* 53 (19): 9983–10002. https://doi.org/10.1021/ic500658x.

56 Savéant, J.M. and Costentin, C. (2019). Enzymatic catalysis of electrochemical reactions. In: *Elements of Molecular and Biomolecular Electrochemistry*, 383–437. Wiley. https://doi.org/10.1002/0471758078.ch5.

57 Wang, V.C.C. and Johnson, B.A. (2019). Interpreting the electrocatalytic voltammetry of homogeneous catalysts by the foot of the wave analysis and its wider implications. *ACS Catal.* 9 (8): 7109–7123. https://doi.org/10.1021/acscatal.9b00850.

58 Marsili, E., Rollefson, J.B., Baron, D.B. et al. (2008). Microbial biofilm voltammetry: direct electrochemical characterization of catalytic electrode-attached biofilms. *Appl. Environ. Microbiol.* 74 (23): 7329–7337. https://doi.org/10.1128/aem.00177-08.

59 Yang, Y., Yu, Y.Y., Wang, Y.Z. et al. (2017). Amplification of electrochemical signal by a whole-cell redox reactivation module for ultrasensitive detection of pyocyanin. *Biosens. Bioelectron.* 98 (7): 338–344. https://doi.org/10.1016/j.bios.2017.07.008.

60 Hamelers, H.V.M., ter Heijne, A., Stein, N. et al. (2011). Butler–Volmer–Monod model for describing bio-anode polarization curves. *Bioresour. Technol.* 102 (1): 381–387. https://doi.org/10.1016/j.biortech.2010.06.156.

61 Srikanth, S., Marsili, E., Flickinger, M.C., and Bond, D.R. (2008). Electrochemical characterization of *Geobacter sulfurreducens* cells immobilized on graphite paper electrodes. *Biotechnol. Bioeng.* 99 (5): 1065–1073. https://doi.org/10.1002/bit.21671.

62 Zheng, T., Xu, Y.S., Yong, X.Y. et al. (2015). Endogenously enhanced biosurfactant production promotes electricity generation from microbial fuel cells. *Bioresour. Technol.* 197: 416–421. https://doi.org/10.1016/j.biortech.2015.08.136.

63 Pérez-García, J.A., Bacame-Valenzuela, F.J., Espejel-Ayala, F. et al. (2023). Effect of adsorption of pyocyanin on the electron transfer rate at the interface of a glassy carbon electrode. *Electrochim. Acta* 445: 142014. https://doi.org/10.1016/j.electacta.2023.142014.

64 Saunders, S.H., Edmund, C.M., Yates, M.D. et al. (2020). Extracellular DNA promotes efficient extracellular electron transfer by pyocyanin in *Pseudomonas aeruginosa* biofilms. *Cell* 182 (4): 919–932.e19. https://doi.org/10.1016/j.cell.2020.07.006.

65 Meirelles, L.A. and Newman, D.K. (2018). Both toxic and beneficial effects of pyocyanin contribute to the lifecycle of *Pseudomonas aeruginosa*. *Mol. Microbiol.* 110 (6): 995–1010. https://doi.org/10.1111/mmi.14132.

66 Kato Marcus, A., Torres, C.I., and Rittmann, B.E. (2007). Conduction-based modeling of the biofilm anode of a microbial fuel cell. *Biotechnol. Bioeng.* 98 (6): 1171–1182. https://doi.org/10.1002/bit.21533.

67 Hamelers, H.V.M., Ter Heijne, A., Sleutels, T. et al. (2010). New applications and performance of bioelectrochemical systems. *Appl. Microbiol. Biotechnol.* 85 (6): 1673–1685. https://doi.org/10.1007/s00253-009-2357-1.

68 Peng, S., Liang, D.W., Diao, P. et al. (2013). Nernst-ping-pong model for evaluating the effects of the substrate concentration and anode potential on the kinetic characteristics of bioanode. *Bioresour. Technol.* 136: 610–616. https://doi.org/10.1016/j.biortech.2013.03.073.

69 Renslow, R., Babauta, J., Kuprat, A. et al. (2013). Modeling biofilms with dual extracellular electron transfer mechanisms. *Phys. Chem. Chem. Phys.* 15 (44): 19262–19283. https://doi.org/10.1039/c3cp53759e.

70 Korth, B., Rosa, L.F.M., Harnisch, F., and Picioreanu, C. (2015). A framework for modeling electroactive microbial biofilms performing direct electron transfer. *Bioelectrochemistry* 106: 194–206. https://doi.org/10.1016/j.bioelechem.2015.03.010.

71 Zacharoff, L., Chan, C.H., and Bond, D.R. (2016). Reduction of low potential electron acceptors requires the CbcL inner membrane cytochrome of *Geobacter sulfurreducens. Bioelectrochemistry* 107: 7–13. https://doi.org/10.1016/j.bioelechem.2015.08.003.

72 Levar, C.E., Chan, C.H., Mehta-Kolte, M.G., and Bond, D.R. (2014). An inner membrane cytochrome required only for reduction of high redox potential extracellular electron acceptors. *MBio* 5 (6): e02034. https://doi.org/10.1128/mBio.02034-14.

73 Yoho, R.A., Popat, S.C., and Torres, C.I. (2014). Dynamic potential-dependent electron transport pathway shifts in anode biofilms of *Geobacter sulfurreducens. ChemSusChem* 7 (12): 3413–3419. https://doi.org/10.1002/cssc.201402589.

74 Howley, E., Krajmalnik-Brown, R., and Torres, C.I. (2023). Cytochrome gene expression shifts in *Geobacter sulfurreducens* to maximize energy conservation in response to changes in redox conditions. *Biosens. Bioelectron.* 237: 115524. https://doi.org/10.1016/j.bios.2023.115524.

75 Edwards, M.J., White, G.F., Butt, J.N. et al. (2020). The crystal structure of a biological insulated transmembrane molecular wire. *Cell* 181 (3): 665–673.e10. https://doi.org/10.1016/j.cell.2020.03.032.

76 Clarke, T.A., Edwards, M.J., Gates, A.J. et al. (2011). Structure of a bacterial cell surface decaheme electron conduit. *Proc. Natl. Acad. Sci.* 108 (23): 9384–9389. https://doi.org/10.1073/pnas.1017200108.

77 Richardson, D.J., Butt, J.N., Fredrickson, J.K. et al. (2012). The 'porin-cytochrome' model for microbe-to-mineral electron transfer. *Mol. Microbiol.* 85 (2): 201–212. https://doi.org/10.1111/j.1365-2958.2012.08088.x.

78 Breuer, M., Rosso, K.M., Blumberger, J., and Butt, J.N. (2015). Multi-haem cytochromes in *Shewanella oneidensis* MR-1: structures, functions and opportunities. *J. R. Soc. Interface* 12 (102): 20141117. https://doi.org/10.1098/rsif.2014.1117.

79 Ross, D.E., Brantley, S.L., and Tien, M. (2009). Kinetic characterization of OmcA and MtrC, terminal reductases involved in respiratory electron transfer for dissimilatory iron reduction in *Shewanella oneidensis* MR-1. *Appl. Environ. Microbiol.* 75 (16): 5218–5226. https://doi.org/10.1128/AEM.00544-09.

80 Wang, Q.Q., Wu, X.Y., Yu, Y.Y. et al. (2017). Facile in-situ fabrication of graphene/riboflavin electrode for microbial fuel cells. *Electrochim. Acta* 232: 439–444. https://doi.org/10.1016/j.electacta.2017.03.008.

81 Zhang, L., Zhang, Y., Liu, Y. et al. (2024). High power density redox-mediated *Shewanella* microbial flow fuel cells. *Nat. Commun.* 15 (1): 8302. https://doi.org/10.1038/s41467-024-52498-w.

82 Baron, D., LaBelle, E., Coursolle, D. et al. (2009). Electrochemical measurement of electron transfer kinetics by *Shewanella oneidensis* MR-1. *J. Biol. Chem.* 284 (42): 28865–28873. https://doi.org/10.1074/jbc.M109.043455.

83 Yong, X.Y., Yan, Z.Y., Shen, H.B. et al. (2017). An integrated aerobic-anaerobic strategy for performance enhancement of *Pseudomonas aeruginosa*-inoculated microbial fuel cell. *Bioresour. Technol.* 241: 1191–1196. https://doi.org/10.1016/j.biortech.2017.06.050.

84 Yong, X.-Y., Feng, J., Chen, Y.-L. et al. (2014). Enhancement of bioelectricity generation by cofactor manipulation in microbial fuel cell. *Biosensors Bioelectron.* 56: 19–25. https://doi.org/10.1016/j.bios.2013.12.058.

85 Glasser, N.R., Wang, B.X., Hoy, J.A., and Newman, D.K. (2017). The pyruvate and alpha-ketoglutarate dehydrogenase complexes of *Pseudomonas aeruginosa* catalyze pyocyanin and phenazine-1-carboxylic acid reduction via the subunit dihydrolipoamide dehydrogenase. *J. Biol. Chem.* 292 (13): 5593–5607. https://doi.org/10.1074/jbc.M116.772848.

86 Chukwubuikem, A., Berger, C., Mady, A., and Rosenbaum, M.A. (2021). Role of phenazine-enzyme physiology for current generation in a bioelectrochemical system. *Microb. Biotechnol.* 14 (4): 1613–1626. https://doi.org/10.1111/1751-7915.13827.

87 Borrero-de Acuña, J.M., Timmis, K.N., Jahn, M., and Jahn, D. (2017). Protein complex formation during denitrification by *Pseudomonas aeruginosa*. *Microb. Biotechnol.* 10 (6): 1523–1534. https://doi.org/10.1111/1751-7915.12851.

88 Das, T., Kutty, S.K., Tavallaie, R. et al. (2015). Phenazine virulence factor binding to extracellular DNA is important for *Pseudomonas aeruginosa* biofilm formation. *Sci. Rep.* 5 (1): 8398. https://doi.org/10.1038/srep08398.

89 Hunter, R.C. and Beveridge, T.J. (2005). Application of a pH-sensitive fluoroprobe (C-SNARF-4) for pH microenvironment analysis in *Pseudomonas aeruginosa* biofilms. *Appl. Environ. Microbiol.* 71 (5): 2501–2510. https://doi.org/10.1128/AEM.71.5.2501-2510.2005.

90 Lawrence, J.R., Wolfaardt, G.M., and Korber, D.R. (1994). Determination of diffusion coefficients in biofilms by confocal laser microscopy. *Appl. Environ. Microbiol.* 60 (4): 1166–1173. https://doi.org/10.1128/aem.60.4.1166-1173.1994.

91 Thalhammer, K.O. and Newman, D.K. (2023). A phenazine-inspired framework for identifying biological functions of microbial redox-active metabolites. *Curr. Opin. Chem. Biol.* 75: 102320. https://doi.org/10.1016/j.cbpa.2023.102320.

92 Yu, Y.-Y., Zhang, Y., and Peng, L. (2022). Investigating the interaction between *Shewanella oneidensis* and phenazine 1-carboxylic acid in the microbial electrochemical processes. *Sci. Total Environ.* 838: 156501. https://doi.org/10.1016/j.scitotenv.2022.156501.

93 Jeuken, L.J.C., Hards, K., and Nakatani, Y. (2020). Extracellular Electron Transfer: Respiratory or Nutrient Homeostasis? *J. Bacteriol.* 202 (7): e00029-20. https://doi.org/10.1128/jb.00029-20.

94 Zhang, J., Liu, H., Zhang, Y. et al. (2023). Enhanced CO_2 reduction by electron shuttle molecules via coupling different electron transport processes in microbial electrosynthesis. *Fermentation* 9 (7): 679. https://doi.org/10.3390/fermentation9070679.

95 Huang, L., Tang, J., Chen, M. et al. (2018). Two modes of riboflavin-mediated extracellular electron transfer in *Geobacter uraniireducens*. *Front. Microbiol.* 9: 2886. https://doi.org/10.3389/fmicb.2018.02886.

96 Kracke, F., Vassilev, I., and Kromer, J.O. (2015). Microbial electron transport and energy conservation - the foundation for optimizing bioelectrochemical systems. *Front. Microbiol.* 6: 575. https://doi.org/10.3389/fmicb.2015.00575.

97 Berry, S., Schneider, D., Vermaas, W.F.J., and Rogner, M. (2002). Electron transport routes in whole cells of *Synechocystis* sp. Strain PCC 6803: The role of the cytochrome *bd*-type oxidase. *Biochemistry* 41 (10): 3422–3429. https://doi.org/10.1021/bi011683d.

98 Yates, M., Strycharz-Glaven, S., Golden, J. et al. (2018). Characterizing electron transport through living biofilms. *J. Vis. Exp.* 136: e54671. https://doi.org/10.3791/54671.

99 Braasch-Turi, M.M., Koehn, J.T., Kostenkova, K. et al. (2022). Electron transport lipids fold within membrane-like interfaces. *Front. Chem.* 10: 827530. https://doi.org/10.3389/fchem.2022.827530.

100 Hsu, L.H., Zhang, Y., Deng, P. et al. (2019). Biosynthetic electronic interfaces for bridging microbial and inorganic electron transport. *Nano Lett.* 19: 8787–8792. https://doi.org/10.1021/acs.nanolett.9b03573.

101 Zhou, S., Chen, S., Yuan, Y., and Lu, Q. (2015). Influence of humic acid complexation with metal ions on extracellular electron transfer activity. *Sci. Rep.* 5: 17067. https://doi.org/10.1038/srep17067.

102 De La Cruz, C., Molina, A., Patil, N. et al. (2020). New insights into phenazine-based organic redox flow batteries by using high-throughput DFT modelling. *Sustainable Energy Fuels* 4 (11): 5513–5521. https://doi.org/10.1039/D0SE00687D.

103 Nomura, Y. and Akashi, R. (2024). Density functional theory☆. In: *Encyclopedia of Condensed Matter Physics (Second Edition)* (ed. T. Chakraborty), 867–878. Oxford: Academic Press.

104 Liu, X., Ye, Y., Yang, N. et al. (2024). Nonelectroactive *Clostridium* obtains extracellular electron transfer-capability after forming chimera with *Geobacter*. *ISME Commun.* 4 (1): ycae058. https://doi.org/10.1093/ismeco/ycae058.

105 Zhang, J., Li, F., Liu, D. et al. (2024). Engineering extracellular electron transfer pathways of electroactive microorganisms by synthetic biology for energy and chemicals production. *Chem. Soc. Rev.* 53 (3): 1375–1446. https://doi.org/10.1039/D3CS00537B.

106 Pluczyk, S., Vasylieva, M., and Data, P. (2018). Using cyclic voltammetry, UV-Vis-NIR, and EPR spectroelectrochemistry to analyze organic compounds. *J. Vis. Exp.: JoVE* 140: e56656. https://doi.org/10.3791/56656.

107 Kornienko, N., Ly, K.H., Robinson, W.E. et al. (2019). Advancing techniques for investigating the enzyme-electrode interface. *Acc. Chem. Res.* 52 (5): 1439–1448. https://doi.org/10.1021/acs.accounts.9b00087.

108 Viehe, H.G., Janousek, Z., Merenyi, R., and Stella, L. (1985). The captodative effect. *Acc. Chem. Res.* 18 (5): 148–154. https://doi.org/10.1021/ar00113a004.

109 Sustmann, R. and Korth, H.-G. (1990). The captodative effect. In: *Advances in Physical Organic Chemistry*, vol. 26 (ed. D. Bethell), 131–178. Academic Press.

110 Blokker, E., ten Brink, M., van der Schuur, J.M. et al. (2023). Origin of the captodative effect: the lone-pair shielded radical. *ChemistryEurope* 1 (1): e202300006. https://doi.org/10.1002/ceur.202300006.

111 Huynh, M.T., Anson, C.W., Cavell, A.C. et al. (2016). Quinone 1 e^- and 2 e^-/2 H^+ reduction potentials: identification and analysis of deviations from systematic scaling relationships. *J. Am. Chem. Soc.* 138 (49): 15903–15910. https://doi.org/10.1021/jacs.6b05797.

112 Ikonen, M., Murtomäki, L., and Kontturi, K. (2007). An electrochemical method for the determination of liposome–water partition coefficients of drugs. *J. Electroanal. Chem.* 602 (2): 189–194. https://doi.org/10.1016/j.jelechem.2006.12.014.

113 Kontturi, K. and Murtomäki, L. (1992). Electrochemical determination of partition coefficients of drugs. *J. Pharm. Sci.* 81 (10): 970–975. https://doi.org/10.1002/jps.2600811003.

114 Chang, H., Zhong, F., Cao, R. et al. (2020). Prediction of the solubilities of water in hydrocarbons with COSMO-RS and interpretation of the solubility characteristics. *J. Solut. Chem.* 49 (3): 365–382. https://doi.org/10.1007/s10953-020-00966-4.

115 Kihara, S., Hartzler, D.A., and Savikhin, S. (2014). Oxygen concentration inside a functioning photosynthetic cell. *Biophys. J.* 106 (9): 1882–1889. https://doi.org/10.1016/j.bpj.2014.03.031.

116 Shang, E., Li, Y., Niu, J. et al. (2017). Relative importance of humic and fulvic acid on ROS generation, dissolution, and toxicity of sulfide nanoparticles. *Water Res.* 124: 595–604. https://doi.org/10.1016/j.watres.2017.08.001.

117 Meyer, J.N., Hartman, J.H., and Mello, D.F. (2018). Mitochondrial Toxicity. *Toxicol. Sci.* 162 (1): 15–23. https://doi.org/10.1093/toxsci/kfy008.

118 Bandyopadhyay, I. (2019), "Synthesis of Redox Active Neuroprotective Therapeutic Agents," Ph.D., Arizona State University, United States -- Arizona, 13861025. [Online]. Available: http://login.ezproxy1.lib.asu.edu/login?url=https://www.proquest.com/dissertations-theses/synthesis-redox-active-neuroprotective/docview/2228126857/se-2?accountid=4485.

119 Pospíšil, P. (2016). Production of reactive oxygen species by photosystem II as a response to light and temperature stress. *Front. Plant Sci.* 7: 1950. https://doi.org/10.3389/fpls.2016.01950.

120 Nevin, K.P. and Lovley, D.R. (2000). Potential for nonenzymatic reduction of Fe(III) via electron shuttling in subsurface sediments. *Environ. Sci. Technol.* 34 (12): 2472–2478. https://doi.org/10.1021/es991181b.

121 Lewis, C.M., Flory, J.D., Moore, T.A. et al. (2022). Electrochemically Driven Photosynthetic Electron Transport in Cyanobacteria Lacking Photosystem II. *J. Am. Chem. Soc.* 144: 2933–2942. https://doi.org/10.1021/jacs.1c09291.

122 Mullineaux, C.W. and Liu, L.N. (2020). Membrane dynamics in phototrophic bacteria. *Ann. Rev. Microbiol.* 74: 633–654. https://doi.org/10.1146/annurev-micro-020518-120134.

123 Peng, X.-X., Gai, S., Cheng, K., and Yang, F. (2022). Roles of humic substances redox activity on environmental remediation. *J. Hazard. Mater.* 435: 129070. https://doi.org/10.1016/j.jhazmat.2022.129070.

124 Suzuki, N. (2023). Fine tuning of ROS, redox and energy regulatory systems associated with the functions of chloroplasts and mitochondria in plants under heat stress. *Int. J. Mol. Sci.* 24 (2): 1356. https://doi.org/10.3390/ijms24021356.

125 Wallerio, J.G. (1761). *Agriculturaea Fundamenta Chemica*, 330. Sweden: Upsala.

126 Kaneko, M., Ishikawa, M., Hashimoto, K., and Nakanishi, S. (2017). Molecular design of cytocompatible amphiphilic redox-active polymers for efficient extracellular electron transfer. *Bioelectrochemistry* 114: 8–12. https://doi.org/10.1016/j.bioelechem.2016.11.001.

127 Lovley, D.R., Coates, J.D., Blunt-Harris, E.L. et al. (1996). Humic substances as electron acceptors for microbial respiration. *Nature* 382: 445–448. https://doi.org/10.1038/382445a0.

128 Kishi, S., Saito, K., Kato, Y., and Ishikita, H. (2017). Redox potentials of ubiquinone, menaquinone, phylloquinone, and plastoquinone in aqueous solution. *Photosynth. Res.* 134 (2): 193–200. https://doi.org/10.1007/s11120-017-0433-4.

129 van Eerden, F.J., de Jong, D.H., de Vries, A.H. et al. (2015). Characterization of thylakoid lipid membranes from cyanobacteria and higher plants by molecular dynamics simulations. *Biochim. Biophys. Acta* 1848 (6): 1319–1330. https://doi.org/10.1016/j.bbamem.2015.02.025.

130 Collins, M.D. (1985). 11 analysis of isoprenoid quinones. *Methods Microbiol.* 329–366. https://doi.org/10.1016/S0580-9517(08)70480-X.

131 Gunner, M.R., Madeo, J., and Zhu, Z. (2008). Modification of quinone electrochemistry by the proteins in the biological electron transfer chains: examples from photosynthetic reaction centers. *J. Bioenerg. Biomembr.* 40 (5): 509–519. https://doi.org/10.1007/s10863-008-9179-1.

132 Uchimiya, M. and Stone, A.T. (2009). Reversible redox chemistry of quinones: impact on biogeochemical cycles. *Chemosphere* 77 (4): 451–458. https://doi.org/10.1016/j.chemosphere.2009.07.025.

133 Chambers, J.Q. (1988). The electrochemistry of quinones. In: *Chemistry of Quinonoid Compounds*, vol. II (ed. S. Patai and Z. Rappoport), 719–757. University of Tennessee, Knoxville, TN: Wiley and Sons Ltd.

134 Gorner, H. (2004). Photoreactions of p-benzo-, p-naphtho- and p-anthraquinones with ascorbic acid. *Photochem. Photobiol. Sci.* 3 (10): 933–938. https://doi.org/10.1039/b410386f.

135 Longatte, G., Rappaport, F., Wollman, F.A. et al. Mechanism and analyses for extracting photosynthetic electrons using exogenous quinones - what makes a good extraction pathway? *Photochem. Photobiol. Sci.* 15 (8): 969–979. https://doi.org/10.1039/c6pp00076b.

136 Gaudin, T. and Aubry, J.-M. (2022). Prediction of pourbaix diagrams of quinones for redox flow battery by COSMO-RS. *J Energy Storage* 49: 104152. https://doi.org/10.1016/j.est.2022.104152.

137 Light, S.H., Su, L., Rivera-Lugo, R. et al. (2018). A flavin-based extracellular electron transfer mechanism in diverse Gram-positive bacteria. *Nature* 562: 140–144. https://doi.org/10.1038/s41586-018-0498-z.

138 Krayz, G.T., Bittner, S., Dhiman, A., and Becker, J.Y. (2021). Electrochemistry of quinones with respect to their role in biomedical chemistry. *Chem. Rec.* 21 (9): 2332–2343. https://doi.org/10.1002/tcr.202100069.

139 Newman, D.K. and Kolter, R. (2000). A role for excreted quinones in extracellular electron transfer. *Nature* 405: 94–97. https://doi.org/10.1038/35011098.

140 Bukhov, N.G., Sridharan, G., Egorova, E.A., and Carpentier, R. (2003). Interaction of exogenous quinones with membranes of higher plant chloroplasts: modulation of quinone capacities as photochemical and non-photochemical quenchers of energy in photosystem II during light-dark transitions. *Biochim. Biophys. Acta* 1604 (2): 115–123. https://doi.org/10.1016/s0005-2728(03)00042-2.

141 Quan, M., Sanchez, D., Wasylkiw, M.F., and Smith, D.K. (2007). Voltammetry of quinones in unbuffered aqueous solution: reassessing the roles of proton transfer and hydrogen bonding in the aqueous electrochemistry of quinones. *J. Am. Chem. Soc.* 129 (42): 12847–12856. https://doi.org/10.1021/ja0743083.

142 Gerhardt, K.L.M.R., Chen, Q., Marshak, M.P. et al. (2015). *Quinone Electrochemistry in Acidic and Alkaline Solutions & Its Application in Large Scale Energy Storage*. American Chemical Society Conference, Boston, Massachusetts, August 16-20. [Online]. Available: http://nrs.harvard.edu/urn-3:HUL.InstRepos:34325476.

143 Franza, T. and Gaudu, P. (2022). Quinones: more than electron shuttles. *Res. Microbiol.* 173 (6–7): 103953. https://doi.org/10.1016/j.resmic.2022.103953.

144 Takeda, K., Birrell, J.A., Kusuoka, R. et al. (2024). Redox properties of pyrroloquinoline quinone in pyranose dehydrogenase measured by direct electron transfer. *ACS Catal.* 14 (16): 12242–12250. https://doi.org/10.1021/acscatal.4c02319.

145 Sandford, C., Edwards, M.A., Klunder, K.J. et al. (2019). A synthetic chemist's guide to electroanalytical tools for studying reaction mechanisms. *Chem. Sci.* 10 (26): 6404–6422. https://doi.org/10.1039/C9SC01545K.

146 Weliwatte, N.S., Grattieri, M., and Minteer, S.D. (2021). Rational design of artificial redox-mediating systems toward upgrading photobioelectrocatalysis. *Photochem. Photobiol. Sci.* 20 (10): 1333–1356. https://doi.org/10.1007/s43630-021-00099-7.

147 Murata, N., Takahashi, S., Nishiyama, Y., and Allakhverdiev, S.I. (2007). Photoinhibition of photosystem II under environmental stress. *Biochim. Biophys. Acta - Bioenerg.* 1767 (6): 414–421. https://doi.org/10.1016/j.bbabio.2006.11.019.

148 Mavrodi, D.V., Peever, T.L., Mavrodi, O.V. et al. (2010). Diversity and evolution of the phenazine biosynthesis pathway. *Appl. Environ. Microbiol.* 76 (3): 866–879. https://doi.org/10.1128/AEM.02009-09.

149 Clifford, E.R., Bradley, R.W., Wey, L.T. et al. (2021). Phenazines as model low-midpoint potential electron shuttles for photosynthetic bioelectrochemical systems. *Chem. Sci.* 12 (9): 3328–3338. https://doi.org/10.1039/d0sc05655c.

150 Zhuang, X., Zhang, Y., Xiao, A.-F. et al. (2022). Applications of synthetic biotechnology on carbon neutrality research: a review on electrically driven microbial and enzyme engineering. *Front. Bioeng. Biotechnol.* 10: 826008. https://doi.org/10.3389/fbioe.2022.826008.

151 Park, D.H. and Zeikus, J.G. (2000). Electricity generation in microbial fuel cells using neutral red as an electronophore. *Appl. Environ. Microbiol.* 66 (4): 1292–1297. https://doi.org/10.1128/AEM.66.4.1292-1297.2000.

152 Sund, C.J., McMasters, S., Crittenden, S.R. et al. (2007). Effect of electron mediators on current generation and fermentation in a microbial fuel cell. *Appl. Microbiol. Biotechnol.* 76 (3): 561–568. https://doi.org/10.1007/s00253-007-1038-1.

153 Simoska, O., Gaffney, E.M., Lim, K. et al. (2021). Understanding the properties of phenazine mediators that promote extracellular electron transfer in *Escherichia coli*. *J. Electrochem. Soc.* 168 (2): 025503. https://doi.org/10.1149/1945-7111/abe52d.

154 Zheng, T., Xu, B., Ji, Y. et al. (2021). Microbial fuel cell-assisted utilization of glycerol for succinate production by mutant of *Actinobacillus succinogenes*. *Biotechnol. Biofuels* 14 (1): 23. https://doi.org/10.1186/s13068-021-01882-5.

155 Ma, J., Feng, Y., Li, L. et al. (2023). Redox mediators stimulated chain elongation process in fluidized cathode electro-fermentation systems for caproate production. *J. Environ. Manag.* 348: 119286. https://doi.org/10.1016/j.jenvman.2023.119286.

156 Gagne, R.R., Koval, C.A., and Lisensky, G.C. (1980). Ferrocene as an internal standard for electrochemical measurements. *Inorg. Chem.* 19: 2854–2855.

157 Rauf, U., Shabir, G., Bukhari, S. et al. (2023). Contemporary developments in ferrocene chemistry: physical, chemical, biological and industrial aspects. *Molecules* 28 (15): 5765. https://doi.org/10.3390/molecules28155765.

158 Fabbrizzi, L. (2020). The ferrocenium/ferrocene couple: a versatile redox switch. *ChemTexts* 6 (4): 22. https://doi.org/10.1007/s40828-020-00119-6.

159 Li, W., Yu, J., Wang, J. et al. (2024). How does ferrocene correlate with ferroptosis? Multiple approaches to explore ferrocene-appended GPX4 inhibitors as anticancer agents. *Chem. Sci.* 15 (27): 10477–10490. https://doi.org/10.1039/D4SC02002B.

160 Brownson, D.A.C. and Banks, C.E. (2014). Interpreting electrochemistry. In: *The Handbook of Graphene Electrochemistry*, 23–77. London: Springer London. https://doi.org/10.1007/978-1-4471-7536-0.

161 Çevik, E., Titiz, M., and Şenel, M. (2018). Light-dependent photocurrent generation: Novel electrochemical communication between biofilm and electrode by ferrocene cored Poly(amidoamine) dendrimers. *Electrochim. Acta* 291: 41–48. https://doi.org/10.1016/j.electacta.2018.08.108.

162 Vassilev, I., Gießelmann, G., Schwechheimer, S.K. et al. (2018). Anodic electro-fermentation: anaerobic production of L-Lysine by recombinant *Corynebacterium glutamicum*. *Biotechnol. Bioeng.* 115 (6): 1499–1508. https://doi.org/10.1002/bit.26562.

163 Kirchhofer, N.D., Rengert, Z.D., Dahlquist, F.W. et al. (2017). A ferrocene-based conjugated oligoelectrolyte catalyzes bacterial electrode respiration. *Chem* 2 (2): 240–257. https://doi.org/10.1016/j.chempr.2017.01.001.

164 Zarenezhad, H., Rezaei, A., Aber, S., and Teimuri-Mofrad, R. (2024). Enhancing extracellular electron transfer and power generation in microbial fuel cell using a ferrocene-based conjugated oligoelectrolyte. *Fuel* 358: 130271. https://doi.org/10.1016/j.fuel.2023.130271.

165 Simoska, O., Cummings, D.A. Jr., Gaffney, E.M. et al. (2023). Enhancing the performance of microbial fuel cells via metabolic engineering of *Escherichia coli* for phenazine production. *ACS Sustain. Chem. Eng.* 11 (32): 11855–11866. https://doi.org/10.1021/acssuschemeng.3c01593.

166 Song, Y.E., Mohamed, A., Kim, C. et al. (2022). Biofilm matrix and artificial mediator for efficient electron transport in CO_2 microbial electrosynthesis. *Chem. Eng. J.* 427: 131885. https://doi.org/10.1016/j.cej.2021.131885.

167 Zhu, X., Dou, F., Long, M. et al. (2023). Electron shuttle-dependent biofilm formation and biocurrent generation: concentration effects and mechanistic insights. *Front. Microbiol.* 14: 1070800. https://doi.org/10.3389/fmicb.2023.1070800.

168 Logan, B.E., Rossi, R., Ragab, A.A., and Saikaly, P.E. (2019). Electroactive microorganisms in bioelectrochemical systems. *Nat. Rev. Microbiol.* 17 (5): 307–319. https://doi.org/10.1038/s41579-019-0173-x.

169 Bond, D.R. and Lovley, D.R. (2003). Electricity production by *Geobacter sulfurreducens* attached to electrodes. *Appl. Environ. Microbiol.* 69 (3): 1548–1555. https://doi.org/10.1128/AEM.69.3.1548-1555.2003.

170 Reguera, G., Nevin, K.P., Nicoll, J.S. et al. (2006). Biofilm and nanowire production leads to increased current in *Geobacter sulfurreducens* fuel cells. *Appl. Environ. Microbiol.* 72 (11): 7345–7348. https://doi.org/10.1128/AEM.01444-06.

171 Nevin, K.P., Richter, H., Covalla, S.F. et al. (2008). Power output and columbic efficiencies from biofilms of *Geobacter sulfurreducens* comparable to mixed community microbial fuel cells. *Environ. Microbiol.* 10 (10): 2505–2514. https://doi.org/10.1111/j.1462-2920.2008.01675.x.

172 Torres, C.I., Krajmalnik-Brown, R., Parameswaran, P. et al. (2009). Selecting anode-respiring bacteria based on anode potential: phylogenetic, electrochemical, and microscopic characterization. *Environ. Sci. Technol.* 43 (24): 9519–9524. https://doi.org/10.1021/es902165y.

173 Ishii, S.I., Logan, B.E., and Sekiguchi, Y. (2012). Enhanced electrode reducing rate during the enrichment process in an air-cathode microbial fuel cell. *Appl. Microbiol. Biotechnol.* 94 (4): 1087–1094. https://doi.org/10.1007/s00253-011-3844-8.

174 Commault, A.S., Lear, G., Packer, M.A., and Weld, R.J. (2013). Influence of anode potentials on selection of *Geobacter* strains in microbial electrolysis cells. *Bioresour. Technol.* 139: 226–234. https://doi.org/10.1016/j.biortech.2013.04.047.

175 Nevin, K.P., Kim, B.C., Glaven, R.H. et al. (2009). Anode biofilm transcriptomics reveals outer surface components essential for high density current production in *Geobacter sulfurreducens* fuel cells. *PLoS One* 4 (5): e5628. https://doi.org/10.1371/journal.pone.0005628.

176 Bond, D.R., Holmes, D.E., Tender, L.M., and Lovley, D.R. (2002). Electrode-reducing microorganisms that harvest energy from marine sediments. *Science* 295 (5554): 483–485. https://doi.org/10.1126/science.1066771.

177 Leang, C., Malvankar, N.S., Franks, A.E. et al. (2013). Engineering *Geobacter sulfurreducens* to produce a highly cohesive conductive matrix with enhanced capacity for current production. *Energy Environ. Sci.* 6 (6): 1901–1908. https://doi.org/10.1039/c3ee40441b.

178 Malvankar, N.S. and Lovley, D.R. (2014). Microbial nanowires for bioenergy applications. *Curr. Opin. Biotechnol.* 27: 88–95. https://doi.org/10.1016/j.copbio.2013.12.003.

179 Feliciano, G.T., Steidl, R.J., and Reguera, G. (2015). Structural and functional insights into the conductive pili of *Geobacter sulfurreducens* revealed in molecular dynamics simulations. *Phys. Chem. Chem. Phys.* 17 (34): 22217–22226. https://doi.org/10.1039/C5CP03432A.

180 Gu, Y., Srikanth, V., Salazar-Morales, A.I. et al. (2021). Structure of *Geobacter* pili reveals secretory rather than nanowire behaviour. *Nature* 597: 430–434. https://doi.org/10.1038/s41586-021-03857-w.

181 Yalcin, S.E., OBrien, J.P., Gu, Y. et al. (2020). Electric field stimulates production of highly conductive microbial OmcZ nanowires. *Nat. Chem. Biol.* 16 (10): 1136–1142. https://doi.org/10.1038/s41589-020-0623-9.

182 Liu, X., Walker, D.J.F., Nonnenmann, S.S. et al. (2021). Direct observation of electrically conductive pili emanating from *Geobacter sulfurreducens*. *MBio* 12 (4): e02209-21. https://doi.org/10.1128/mBio.02209-21.

183 Wang, F., Gu, Y., OBrien, J.P. et al. (2019). Structure of microbial nanowires reveals stacked hemes that transport electrons over micrometers. *Cell* 177 (2): 361–369.e10. https://doi.org/10.1016/j.cell.2019.03.029.

184 Wang, F., Chan, C.H., Suciu, V. et al. (2022). Structure of Geobacter OmcZ filaments suggests extracellular cytochrome polymers evolved independently multiple times. *elife* 11: e81551. https://doi.org/10.7554/eLife.81551.

185 Berman, H.M., Westbrook, J., Feng, Z. et al. (2000). The protein data bank. *Nucleic Acids Res.* 28 (1): 235–242. https://doi.org/10.1093/nar/28.1.235.

186 Sehnal, D., Bittrich, S., Deshpande, M. et al. (2021). Mol* Viewer: modern web app for 3D visualization and analysis of large biomolecular structures. *Nucleic Acids Res.* 49 (W1): W431–W437. https://doi.org/10.1093/nar/gkab314.

187 Ye, Y., Liu, X., Nealson, K.H. et al. (2022). Dissecting the structural and conductive functions of nanowires in *Geobacter sulfurreducens* electroactive biofilms. *MBio* 13 (1): e03822-21. https://doi.org/10.1128/mbio.03822-21.

188 Chen, X., Yang, G., Quan, X. et al. (2024). Significance of a minor pilin PilV in biofilm cohesion of *Geobacter sulfurreducens*. *Sci. Total Environ.* 927: 172242. https://doi.org/10.1016/j.scitotenv.2024.172242.

189 Portela, P.C., Shipps, C.C., Shen, C. et al. (2024). Widespread extracellular electron transfer pathways for charging microbial cytochrome OmcS nanowires via periplasmic cytochromes PpcABCDE. *Nat. Commun.* 15 (1): 2434–2434. https://doi.org/10.1038/s41467-024-46192-0.

190 Malvankar, N.S., Tuominen, M.T., and Lovley, D.R. (2012). Lack of cytochrome involvement in long-range electron transport through conductive biofilms and nanowires of *Geobacter sulfurreducens*. *Energy Environ. Sci.* 5 (9): 8651–8659. https://doi.org/10.1039/c2ee22330a.

191 Wang, F., Mustafa, K., Suciu, V. et al. (2022). Cryo-EM structure of an extracellular Geobacter OmcE cytochrome filament reveals tetrahaem packing. *Nat. Microbiol.* 7 (8): 1291–1300. https://doi.org/10.1038/s41564-022-01159-z.

192 Jiang, J., He, P., Luo, Y. et al. (2023). The varied roles of pilA-N, omcE, omcS, omcT, and omcZ in extracellular electron transfer by *Geobacter sulfurreducens*. *Front. Microbiol.* 14: 1251346. https://doi.org/10.3389/fmicb.2023.1251346.

193 Inoue, K., Leang, C., Franks, A.E. et al. (2011). Specific localization of the *c*-type cytochrome OmcZ at the anode surface in current-producing biofilms of *Geobacter sulfurreducens*. *Environ. Microbiol. Rep.* 3 (2): 211–217. https://doi.org/10.1111/j.1758-2229.2010.00210.x.

194 Rollefson, J.B., Stephen, C.S., Tien, M., and Bond, D.R. (2011). Identification of an extracellular polysaccharide network essential for cytochrome anchoring and biofilm formation in *Geobacter sulfurreducens*. *J. Bacteriol.* 193 (5): 1023–1033. https://doi.org/10.1128/jb.01092-10.

195 Clark, M.M., Paxhia, M.D., Young, J.M. et al. (2021). Adaptive synthesis of a rough lipopolysaccharide in *Geobacter sulfurreducens* for metal reduction and detoxification. *Appl. Environ. Microbiol.* 87 (20): e00964-21. https://doi.org/10.1128/AEM.00964-21.

196 Logan, B.E., Call, D., Cheng, S. et al. (2008). Microbial electrolysis cells for high yield hydrogen gas production from organic matter. *Environ. Sci. Technol.* 42 (23): 8630–8640. https://doi.org/10.1021/es801553z.

197 Peng, L., Zhang, X.-T., Kawaichi, S. et al. (2015). Using acetate and formate as the substrates for *Geobacter sulfurreducens* exoelectrogenesis resulted in different half-saturation potentials. *Electrochemistry* 83 (8): 600–604. https://doi.org/10.5796/electrochemistry.83.600.

198 Malvankar, N.S., Vargas, M., Nevin, K.P. et al. (2011). Tunable metallic-like conductivity in microbial nanowire networks. *Nat. Nanotechnol.* 6 (9): 573–579. https://doi.org/10.1038/nnano.2011.119.

199 Lampa-Pastirk, S., Veazey, J.P., Walsh, K.A. et al. (2016). Thermally activated charge transport in microbial protein nanowires. *Sci. Rep.* 6 (1): 23517. https://doi.org/10.1038/srep23517.

200 Adhikari, R.Y., Malvankar, N.S., Tuominen, M.T., and Lovley, D.R. (2016). Conductivity of individual *Geobacter* pili. *RSC Adv.* 6 (10): 8363–8366. https://doi.org/10.1039/c5ra28092c.

201 Tan, Y., Adhikari, R.Y., Malvankar, N.S. et al. (2016). Synthetic biological protein nanowires with high conductivity. *Small* 12 (33): 4481–4485. https://doi.org/10.1002/smll.201601112.

202 Liu, X., Nealson, K.H., Zhou, S., and Rensing, C. (2022). Reply to Lovley, "Untangling *Geobacter sulfurreducens* nanowires". *MBio* 13 (3): e01041-22. https://doi.org/10.1128/mbio.01041-22.

203 Guberman-Pfeffer, M.J. (2023). Structural determinants of redox conduction favor robustness over tunability in microbial cytochrome nanowires. *J. Phys. Chem. B* 127 (32): 7148–7161. https://doi.org/10.1021/acs.jpcb.3c02912.

204 Yi, H., Nevin, K.P., Kim, B.C. et al. (2009). Selection of a variant of *Geobacter sulfurreducens* with enhanced capacity for current production in microbial fuel cells. *Biosens. Bioelectron.* 24 (12): 3498–3503. https://doi.org/10.1016/j.bios.2009.05.004.

205 Chadwick, G.L., Otero, F.J., Gralnick, J.A. et al. (2019). NanoSIMS imaging reveals metabolic stratification within current-producing biofilms. *Proc. Natl. Acad. Sci. USA* 116 (41): 20716–20724. https://doi.org/10.1073/pnas.1912498116.

206 Jiménez Otero, F., Chadwick, G.L., Yates, M.D. et al. (2021). Evidence of a streamlined extracellular electron transfer pathway from biofilm structure, metabolic stratification, and long-range electron transfer parameters. *Appl. Environ. Microbiol.* 87 (17): e00706-21. https://doi.org/10.1128/AEM.00706-21.

207 Sun, D., Chen, J., Huang, H. et al. (2016). The effect of biofilm thickness on electrochemical activity of *Geobacter sulfurreducens*. *Int. J. Hydrog. Energy* 41 (37): 16523–16528. https://doi.org/10.1016/j.ijhydene.2016.04.163.

208 Methe, B.A., Nelson, K.E., Eisen, J.A. et al. (2003). Genome of *Geobacter sulfurreducens*: metal reduction in subsurface environments. *Science* 302 (5652): 1967–1969. https://doi.org/10.1126/science.1088727.

209 Coppi, M.V., Leang, C., Sandler, S.J., and Lovley, D.R. (2001). Development of a genetic system for *Geobacter sulfurreducens*. *Appl. Environ. Microbiol.* 67 (7): 3180–3187. https://doi.org/10.1128/AEM.67.7.3180-3187.2001.

210 Miceli, J.F., Parameswaran, P., Kang, D.-W. et al. (2012). Enrichment and analysis of anode-respiring bacteria from diverse anaerobic inocula. *Environ. Sci. Technol.* 46 (18): 10349–10355. https://doi.org/10.1021/es301902h.

211 Marshall, C.W. and May, H.D. (2009). Electrochemical evidence of direct electrode reduction by a thermophilic Gram-positive bacterium, *Thermincola ferriacetica*. *Energy Environ. Sci.* 2 (6): 699–705. https://doi.org/10.1039/b823237g.

212 Imran, M., Crowley, D.E., Khalid, A. et al. (2015). Microbial biotechnology for decolorization of textile wastewaters. *Rev. Environ. Sci. Biotechnol.* 14 (1): 73–92. https://doi.org/10.1007/s11157-014-9344-4.

213 Speers, A.M. and Reguera, G. (2012). Electron donors supporting growth and electroactivity of *Geobacter sulfurreducens* anode biofilms. *Appl. Environ. Microbiol.* 78 (2): 437–444. https://doi.org/10.1128/AEM.06782-11.

214 Lin, T., Ding, W., Zhang, D. et al. (2024). Expression of filaments of the *Geobacter* extracellular cytochrome OmcS in *Shewanella oneidensis*. *Biotechnol. Bioeng.* 121 (6): 2002–2012. https://doi.org/10.1002/bit.28702.

215 Jensen, H.M., Albers, A.E., Malley, K.R. et al. (2010). Engineering of a synthetic electron conduit in living cells. *Proc. Natl. Acad. Sci.* 107 (45): 19213–19218. https://doi.org/10.1073/pnas.1009645107.

216 Howley, E., Mangus, A., Williams, D., and Torres, C.I. (2023). Intracytoplasmic membranes develop in *Geobacter sulfurreducens* under thermodynamically limiting conditions. *npj Biofilms Microbiomes* 9 (1): 18. https://doi.org/10.1038/s41522-023-00384-6.

217 Bosch, J., Lee, K.-Y., Hong, S.-F. et al. (2014). Metabolic efficiency of *Geobacter sulfurreducens* growing on anodes with different redox potentials. *Curr. Microbiol.* 68 (6): 763–768. https://doi.org/10.1007/s00284-014-0539-2.

218 Strycharz, S.M., Malanoski, A.P., Snider, R.M. et al. (2011). Application of cyclic voltammetry to investigate enhanced catalytic current generation by biofilm-modified anodes of *Geobacter sulfurreducens* strain DL1vs. variant strain KN400. *Energy Environ. Sci.* 4 (3): 896–913. https://doi.org/10.1039/C0EE00260G.

219 Guberman-Pfeffer, M.J. (2022). Assessing thermal response of redox conduction for *Anti*-Arrhenius kinetics in a microbial cytochrome nanowire. *J. Phys. Chem. B* 126 (48): 10083–10097. https://doi.org/10.1021/acs.jpcb.2c06822.

220 Wang, H., Jing, X., Feng, K. et al. (2024). In situ tuning the structure of *Geobacter* biofilm for bioelectricity enhancement. *Environ. Sci. Technol. Lett.* 11 (2): 106–113. https://doi.org/10.1021/acs.estlett.3c00783.

221 Scarabotti, F., Rago, L., Bühler, K., and Harnisch, F. (2021). The electrode potential determines the yield coefficients of early-stage *Geobacter sulfurreducens* biofilm anodes. *Bioelectrochemistry* 140: 107752. https://doi.org/10.1016/j.bioelechem.2021.107752.

222 Kondaveeti, S., Lee, S.-H., Park, H.-D., and Min, B. (2020). Specific enrichment of different *Geobacter* sp. in anode biofilm by varying interspatial distance of electrodes in air-cathode microbial fuel cell (MFC). *Electrochim. Acta* 331: 135388. https://doi.org/10.1016/j.electacta.2019.135388.

223 Härrer, D., Elreedy, A., Ali, R. et al. (2023). Probing the robustness of *Geobacter sulfurreducens* against fermentation hydrolysate for uses in bioelectrochemical systems. *Bioresour. Technol.* 369: 128363. https://doi.org/10.1016/j.biortech.2022.128363.

224 Ramírez-Moreno, M., Berenguer, R., Ortiz, J.M., and Esteve-Núñez, A. (2024). Study of the influence of nanoscale porosity on the microbial electroactivity between expanded graphite electrodes and *Geobacter sulfurreducens* biofilms. *Microb. Biotechnol.* 17 (1): e14357. https://doi.org/10.1111/1751-7915.14357.

4

Glucometers and Continuous Glucose Monitors

Nunzio Giorgio G. Carducci and David P. Hickey

Department of Chemical Engineering and Material Science, Michigan State University, East Lansing, MI, USA

4.1 Introduction

Electrochemical glucose sensors are among the most widely used point-of-care (POC) medical devices in the world and are essential for the estimated 536.6 million people living with diabetes mellitus worldwide to manage their blood-glucose levels [1]. Electrochemical glucose sensors incorporate a glucose-oxidizing enzyme onto an electrode interface to generate an electrochemical signal proportional to the bulk concentration of glucose in a sample. The first glucose sensors, reported in the early 1960s, were integrated into large instruments that had to be run by trained technicians. Since then, electrochemical glucose sensors have matured into exquisitely engineered commercial devices that can be worn and operated by the patient to continuously monitor blood-glucose levels [2–4]. Moreover, the need for accurate and reliable glucose monitoring in the 1970s helped to spark the modern field of bioelectrocatalysis [5–10]. Subsequently, several unique and innovative strategies have been developed in the field of electrochemical glucose sensors (including enzyme engineering, biocompatible electrode coatings, flexible electronics, electroactive hydrogels, 3D-printing, microelectrode array design, and microelectronics fabrication) [11–17] to incrementally address challenges on the way to commercialization and expansion into a multibillion-dollar industry. In this way, electrochemical glucose sensors have established the foundation for emerging bioelectrode interfaces while also providing a potential blueprint for scaling and commercialization of emerging bioelectrocatalysis applications. In this chapter, we describe the operating principles, components, and design of electrochemical glucometers. Additionally, we provide an overview of modern applications ranging from single-point testing to continuous glucose monitoring and highlight the remaining challenges for future sensor development. Given the significance of glucose sensors in the context of advancing the field of bioelectrocatalysis, this chapter includes a brief history of glucose sensors from both scientific development and commercialization perspectives.

Electrochemical glucose sensors consist of three primary components: (i) a recognition element to differentiate glucose from any other chemicals present in the analytical sample, (ii) a transduction element that produces and propagates an electrochemical signal to an electrode detector, and (iii) a detector/signal processing element (i.e. a potentiostat) that converts the signal into a readable format (Figure 4.1).

Glucometers are broadly categorized into (i) clinical glucometers, (ii) self-monitoring single-measurement glucometers, and (iii) continuous glucose sensors. Clinical glucose sensors are exclusively employed in a clinical setting – utilized both in central laboratory and

Bioelectrocatalysis: From Electron Transfer Processes to Emerging Technological Applications, First Edition.
Edited by Shelley D. Minteer and Matteo Grattieri.

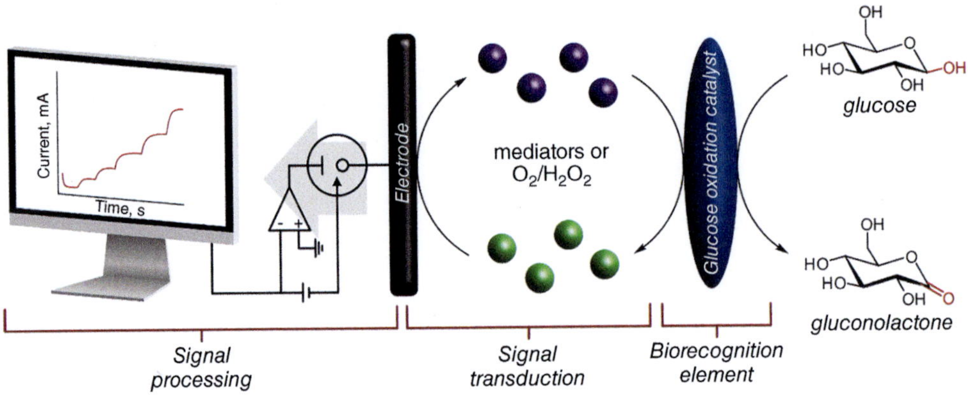

Figure 4.1 General schematic of an electrochemical glucose sensor.

inpatient POC settings – and are often integrated into larger blood-gas analyzers [18]. In contrast to clinical glucometers, self-monitoring single-measurement glucometers (i.e. glucose test strips) [19] and continuous glucose monitors (e.g. worn glucose sensing patches and implanted glucose sensors) [20] are most often used for in-home diabetes management. For all three classes of glucometers, the detection principles are largely the same: glucose is oxidized by a catalyst at an electrode surface, and a co-product of the reaction (e.g. hydrogen peroxide or a small molecule redox mediator) diffuses to the electrode where it is detected. In this way, the signal strength is proportional to (and can thus be used to calculate) glucose concentration in the parent sample solution. From these simple detection principles, glucometers have evolved significantly over the last 60 years to incorporate advanced technologies and refined methodologies that have greatly enhanced their accuracy, reliability, and user-friendliness.

4.1.1 Evolution of Modern Glucometers

Much of the early work on glucose detection was focused on the use of an enzyme, glucose oxidase (GOx), as a highly selective glucose oxidation catalyst. In the presence of O_2, the flavin adenine dinucleotide ($FAD/FADH_2$) cofactor of GOx oxidizes glucose to gluconolactone while producing hydrogen peroxide (H_2O_2) via the following reaction sequence [21–25]:

$$\text{glucose} + \text{GOx}_{\text{FAD}} \rightarrow \text{gluconolactone} + \text{GOx}_{\text{FADH}_2} \tag{4.1}$$

$$\text{GOx}_{\text{FADH}_2} + O_2 \rightarrow \text{GOx}_{\text{FAD}} + H_2O_2 \tag{4.2}$$

Monitoring the conversion of O_2 to H_2O_2 (reaction 4.2) forms the foundation of the first generation of electrochemical glucose sensors. The first electrochemical glucose assay – described by Malmstadt and Pardue in 1961 – used the H_2O_2 produced during glucose oxidation (along with a molybdenum catalyst) to oxidize potassium iodide to elemental iodine [26]. The resultant change in iodine concentration (relative to iodide) induces a change in the open circuit potential of the electrochemical cell, which was then correlated back to the glucose concentration. Unfortunately, this approach required complete (or near-complete) oxidation of glucose in the sample. Therefore, measurement times heavily influenced the accuracy of measured glucose concentrations.

The first practical electrochemical glucose sensor was proposed by Clark and Lyons in 1962 and expanded on an earlier design for measuring dissolved oxygen concentrations (now referred to as a Clark electrode, where O_2 is electrochemically reduced at a Pt cathode) [27, 28]. The sensor utilized a layer of GOx sandwiched between two cellulose dialysis membranes stretched over a Pt electrode. As glucose and O_2 diffuse from the bulk solution through the first membrane, O_2 is consumed while gluconolactone is produced. Under mild aqueous conditions, gluconolactone is spontaneously hydrolyzed to gluconic acid, thereby causing a change in local pH values.

$$\text{gluconolactone} + H_2O \rightarrow \text{gluconate} + H^+ \tag{4.3}$$

In this way, the proposed sensor could use the working electrode to either monitor consumption of O_2 (via a Clark electrode) or a shift in local pH (via a pH probe). While many early glucose sensors followed this approach, several factors – including the baseline oxygen levels and glucose permeability through the membrane – were shown to significantly affect sensor performance. Updike and Hicks improved upon this by utilizing GOx entrapped in a polyacrylamide membrane to increase glucose permeability, while issues with background oxygen levels were addressed by using an electrochemical cell containing two multiplexed Pt cathodes [5]. In this configuration, both cathodes were poised to reduce O_2; however, the primary cathode contained a layer of active enzyme, while the second cathode included inactive enzyme to establish a baseline current and serve as a negative control for the active GOx electrode.

In 1972, Guilbault and Lubrano introduced the first glucose biosensor that operated by measuring hydrogen peroxide generated by glucose oxidation of GOx (reaction 4.2) [29, 30]. Reliance on peroxide for signal transduction presented several drawbacks due to undesired reactivity as a mild oxidant (leading to sensor inaccuracy and instability) and overlapping oxidation potential with other compounds in blood (leading to an overestimation of glucose in the sample) [31, 32]. Nevertheless, peroxide detection-based approaches have been widely studied and were used for the first direct measurement of blood-glucose levels. In 1975, this approach was utilized for the first commercially successful glucometer when the Yellow Springs Instrument Company's Model 23 whole blood analyzer was introduced [33].

The need to improve sensor response time and accuracy led to the utilization of redox mediators – compounds that behave as electron shuttles between an electrode and the cofactor of a glucose-oxidizing enzyme. Redox mediators can exhibit much higher electron transfer rates compared to oxygen, which helps mitigate accuracy issues due to variation of the background oxygen level. One of the earliest glucose biosensors incorporating a redox mediator – ferricyanide – appeared in a 1968 patent filed by Blaedel and Olson. From this seminal work, ferricyanide later became widely adopted in commercial blood-glucose monitoring strips [34, 35]. Another early study by Williams *et al.* employed benzoquinone as a redox mediator, noting its competition with oxygen, which was overcome by increasing quinone concentrations [36]. Further simplification came with immobilizing the mediator and enzyme directly on the electrode surface. In 1984, Cass *et al.* reported graphite electrodes with bound ferrocene and GOx [37]. This approach continued with Degani and Heller, who covalently attached GOx to electrodes using ferrocene derivatives in 1987 [38]. In 1990, Heller's group interfaced GOx using osmium-containing redox hydrogels that are now widely used [39].

Commercialization began in 1975 with Yellow Spring Instrument Company's Model 23, but their large size and high costs due to the platinum electrodes largely restricted them to clinical use settings [33]. The first portable electrochemical blood-glucose monitor was patented in 1985, leading to the development of the first portable home-use disposable test strips known as ExacTech [40–43]. The device employed a ferrocene redox mediator initially described by Cass et al. and was developed through a collaboration between Genetics International (later known as Medisense) and the Universities of Cranfield and Oxford [33, 37, 44]. Newer models only need several seconds and sub-microliter sample amounts to provide accurate results [45–47]. These advancements

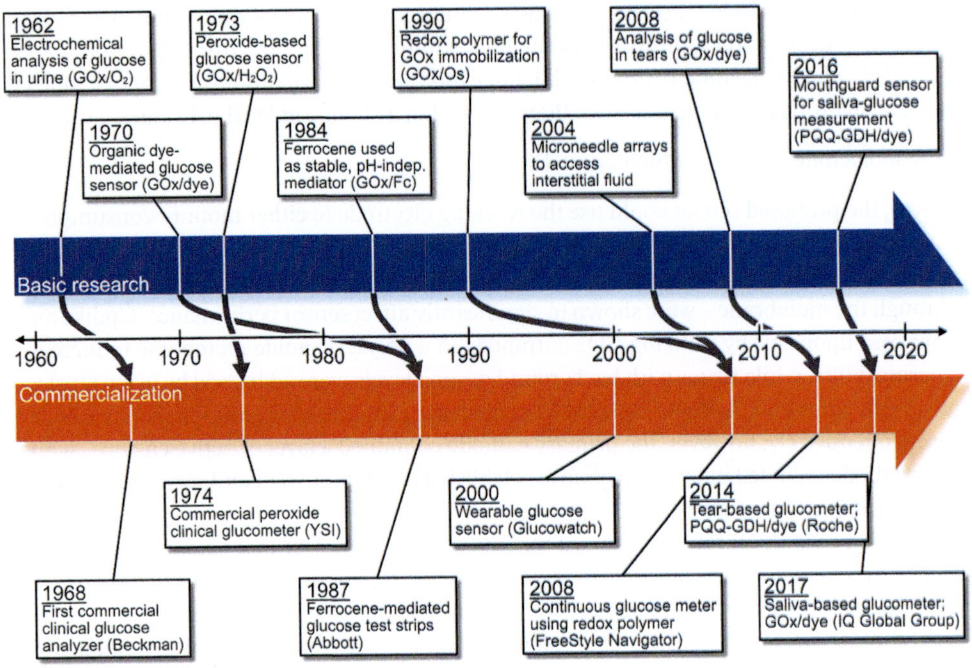

Figure 4.2 Parallel timelines of the basic research and commercial developments of electrochemical glucometers.

have been driven by key innovations in sensor technology, particularly in the elements of the bioelectrochemical interface. A summary of glucometer developments is illustrated in Figure 4.2.

4.2 Glucose Oxidation Catalysts in Glucometers

The design of a glucose biosensor starts with the selection of a glucose oxidation catalyst, because all other elements of the bioelectrochemical interface (e.g. electrode material, medium, pH, signal transduction) must be selected to ensure catalytic activity under the desired operating conditions. Electrochemical glucose sensors operate by catalyzing the oxidation of glucose to gluconolactone and use either potentiometric (measuring the electrochemical potential of the working electrode), amperometric (measuring the current response at a constant applied potential), or conductometric (measuring the ionic conductivity at the working electrode) techniques to measure the rate of glucose oxidation (and thus, the concentration of glucose in a sample). Recently, several nonenzymatic glucose oxidation electrocatalysts have been described for glucose sensors [48–50]; however, the majority of glucometers utilize one of two types of enzymatic catalysts – GOx or glucose dehydrogenase (GDH) – to oxidize glucose and generate an electrochemical signal. GOx contains a non-dissociable FAD cofactor that catalyzes glucose oxidation with molecular oxygen as a native terminal electron acceptor (Figure 4.3a), while GDHs use either (i) a dissociable nicotinamide adenine dinucleotide (NAD^+) cofactor, (ii) a non-dissociable pyrroloquinoline (PQQ) cofactor, or (iii) a non-dissociable FAD cofactor (Figure 4.3b). Apart from their cofactors, these two families of enzymes differ in their turnover rates, specificity to glucose, binding affinity to glucose (as measured by the respective Michaelis constant, K_M), redox potential, and their co-substrates (i.e. native

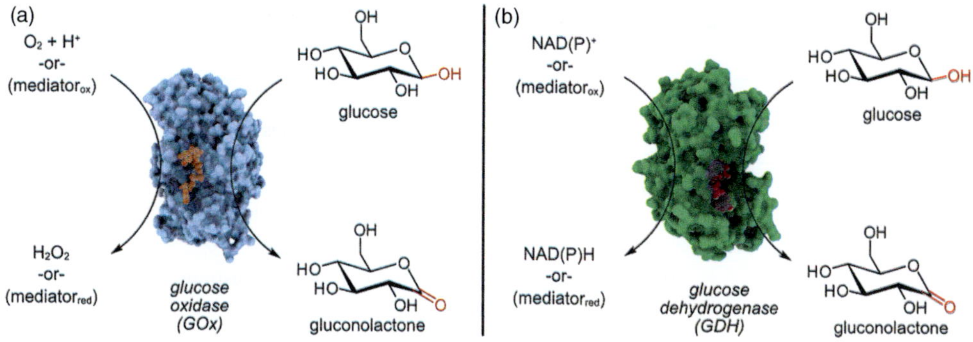

Figure 4.3 Reaction scheme of (a) FAD-dependent glucose oxidase (GOx) and (b) glucose dehydrogenase (GDH) with glucose and an electron acceptor. PDB 2GAL (left) and 2VWH (right).

electron acceptors). In the context of glucose sensors, each of these differences offers a balance of advantages and drawbacks.

4.2.1 Glucose Oxidase

GOx is the most well-studied enzyme in the context of bioelectrocatalysis due to its high substrate specificity for β-D-glucose and moderate specific activity (K_M = 26 mM; specific activity = 132 μmol min^{-1} mg$_{GOx}$$^{-1}$) [51]. Glucose is oxidized by GOx via sequential hydride and proton transfers from the oxidized form of the enzymatic cofactor, FAD^+, to produce gluconolactone and the reduced form of the cofactor, $FADH_2$. Wild-type GOx from *Aspergillus niger* is active from pH 3 to pH 8 (maintaining 90% of its maximum activity); however, low concentrations of chloride anions ([Cl^-] > 5 mM) have an inhibitory effect at low pH and result in a truncated pH range (pH 5–8) [21]. The non-dissociable FAD/$FADH_2$ cofactor has a pH-dependent redox potential of −97 mV vs. the normal hydrogen electrode (NHE) (−338 mV vs. a saturated calomel electrode [SCE] at pH 7.4) that, once reduced, reacts with O_2 to produce FAD^+ and hydrogen peroxide (H_2O_2) [52]. As described above, H_2O_2 concentration can be quantified electrochemically as an indirect measure of glucose concentration; however, H_2O_2 acts as an inhibitor to GOx at low concentrations and can denature GOx at high concentrations [53]. A wide range of chemical oxidants can act as alternative electron acceptors for GOx to avoid the undesired production of H_2O_2, but it should be noted that the primary drawback of GOx arises from reactivity with O_2 as a co-substrate. In the context of glucose sensor engineering, the presence of O_2 in GOx-based sensors can result in shorter device lifespan due to the buildup of H_2O_2 and can result in erroneous glucose measurements in second-generation biosensors [54]. The latter arises due to "leeching" of electrons from the desired mediated signal transduction pathway by O_2.

4.2.2 Glucose Dehydrogenase

The use of GDHs in electrochemical glucose monitoring technologies has grown dramatically in the last 10 years, largely due to a lack of undesired reactivity with O_2 [55–59]. Unlike GOx, GDHs cannot effectively utilize O_2 as a terminal electron acceptor. Instead, the native terminal oxidant depends on the type of GDHs being used. Both FAD-GDH and PQQ-GDH utilize quinone of quinoid redox dyes as co-substrates during glucose oxidation, while the dissociable nicotinamide cofactor of NAD-GDH can be reoxidized by numerous artificial catalysts or natural metabolic

Table 4.1 Non-comprehensive summary of glucose-oxidizing enzymes used in electrochemical glucose biosensors [65].

Enzyme (cofactor)	EC number	Advantages	Disadvantages
GOx (FAD)	1.1.3.4	High glucose specificity	Relatively low specific activity; reacts with O_2 (variable O_2 may lead to erroneous readings)
GDH (PQQ)	1.1.5.2	High specific activity	Promiscuous substrate scope; high reactivity to maltose may lead to erroneous readings
GDH (NAD[P])	1.1.1.118 (NAD) 1.1.1.119 (NADP)	Moderate glucose specificity; moderate specific activity	Dissociable cofactor must be supplied
Bacterial GDH (FAD)	1.1.5.9	Moderate glucose specificity	Promiscuous substrate scope; high reactivity to maltose may lead to erroneous readings
Fungal GDH (FAD)	1.1.5.9	Moderate glucose specificity	Promiscuous substrate scope; high reactivity to xylose may lead to erroneous readings

pathways [60]. The primary drawback to GDHs in the context of glucose sensors is their promiscuity toward non-glucose substrates (i.e. other monosaccharides), which causes poor selectivity in sensors that incorporate them into their design (relative to GOx). Specifically, PQQ-GDH exhibits high relative activity toward 3-O-methyl-glucose (81%), maltose (61%), lactose (61%), allose (47%), and galactose (11%), while FAD-GDH exhibits high relative activity toward maltose (26%). It should be noted that NAD-GDH has fallen out of favor in the field of glucose sensors because of the need to supply the dissociable NAD^+ cofactor – resulting in high cost, inefficient electrochemical regeneration, and leaching of cofactor in implantable sensors – however, substantial interest remains for both PQQ-GDH and FAD-GDH [61–64]. A list of enzymes commonly used in glucose biosensors is included in Table 4.1.

4.2.3 Engineering Enzymes for Improved Glucometers

While the wild-type GOx and PQQ-GDH were originally expressed from *A. niger* and *Acinetobacter calcoaceticus*, respectively, engineered derivatives of both enzymes have been expressed in several different organisms to improve their efficacy in glucose meters. Among these efforts, most have been focused on using site-directed mutagenesis to improve the substrate specificity of FAD-GDH (to favor glucose over maltose), as well as to improve both specificity and solubility of PQQ-GDH [66, 67]. In addition to improving substrate specificity, FAD-GDH has been bioengineered to (i) improve heterogeneous electron transfer kinetics, (ii) enhance enzyme stability, and (iii) increase enzyme expression. To improve electron transfer kinetics between the flavin cofactor and either a mediator or a heterogeneous electrode surface, several groups have utilized enzyme engineering to fuse a heme-containing cytochrome domain to the N-terminus of FAD-GDH [68, 69]. The cytochrome domain – which itself is sufficiently small to undergo moderately fast direct electron transfer (DET) – was shown to act as a fused mediator for the parent FAD-GDH enzyme. Improved enzyme stability has been achieved through the utilization of FAD-GDH homologs identified in thermophilic filamentous fungi. While GOx exhibits decreased activity above ~40 °C [68], gene homologs cloned from thermophilic *Talaromyces emersonii* and expressed in *Escherichia coli* produced FAD-GDH that exhibit maximum activity at ~65 °C (400% relative activity compared to that at 37 °C) [70]. Stability studies suggest that the increased activity at elevated temperatures of *T. emersonii* GDH is accompanied by an enhanced stability up to 60 °C. Finally, the utilization of glucose-oxidizing enzymes in disposable glucose sensors necessitates production on an industrial scale. The most widely employed enzyme is *A. niger* GOx; however, the expression of fungal GOx is relatively low. Consequently, substantial efforts have been made to improve the production of GOx by expressing it in a variety of different organisms [71]. While many expression systems, such as *E. coli* and *Saccharomyces cerevisiae*, enable rapid growth and production of GOx, the resulting recombinant proteins often exhibit significantly lower specific activity due to hyperglycosylation (which may interfere with glucose transport to the FAD cofactor) [72]. To overcome this challenge, several variants of GOx have been expressed in yeast *Pichia pastoris*, which results in less glycosylation and thus specific activity that is three times higher than the same variants expressed in *S. cerevisiae* [72, 73].

4.3 Operating Principles of Glucometers

Regardless of the enzyme employed, glucose sensors are categorized by the type of signal transduction mechanism used in their design. Glucometers, whose electrochemical signal is propagated either (i) indirectly via hydrogen peroxide, (ii) indirectly via a chemical redox mediator, or (iii) via DET between the enzymatic cofactor and electrode surface, are defined as either first-, second-, or third-generation sensors, respectively (Figure 4.4). Recent reports have described a "fourth-generation" glucose sensors that use nonenzymatic electrocatalysts for glucose oxidation; however, these sensors often require alkaline conditions to operate and lack the selectivity and sensitivity of enzymatic sensors [48–50]. Consequently, many such sensors are not yet practical for POC devices, and the remainder of this chapter is focused on enzymatic glucose sensors.

4.3.1 Detection of H_2O_2 from GOx/O_2 Glucose Oxidation

Indirect glucose assays that measure the peroxide generated during the oxidation of glucose by GOx and O_2 (or first-generation glucose biosensors) can work either by electrooxidation of H_2O_2 to O_2 or electroreduction of H_2O_2 to H_2O. Both measurements are done amperometrically using a constant potential at the working electrode while monitoring the current response. Electrooxidation of H_2O_2 is accomplished at 0.3–0.8 V vs. SCE using a catalytic working electrode, such as Pt on graphite, functionalized carbon nanotubes, palladium, iridium dioxide, and ruthenium–platinum alloys, among many others. The majority of catalyst research is focused on selectively lowering the overpotential required for H_2O_2 oxidation. As described earlier, several common interferants (including ascorbate and acetaminophen) are oxidized in the same electrochemical window as H_2O_2 and, therefore, must be either removed from the sample (via permselective membranes) or measured using a secondary method for proper signal subtraction. Electrocatalytic reduction is considerably more common and is accomplished at −0.1 V vs. SCE using a catalytic electrode, such as noble metals (e.g. Pd, Pd-Pt, Pt nanowires, Au nanoparticles), metal–DNA hybrid nanoparticles, and metal-modified silicate xerogels [74–77].

As an alternative to transition metal electrocatalysts, the electroreduction of H_2O_2 for the indirect measurement of glucose can be catalyzed by a peroxidase enzyme. Horseradish peroxidase (HRP) – the most widely used enzyme for this purpose – contains a tetravalent Fe cofactor that

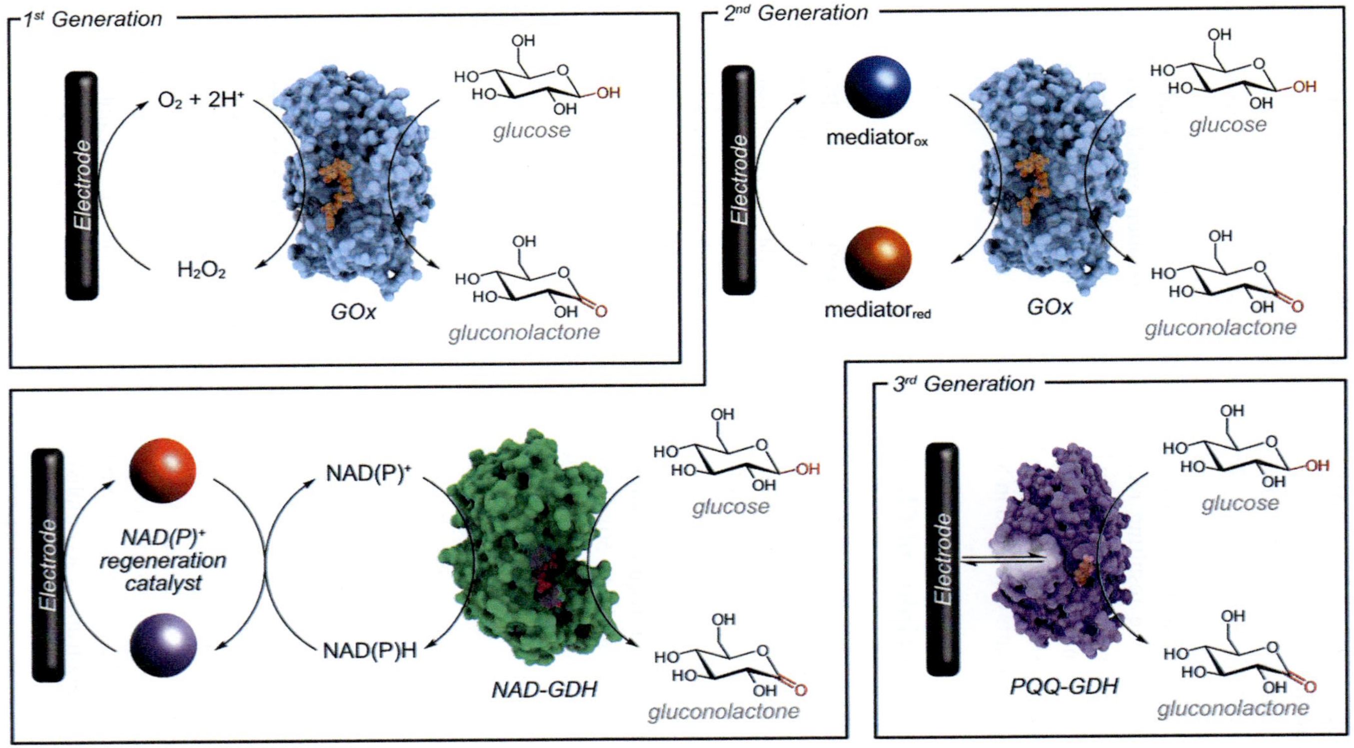

Figure 4.4 Schemes of first-, second-, and third-generation electrochemical enzymatic glucose sensors.

reduces H_2O_2 to hydroxide ions [78–80]. The resulting oxidized HRP can then be electrochemically reduced either by DET or with the use of an electrochemical mediator to generate the current measured for glucose detection. The primary benefit of using HRP for the detection of H_2O_2 is that it maintains high catalytic activity and selectivity under physiological conditions. However, unlike its transition metal counterparts, HRP (along with any redox enzyme) must be integrated into a bioelectrode interface to enable efficient electrochemical regeneration of the redox cofactor.

4.3.2 Mediated Bioelectrocatalysis for Glucose Sensing

The amperometric glucose assay with GOx can avoid the production of H_2O_2 altogether by electrochemically regenerating the FAD^+ cofactor using an exogenous redox mediator (or second-generation glucose biosensor). A redox mediator is any organic, organometallic, or inorganic molecule capable of shuttling electrons between the $FAD^+/FADH_2$ active site and an electrode surface. This process of mediated electron transfer (MET)-based bioelectrocatalysis was first demonstrated in 1970, where glucose oxidation by GOx was electrochemically mediated using several electrochemically active small molecules, including phenothiazine dyes (e.g. methylene green, phenazine methosulfate), phenazine dyes (e.g. phenosafranin), and quinoid dyes (e.g. 2,6-dichloroindophenol) [81–83]. Since then, several classes of redox-active molecules have been reported as mediators for GOx as well as PQQ-GDH and FAD-GDH, including 5,10-dihydro-5,10-dimethylphenazine [83], 1,6-dithispyrene [84], 1,4-benzoquinone [85], potassium hexacyanoferrate [86], and ferrocenylmethanol [87]. Additionally, numerous quinones, ferrocene derivatives, osmium complexes, cobalt complexes, and phenazine derivatives have been extensively studied as mediators for enzymatic glucose oxidation (Figure 4.5) [82, 89–91].

Selection of an effective mediator is traditionally based on three criteria: stability in the reduced and oxidized form, diffusivity, and redox potential. In the context of glucose sensors, mediators must not chemically decompose on the timescale of the enzymatic reaction that is being mediated

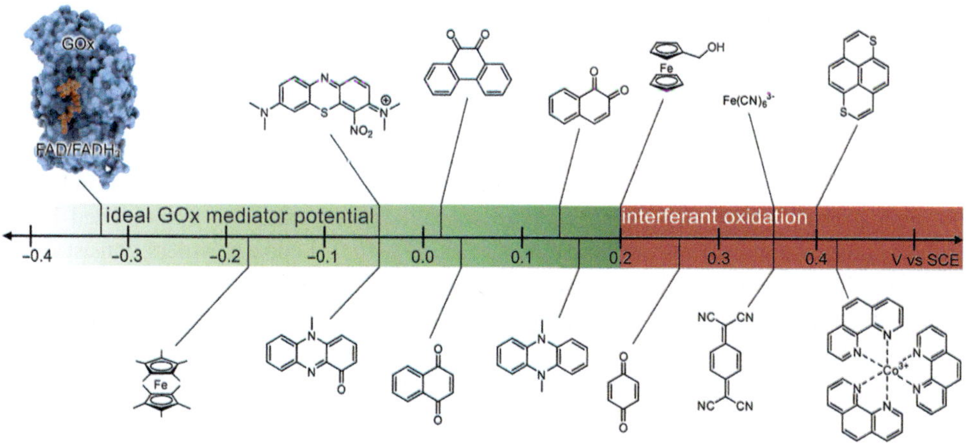

Figure 4.5 A non-comprehensive overview of mediators that have been used to mediate GOx for electrochemical glucose sensors. The formal potential of each mediator (and the FAD cofactor of GOx) is reported at pH 7. Mediators include (from right to left) $Co(phen)_3^{3+}$, 1,6-dithiapyrene, tetracyanoquinodimethane, ferricyanide, 1,4-benzoquinone, ferrocenemethanol, 5,10-dihydro-5,10-dimethylphenazine, 1,2-napthoquinone, 1,4-naphthoquinone, 9,10-phenanthrenequinone, methylene green, pyocyanin, decamethylferrocene, and the FAD cofactor of GOx [52, 82–86, 88].

[92]. In this way, undesired chemical reactions of the mediator must be slower than the bimolecular electron transfer between it and the redox cofactor. Diffusivity refers either to the rate of mass transport for the mediator (for freely dissolved mediators) or the apparent electron diffusivity through a redox polymer (for immobilized mediators/enzymes), as discussed in more detail below. Faster diffusivity is preferred, but the range of diffusivities for small molecule mediators is relatively narrow (generally 5×10^{-6} – 2×10^{-5} cm^2 s^{-1}) so it can often be neglected. The redox potential of a mediator relative to the redox cofactor of the glucose-oxidizing enzyme is also an important property. The mediator must have a potential that is more oxidizing than the redox cofactor to provide a thermodynamic driving force for electron transfer to occur. The difference in redox potential of the cofactor and mediator is related to the rate of electron transfer between them [92]. A higher redox potential results in faster bimolecular rate constant for cofactor regeneration; however, mediator potentials above ~0.2 V vs. SCE result in undesired oxidation of common interferants and thus lead to erroneously high glucose readings [93–96]. In accordance with Marcus theory, it is possible for a mediator potential to be high enough to be in the inverted region (i.e. the driving force beyond which electron transfer rates decrease with increasing mediator potential); however, this is rarely observed due to diffusion limitations.

These mediator selection criteria generally assume that the chemical structure of a mediator does not influence the corresponding mediator–enzyme bimolecular rate constant (i.e. mediator is assumed to behave as a sphere); however, there is growing evidence that the charge and shape of a mediator can play a significant role in mediating glucose bioelectrocatalysis [90, 97]. Recent work used parameterized modelling to identify quantitative structure–activity relationships for quinone mediators with GOx [90]. The combination of redox potential and mediator size was found to correlate strongly with bimolecular rate constant (and, thus, the maximum catalytic current density and glucose sensitivity), thereby suggesting that small mediators with lower redox potentials may achieve the same current density and sensitivity as large mediators with higher redox potentials.

It should be noted that there are specific circumstances where additional mediator selection criteria should be considered. An example of this is discussed below for single-point test glucometers, where mediator solubility and solvation rate are important parameters to reduce measurement times.

4.3.3 Electrical "Wiring" of GOx or GDH in Redox Hydrogels

For applications like implantable glucose sensors, the use of dissolved redox mediators is impractical. In such devices, redox mediators and glucose-oxidizing enzymes are immobilized in a cross-linked polymer hydrogel on the electrode surface [98–105]. Redox polymers consist of a parent polymer backbone to which redox mediators are covalently attached. Hydrogel films are then prepared by cross-linking the redox polymer in the presence of enzyme on an electrode surface. Unlike freely dissolved mediators, which transport charges to/from the electrode via diffusion of the oxidized mediator, immobilized mediators facilitate charge transfer via self-exchange (i.e. electron hopping between immobilized mediators). This enables efficient electron transfer to any enzyme in the redox hydrogel. In the context of glucose biosensors, higher loading of glucose-oxidizing enzyme results in higher sensitivity and limit of detection for the resulting device. Furthermore, because the mediator and enzyme are enveloped in the hydrogel matrix, leaching of potentially toxic sensor components (e.g. heavy metal- or viologen-based mediators) is not observed.

4.3.4 DET Bioelectrocatalysis for Glucose Sensing

Despite the efficacy of mediated glucometers, the use of a redox shuttle inherently increases the overpotential required for electrochemical detection. Furthermore, for non-covalently immobilized mediators, leeching of potentially toxic mediators from the glucose sensor interface is a concern for implanted continuous glucose monitor. Consequently, there are potential advantages to utilizing an enzyme–electrode interface that enables direct electrochemical oxidation of the cofactor. When possible, DET between an electrode and enzymatic cofactor avoids the drawbacks of MET; however, electron tunneling rates decrease logarithmically with distance between the electrode and redox cofactor [106–108]. Because of this distance dependence, redox enzymes often need to be properly oriented on the electrode surface to observe DET. It should be noted that even with optimal orientation, DET is not always possible [109].

Notably, the FAD cofactor of GOx is buried in the protein matrix with an average distance to the outer shell of ~22 Å, which is considerably greater than the maximum distance of ~12 Å necessary for efficient electron transfer [110]. It is worth noting that this maximum distance value is based on achieving a transmission coefficient (i.e. probability that electron transfer will take place) that, in turn, would be sufficient to enable heterogeneous electron transfer rates that are comparable to the activity of (e.g.) GOx activity (~10^6 M^{-1} s^{-1}). Because the maximum rate of DET at distances greater than 12 Å is much slower than the rate of glucose oxidation by GOx, any signal would be indistinguishable from the background current. Despite this, most of the early DET-based glucose biosensor studies utilized GOx, leading to controversial results because the use of O_2 as a native co-substrate can be misinterpreted as DET [109].

DET-type (or third-generation) glucose sensors have been successfully demonstrated with a few notable enzymes. In early applications of DET, type I HRP was effectively utilized in several GOx/H_2O_2-based glucose sensors [78, 79, 111]. Due to the close proximity of the Fe-heme cofactor in HRP to the protein surface, DET can be achieved by adsorption onto CNT-modified glassy carbon electrodes while using a hydrogel coating (such as Nafion or polypyrrole) to prevent desorption. Additionally, cellobiose dehydrogenase (CDH) – an O_2-independent enzyme that catalyzes the oxidation of glucose to gluconolactone – contains a flexible cytochrome domain in addition to the glucose-oxidizing flavin domain. When adsorbed onto an electrode surface, this cytochrome domain acts as a built-in mediator to enable DET of CDH in glucose biosensors. Electrode immobilization has been accomplished via either covalent attachment to a self-assembled monolayer, chemical cross-linking of physically adsorbed CDH, or adsorbing onto a poly(ethylenimine)-coated graphite electrode [112–115]. The use of CDH in glucose sensors is limited because the wild-type CDH has a much higher K_M for glucose (relative to GOx and GDH); however, its bioelectrocatalytic response to glucose is linear over the physiologically relevant range, and the low potential of the cytochrome domain (~0.1 V vs. NHE) allows DET-type sensors to operate without significant background signal of many electrochemically active interferants [113].

4.3.5 Glucose Concentrations and Forms in Clinically Relevant Conditions

The effectiveness of a glucometer hinges on the performance of its biorecognition and signal transduction elements within the chosen analytical medium. Venous blood-glucose measurements provide the most accurate determination of an individual's glycemic status; however, accessing blood samples (either through finger pricking or blood draws) causes a significant amount of discomfort to patients. Alternatively, human blood-glucose concentrations can be estimated by measuring glucose levels in various biological fluids, including excreted urine, sweat, saliva, tears, or interstitial fluid (ISF). Each of these media presents its own unique array of chemical interferants and glucose concentration ranges that must be considered.

Under fasting conditions, venous blood-glucose concentrations range from 3.9 to 5.6 mM (70–100 mg dL^{-1}), while non-fasting blood-glucose concentrations typically vary from 3.9 to 7.8 mM (70–140 mg dL^{-1}) in individuals without diabetes [116, 117]. The concentration range of glucose in ISF (i.e. fluid between blood vessels and cells) is approximately the same as that in venous blood samples. In contrast to the relatively high glucose concentration in ISF, the concentration of glucose in saliva, sweat, and tears is 1–2 orders of magnitude smaller than that in the blood. For healthy patients, salivary glucose levels are in the range of 0.2–0.4 mM, sweat-glucose levels are in the range of 0.06–0.1 mM, and glucose concentrations in ocular fluids (i.e. tears) are in the range of 0.05–0.5 mM [118–121]. It should be noted that the upper limit of glucose concentrations in diabetic patients is significantly higher for all biological fluids by an order of magnitude (a summary of typical glucose concentrations in various biological media is provided in Table 4.2).

The efficacy of glucose as a biomarker for a patient's glycemic state depends heavily on the biological fluid. There is generally a very good correlation between glucose levels in ISF and blood glucose – making it (currently) the most promising blood-free media for diabetes management. The primary drawback to using ISF is that changes in physiological ISF-glucose levels lag behind changes in blood-glucose levels by several minutes due to mass transport limitations of glucose from blood vessels into ISF beneath the skin [122]. In contrast to ISF-glucose, there is no consensus as to whether correlations between glucose concentrations in sweat or saliva and blood-glucose levels are sufficient to be diagnostically relevant [121, 123, 124]. In particular, salivary glucose measurements may be contaminated by metabolites or impurities from ingested food, while sweat-glucose measurements must account for variable pH, temperature, dilution effects (from variable sweat rates), and ionic strength. The use of ocular fluid as a blood substitute for monitoring

Table 4.2 A summary of glucose concentrations in clinically relevant biological media.

	Glucose concentration[a]/mM			
Human biofluid	**Healthy patients**	**Diabetic patients**	**Advantages**	**Disadvantages**
Blood (venous)	4.9–6.9	2–40	Gold standard for determining glycemic state	Poor accessibility; contains biofouling agents
Interstitial fluid	3.9–6.6	2–22	Excellent correlation to glycemic state; accessible with the use of microneedles	Significant lag time for changes in glycemic state
Urine	2.8–5.6	>5.6	Easily accessible	Poor sensitivity to changes in glycemic state; poor correlation to glycemic state
Saliva	0.23–0.38	0.6–1.8	Reasonable correlation to glycemic state; easily accessible	Mild sensitivity to changes in glycemic state
Sweat	0.06–0.11	0.01–1	Easily accessible; enables wearable sensors	Variable pH, ionic strength; poor correlation to glycemic state
Tears	0.05–0.5	0.5–5	Readily accessible; consistent pH, ionic strength	Highly variable; poor correlation to glycemic state

a) All values taken from a review by Bruen et al. [120].

glucose may eventually show promise, but there are currently several limitations. Specifically, glucose levels in ocular fluids have not yet been well-correlated to changes in blood glucose, and baseline glucose levels are 50–100 times lower than those in blood (thus requiring very low detection limits) [125]. Finally, urine is a noninvasive fluid for glucose sensing and has been used historically for diagnosing diabetes (e.g. diabetes-caused glycosuria results in glucose concentrations exceeding 5.55 mM in excreted urine); however, glucose levels in urine vary wildly as a function of time due to reabsorption by the kidneys [126, 127]. Consequently, glucose in the urine is a highly volatile biomarker that cannot be effectively used to monitor a patient's glycemic state.

4.3.6 Error Analysis and Sampling of Glucometers

Several electrochemical interferants are present in blood samples and cause erroneously high glucose readings. Both the National Center for Clinical Laboratory Standards and the U.S. Food and Drug Administration (FDA) have developed a list of standard interferants that should be screened when developing new sensors (Table 4.3) [93]. Among the interferants listed, ascorbate, acetaminophen, and urate are the most likely to cause electrochemical interference. These interferants are readily oxidized within a potential window of −0.2 to 0.1 V vs. SCE, which is lower than the oxidation potential of several common redox mediators for GOx. This interference is most often overcome with the use of low-potential redox mediators; however, sufficiently low concentrations of interferants and the use of mediators with rapid electron transfer kinetics help minimize the impact of excess current due to interferant oxidation. Typically, an error of less than 10% of the upper current range is considered acceptable.

Table 4.3 List of standard interferants defined by the FDA and their suggested testing level for glucose biosensors [128].

Interferant	Suggested test level/mg dL^{-1}
Acetaminophen[a)]	20
Salicylic acid	50
Tetracycline	4
Dopamine[a)]	13
Ibuprofen	40
Methyl-DOPA[a)]	2.5
Tolbutamide	100
Ascorbic acid[a)]	3
Bilirubin	20
Cholesterol	500
Creatine	30
Triglycerides	3000
Urate[a)]	20

a) Electrochemically active interferents within the potential window of common redox mediators.

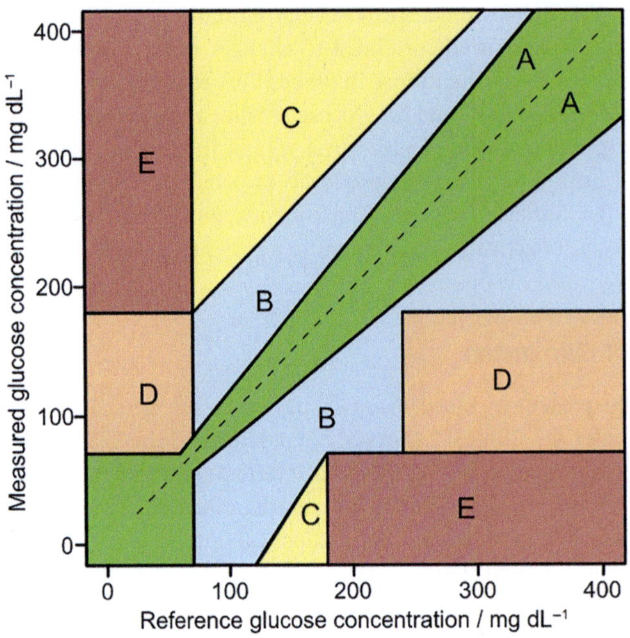

Figure 4.6 Clark error grid for analysis of the accuracy of glucose sensors compared to an established reference sensor.

Accuracy of a glucometer is most often determined using a Clark error grid analysis to compare the glucometer glucose measurements with a blood-glucose measurement from a central laboratory technique (such as a blood-gas analyzer) [129]. The Clark error grid analysis compares blood glucose measured by a reference technique (x-axis) to that measured by the glucometer where the graph is divided into zones of clinical relevance (Figure 4.6). The zones are designated as A–E, where measurements in zone A are clinically accurate, measurements in zone B are acceptable, but measurements in zones C, D, and E are erroneous such that clinical actions taken based on these measurements (i.e. inject insulin, ingest glucose, or do nothing) would result in increasingly dire outcomes. For an ideal sensor, all measured points should fall on a correlation line for which the slope = 1 and the y-intercept = 0 (zone A). Zone A becomes narrower under conditions where higher accuracy is required to take the correct clinical action. Variations on Clark's original error grid have been introduced to provide additional accuracy for patients with type 1 and type 2 diabetes [130]. In this way, it is important to distinguish "clinical accuracy," for which the goal is making the correct therapeutic decision, from analytical accuracy, which only compares the device-measured value with that of a gold standard technique.

4.4 From Single-Point Testing to Continuous Monitoring

4.4.1 POC Glucometers

POC glucose sensors – such as those used in emergency units and intensive care units – are often integrated into arterial blood-gas analyzers, such as the Radiometer ABL90 FLEX [131]. As their name implies, these devices typically measure blood gases (such as partial pressure of O_2 and CO_2) in addition to several electrolytes (e.g. Na^+, K^+, and pH) and metabolites (e.g. glucose, lactate,

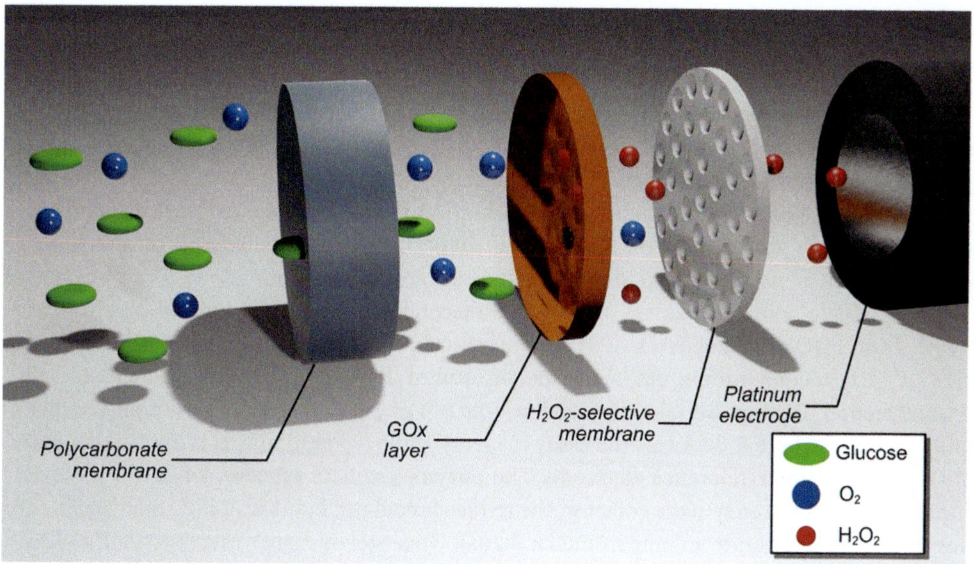

Figure 4.7 Illustration of a YSI-type glucose sensor containing a polycarbonate membrane to limit the flux of glucose, a reactive GOx layer to convert glucose to gluconolactone while generating peroxide, and a peroxide-selective layer to enable transport of peroxide to the electrode surface while limiting the flux of oxygen.

and total hemoglobin) using 45–65 μL samples of intravenous blood. Blood-gas analyzers are benchtop instruments that use potentiometry to measure electrolyte concentrations, spectroscopy to measure hemoglobin, and amperometry to measure glucose and lactate concentrations. Specifically, the Radiometer ABL90 FLEX uses a GOx/H_2O_2 assay with a Pt anode for peroxide oxidation. Instruments like this are modeled after the Yellow Spring Instruments (YSI) Model 23 glucose analyzer, which uses a glucose monitoring probe containing two polymer layers around the inner Pt or Au electrode (Figure 4.7). The outer membrane is a polycarbonate layer to reduce the influx of glucose and prevent electrode fouling. The inner membrane consists of cellulose acetate and GOx cross-linked with glutaraldehyde and is used for glucose oxidation/H_2O_2 generation. Importantly, the outer membrane slows the influx of glucose to ensure that the concentration of glucose arriving at the inner membrane is close to the concentration of saturated aqueous O_2 at 25 °C (approximately 0.2 mM), which, in turn, ensures complete conversion of all the glucose passing through the outer membrane [93]. This has the effect of extending the linear detection range and (coupled with using an initially large excess of GOx) ensuring a constant signal even after much of the initial GOx activity is lost. These design principles helped establish YSI-type analyzers as the gold standard for accuracy.

4.4.2 Single-Point Test Glucometers

Since the first blood-glucose monitor for in-home self-monitoring was launched in 1987, single-point glucose test strips have become integral for diabetes management, with greater than six billion test strips being used per year globally [93, 132]. Commercial glucose test strips are single-use probes for which the electrodes are often screen-printed (or vapor-deposited in the case of Au or Pd electrodes) onto a plastic substrate using a series of patterned masks. Once printed, the working electrode is coated with the assay reagents (enzyme, mediator, surfactants, stabilizer, etc.)

via ink-jet printing technology and stored in the dry form [133, 134]. Test strip size (approximately 1-inch long) is designed for ergonomic use; however, the actual sample chamber size constitutes a small fraction of the test strip. The bottom of each test strip contains electrical contact pads to connect to a specialized meter, which converts the electrical signal into a measurement of glucose concentration. All test strips consist of six basic elements: a plastic substrate, a working electrode, a counter/reference electrode, a capillary chamber, the assay reagents, and a fill detection electrode. Additionally, many sensors have an extra set of working and counter/reference electrodes that are housed in a separate compartment on the test strip and do not contain any assay reagents; their purpose is to obtain an accurate measurement of the background current.

Commercial single-point glucose test strips utilize a second-generation (mediated) amperometric assay with GOx, PQQ-GDH, or FAD-GDH as glucose-oxidation catalysts, and (e.g.) ferricyanide ($Fe(CN)_6^{3+/4+}$), ferrocene, ruthenium hexamine, or osmium tris(2,2-bispyridine) as mediators for signal transduction. During use, a small blood sample (~1 μL) is pulled into the sample chamber by capillary action, where it dissolves the assay reagents while a potential bias is applied between the working and counter/reference electrode. The enzyme oxidizes glucose, while the oxidized mediator regenerates the enzymatic cofactor; the reduced mediator is subsequently oxidized at the working electrode to generate an amperometric signal. Once steady-state turnover conditions are reached (the timing of which is calibrated based on the specific probe design), the strip response is measured (as either current or charge) and compared to a calibration curve stored in the meter.

There are several additional variables that uniquely affect measurements of home blood-glucose test strips, including 1) manufacturing variability, 2) mediator physical properties, and 3) characteristics of a given blood sample.

1) To ensure that manufactured test strips perform accurately over the entire clinically relevant range, several sample strips from each batch are tested at multiple glucose concentrations and compared to a reference analyzer. Specifically, these comparative tests ensure that the current response is linear with glucose concentration and that the coefficient of variation – or the amount of scattering in the data – is sufficiently low (for commercial sensors, the coefficient of variation must not exceed 2–5%) [93].
2) Because the assay components are stored in the dry state, dissolution time and solubility (especially of the mediator) can limit the reading time of single-point measurement glucose test strips.
3) Both pH and hematocrit – or the volume percent of red blood cells in the blood – are known to significantly impact the accuracy of home blood-glucose test strips [135, 136]. Consequently, glucose test strips are commonly calibrated to operate within a pH range of 7.00–7.80 (assuming a normal glucose level) and a range of 25–55% hematocrit. The latter challenge comes from undesired reactivity between the dissolved mediator and erythrocytes; however, this can be overcome by performing impedance measurements at a secondary electrode to enable appropriate electrochemical compensation.

4.4.3 Continuous Glucose Monitoring

Continuous glucose monitoring (CGM) is accomplished by measuring the quantity of glucose in either blood, ISF (i.e. the fluid between cells), sweat, tears, or saliva. The first continuous glucose monitoring (or semicontinuous glucose monitoring) was developed by Cygnus, Inc. as the GlucoWatch automatic glucose biographer, and was released in 1999 [137]. This device employed the first-generation electrochemical glucose sensor with GOx and a Pt electrode to monitor H_2O_2 oxidation and monitored glucose concentrations in the ISF, which was extracted using reverse iontophoresis. Reverse iontophoresis operates by electroosmotically drawing cations from the ISF to the cathode that is placed on the skin. The resulting electroosmotic flow enables the extraction of neutral compounds (including glucose) without piercing the skin [138]. Notably, the GlucoWatch was programmed to take measurements and contained several miniaturized electrical devices to monitor temperature, pH, and ionic conductivity of the skin, to enable real-time background correction of the electrochemical data as well as criteria for rejecting data points altogether if (e.g.) the wearer were sweating excessively (a condition known to confound the glucose sensor). Despite its novelty, the GlucoWatch suffered from a long lag time – time between a change in blood-glucose level and the change in glucose concentration at the sensor – and patient reports of skin irritation and pain while wearing the device (ultimately leading to the removal of the product from the market in 2007) [139]. While significant improvements have been made in continuous glucose monitoring devices, most of the existing devices are only approved as adjunct devices that supplement glucose monitoring by single-point testing methods.

4.4.4 Interstitial Fluid

For long-term continuous glucose monitoring, early work focused on fully implantable glucose sensors that contain a first-generation GOx-based electrode interface (glucose-limiting membrane, GOx-containing polymer membrane, and a Pt electrode) with battery-powered electronics for storing and transmitting the glucose signal. Unfortunately, these devices have suffered from poor accuracy and limited lifetime. The device sensitivity is dependent on a fixed glucose-permeability across the outer membrane, but, upon implantation, sensors are encapsulated by glucose-metabolizing tissue that eventually depletes the local glucose concentration (rendering the sensor useless). Much of the research on implantable electrochemical glucose sensors has aimed at utilizing outer coatings or external membranes (such as cellulose acetate dialysis membranes) to prevent glucose depletion and maintain a constant vascularization (formation and maintenance of blood vessels around the implanted sensor) while avoiding inflammation around the device and implant site [140, 141]. Significant progress has been made toward extending the lifetime and accuracy of implantable electrochemical glucose sensors, and the first fully implantable glucose sensor approved by the FDA in 2018 (Eversense CGM) is a fluorescence-based device that is approved for CGM over 90 days. It should be noted that, unlike indwelling sensors (discussed below), the Eversense device must be surgically implanted and removed.

Indwelling sensors are a well-established alternative to fully implantable devices. Indwelling glucose sensors integrate a three-electrode electrochemical cell into a needle (0.45 mm × 5 mm) that can be placed into the subcutaneous fat of a patient's upper arm by a spring-loaded injector (leaving the electrodes accessible for an external reader). Indwelling glucose sensors, such as the FreeStyle Libre Flash Glucose Monitoring Kit, provide accurate measurements of glucose in the ISF without the need for external calibration and can be used to measure glucose levels for 14 days. The primary limitations of indwelling sensors for CGM are the need for a 1- to 10-hour break-in period to equilibrate with the biological environment and the short lifetime of the sensor. A summary of commercial indwelling glucose sensors is shown in Table 4.4.

The primary limitation of implantable glucose sensors is the need (either by the patient or a surgeon) to implant and remove a physical device periodically. An alternative approach uses hollow microneedle arrays to puncture the skin and passively extract samples from the ISF for glucose analysis. In this approach, an array of 200 hollow silicon microneedles extracts ISF via capillary action into a sensing chamber that contains a screen-printed three-electrode amperometric glucose sensor (with GOx immobilized in cross-linked bovine serum albumin and a Pt/C working electrode to detect the resulting H_2O_2) [143]. While the use of hollow microneedles has been

Table 4.4 Summary of the lifetime and limitations of commercially available continuous glucose monitors [142].

Manufacturer	CGM model	Sensor lifetime/days	Limitations
Medtronic	Enlite Sensor	6	Adjunctive device only, acetaminophen interference
	Guardian Sensor 3	7	
Abbott	Freestyle Navigator II	5	Adjunctive device only
	Freestyle Libre	14	Not recommended for patients with hypoglycemic unawareness, single-point test confirmation recommended
	Freestyle Libre 2	14	
Dexcom	G4 Platinum	7	Adjunctive device only
	G5 Mobile	7	Single-point test confirmation recommended, acetaminophen interference
	G6	10	

demonstrated several times for continuous glucose monitoring on the lab scale, the system suffers from a limited sample volume collection, clogging, and a substantial lag time between application on the skin and readout of glucose levels [144, 145]. Alternatively, electrodes have been fabricated at the end of microneedles (on-tip microneedle glucose sensor) to directly measure glucose in the ISF. In the earliest fabrication of these bioelectrode arrays, GOx was entrapped in a conducting hydrogel film at the electrode surface to generate and detect H_2O_2 [146]. Subsequent work has significantly improved the signal transduction by introducing a redox mediator into the electrode interface and improving fabrication techniques to demonstrate operational lifetimes exceeding 24 hours [147–149].

4.4.5 Tears, Saliva, and Sweat

Significant work has been done developing continuous glucose sensors to measure glucose concentrations in a variety of additional biological fluids, including tears, sweat, saliva, and urine. First-, second-, and third-generation glucose sensor interfaces have been successfully integrated into (i) wearable patches for monitoring glucose in sweat, (ii) contact lenses to measure glucose in tears, and (iii) mouthguards, infant pacifiers, and tooth inserts to measure the concentration of glucose in the saliva. While detection of glucose in these biological fluids has been demonstrated successfully in lab environments, significant lag times and poor correlations between physiologically relevant blood-glucose levels and glucose concentrations in these alternative biological fluids have limited their development and study in clinical trials. Consequently, much of the commercial interest in glucose sensors for sweat, tears, and saliva has weakened substantially. It should be noted that bioelectrochemical sensors are being actively developed for monitoring a variety of other clinically relevant biomarkers in these biological fluids in which the device is identical except for the enzyme that is used for biorecognition.

4.5 Ongoing Challenges

Since their inception, electrochemical glucose sensors have developed from a benchtop experiment into a billion-dollar industry through a combination of analytical electrochemistry, materials science, and biomedical engineering. From the foundational principles established by early glucose sensors to modern innovations in continuous glucose monitoring, these advancements have transformed diabetes management and clinical diagnostics. While significant progress has been made to improve sensor sensitivity, selectivity/accuracy, and practical continuous monitoring capabilities, there are several persistent challenges in the field related to sensor lifetime and correlating glucose levels across different biological media. Future developments will need to focus on further enhancing the precision and usability of these devices, with an eye toward improving patient outcomes and quality of life. As we look forward, the integration of new biorecognition elements, nanomaterials, and advanced data analytics will undoubtedly play a critical role in shaping the next generation of glucose sensors and beyond.

References

1 Sun, H., Saeedi, P., Karuranga, S. et al. (2022). IDF diabetes atlas: global, regional and country-level diabetes prevalence estimates for 2021 and projections for 2045. *Diabetes Res. Clin. Pract.* 183: 109119.

2 Mano, N., Mao, F., and Heller, A. (2003). Characteristics of a miniature compartment-less glucose–O_2 biofuel cell and its operation in a living plant. *JACS* 125 (21): 6588–6594.

3 Bandodkar, A.J. and Wang, J. (2014). Non-invasive wearable electrochemical sensors: a review. *Trends Biotechnol.* 32 (7): 363–371.

4 Jina, A., Tierney, M.J., Tamada, J.A. et al. (2014). Design, development, and evaluation of a novel microneedle array-based continuous glucose monitor. *J. Diabetes Sci. Technol.* 8 (3): 483–487.

5 Updike, S.J. and Hicks, G.P. (1967). The enzyme electrode. *Nature* 214: 986–988.

6 Williams, D.L., Doig, A.R., and Korosi, A. (1970). Electrochemical-enzymatic analysis of blood glucose and lactate. *Anal. Chem.* 42 (1): 118–121.

7 Guilbault, G.G. and Lubrano, G.J. (1972). Enzyme electrode for glucose. *Anal. Chim. Acta* 60 (1): 254–256.

8 Nagy, G., Von Storp, L.H., and Guilbault, G.G. (1973). Enzyme electrode for glucose based on an iodide membrane sensor. *Anal. Chim. Acta* 66 (3): 443–445.

9 Guilbault, G.G. and Lubrano, G.J. (1973). An enzyme electrode for the amperometric determination of glucose. *Anal. Chim. Acta* 64 (3): 439–455.

10 Koyama, M., Sato, Y., Aizawa, M., and Suzuki, S. (1980). Improved enzyme sensor for glucose with an ultrafiltration membrane and immobilized glucose oxidase. *Anal. Chim. Acta* 116 (2): 307–314.

11 Gao, W., Emaminejad, S., Nyein, H.Y.Y. et al. (2016). Fully integrated wearable sensor arrays for multiplexed in situ perspiration analysis. *Nature* 529 (7587): 509–514.

12 Kalkal, A., Kumar, S., Kumar, P. et al. (2021). Recent advances in 3D printing technologies for wearable (bio)sensors. *Addit. Manuf.* 46: 102088.

13 Sharma, S., Huang, Z., Rogers, M. et al. (2016). Evaluation of a minimally invasive glucose biosensor for continuous tissue monitoring. *Anal. Bioanal.Chem.* 408 (29): 8427–8435.

14 Mano, N. (2019). Engineering glucose oxidase for bioelectrochemical applications. *Bioelectrochemistry* 128: 218–240.

15 Cai, R., Ngwadom, C., Saxena, R. et al. (2024). Creation of a point-of-care therapeutics sensor using protein engineering, electrochemical sensing and electronic integration. *Nat. Commun.* 15 (1): 1–11.

16 Gregg, B. and a, and Heller, A. (1990). Cross-linked redox gels containing glucose oxidase for amperometric biosensor applications. *Anal. Chem.* 62 (3): 258–263.

17 Hickey, D.P., Halmes, A.J., Schmidtke, D.W., and Glatzhofer, D.T. (2014). Electrochemical characterization of glucose bioanodes based on tetramethylferrocene-modified linear poly(ethylenimine). *Electrochim. Acta* 149: 252–257.

18 Inoue, S., Egi, M., Kotani, J., and Morita, K. (2013). Accuracy of blood-glucose measurements using glucose meters and arterial blood gas analyzers in critically ill adult patients: systematic review. *Crit. Care* 17 (2): 1–13.

19 Heller, A. and Feldman, B. (2010). Electrochemistry in diabetes management. *Acc. Chem. Res.* 43 (7): 963–973.

20 Saha, T., Del Caño, R., Mahato, K. et al. (2023). Wearable electrochemical glucose sensors in diabetes management: a comprehensive review. *Chem. Rev.* 123 (12): 7854–7889.

21 Weibel, M.K. and Bright, H.J. (1971). The glucose oxidase mechanism. *J. Biol. Chem.* 246 (9): 2734–2744.

22 Meyer, M., Wohlfahrt, G., Knäblein, J., and Schomburg, D. (1998). Aspects of the mechanism of catalysis of glucose oxidase: a docking, molecular mechanics and quantum chemical study. *J. Comput.-Aided Mol. Des.* 12 (5): 425–440.

23 Bauer, J.A., Zámocká, M., Majtán, J., and Bauerová-Hlinková, V. (2022). Glucose oxidase, an enzyme "Ferrari": its structure, function, production and properties in the light of various industrial and biotechnological applications. *Biomolecules* 12 (3): 472.

24 Gibson, Q.H., Swoboda, B.E.P., and Massey, V. (1964). Kinetics and mechanism of action of glucose oxidase. *J. Biol. Chem.* 239: 3927–3934.

25 Pazur, J.H. and Kleppe, K. (1964). The oxidation of glucose and related compounds by glucose oxidase from Aspergillus niger. *Biochemistry* 3 (4): 578–583.

26 Malmstadt, H.V. and Pardue, H.L. (1961). Quantitative analysis by an automatic potentiometric reaction rate method: specific enzymatic determination of glucose. *Anal. Chem.* 33 (8): 1040–1047.

27 Clark, L.C. and Lyons, C. (1962). Electrode systems for continuous monitoring in cardiovascular surgery. *Ann. N. Y. Acad. Sci.* 102 (1): 29–45.

28 Clark, L.C., Wolf, R., Granger, D., and Taylor, Z. (1953). Continuous recording of blood oxygen tensions by polarography. *J. Appl. Physiol.* 6 (3): 189–193.

29 Guilbault, G.G. and Lubrano, G.J. (1973). An enzyme electrode for the amperometric determination of glucose. *Anal. Chim. Acta* 64 (3): 439–455.

30 Guilbault, G.G. and Lubrano, G.J. (1972). Enzyme electrode for glucose. *Anal. Chim. Acta* 60 (1): 254–256.

31 Nanjo, M. and Guilbault, G.G. (1974). Enzyme electrode for l-amino acids and glucose. *Anal. Chim. Acta* 73 (2): 367–373.

32 Karyakin, A.A. (2021). Glucose biosensors for clinical and personal use. *Electrochem. Commun.* 125: 106973.

33 Newman, J.D. and Turner, A.P.F. (2004). Biosensors for monitoring glucose. In: *Sensors in Medicine And Health Care*, 45–78. Weinheim, Germany: Wiley.

34 Blaedel, W.J., and Olson, C.L. (1968) Amperometric Determination of Glucose. US3367849, issued Feb. 6, 1968.

35 Heller, A. and Feldman, B. (2008). Electrochemical glucose sensors and their applications in diabetes management. *Chem. Rev.* 108 (7): 2482–2505.

36 Williams, D.L., Doig, A.R., and Korosi, A. (1970). Electrochemical-enzymatic analysis of blood glucose and lactate. *Anal. Chem.* 42 (1): 118–121.

37 Cass, A.E.G., Davis, G., Francis, G.D. et al. (1984). Ferrocene-mediated enzyme electrode for amperometric determination of glucose. *Anal. Chem.* 56 (4): 667–671.

38 Degani, Y. and Heller, A. (1987). Direct electrical communication between chemically modified enzymes and metal electrodes. I. Electron transfer from glucose oxidase to metal electrodes via electron relays, bound covalently to the enzyme. *J. Phys. Chem.* 91 (6): 1285–1289.

39 Gregg, B.A. and Heller, A. (1990). Cross-linked redox gels containing glucose oxidase for amperometric biosensor applications. *Anal. Chem.* 62 (3): 258–263.

40 Higgins, I.J., Hill, H.A.O., and Plotkin, E. V. (1987) Sensor for Components of a Liquid Mixture. 4,711,245, issued Dec. 8, 1987.

41 Higgins, I.J., Hill, H.A.O., and Plotkin, E. V. (1985) Sensor for Components of a Liquid Mixture. 4,545,382, issued Oct. 8, 1985.

42 Kyvik, K.O., Traulsen, J., Reinholdt, B., and Frøland, A. (1990). The ExacTech blood glucose testing system. *Diabetes Res. Clin. Pract.* 10 (1): 85–90.

43 Ross, D., Heinemann, L., and Chantelau, E.A. (1990). Short-term evaluation of an electro-chemical system (ExacTech™) for blood glucose monitoring. *Diabetes Res. Clin. Pract.* 10 (3): 281–285.

44 Matthews, D.R., Bown, E., Watson, A. et al. (1987). Pen-sized digital 30-second blood glucose meter. *The Lancet* 329 (8536): 778–779.

45 Vashist, S.K., Zheng, D., Al-Rubeaan, K. et al. (2011). Technology behind commercial devices for blood glucose monitoring in diabetes management: a review. *Anal. Chim. Acta* 703 (2): 124–136.

46 Pullano, S.A., Greco, M., Bianco, M.G. et al. (2022). Glucose biosensors in clinical practice: principles, limits and perspectives of currently used devices. *Theranostics* 12 (2): 493–511.

47 Saha, T., Del Caño, R., Mahato, K. et al. (2023). Wearable electrochemical glucose sensors in diabetes management: a comprehensive review. *Chem. Rev.* 123 (12): 7854–7889.

48 Wei, M., Qiao, Y., Zhao, H. et al. (2020). Electrochemical non-enzymatic glucose sensors: recent progress and perspectives. *Chem. Commun.* 56 (93): 14553–14569.

49 Hwang, D.W., Lee, S., Seo, M., and Chung, T.D. (2018). Recent advances in electrochemical non-enzymatic glucose sensors – a review. *Anal. Chim. Acta* 1033: 1–34.

50 Park, S., Boo, H., and Chung, T.D. (2006). Electrochemical non-enzymatic glucose sensors. *Anal. Chim. Acta* 556 (1): 46–57.

51 Nakamura, S., Hayashi, S., and Koga, K. (1976). Effect of periodate oxidation on the structure and properties of glucose oxidase. *Biochim. Biophys. Acta Biomembr. Enzymol.* 445 (2): 294–308.

52 Vogt, S., Schneider, M., Schäfer-Eberwein, H., and Nöll, G. (2014). Determination of the pH dependent redox potential of glucose oxidase by spectroelectrochemistry. *Anal. Chem.* 86 (15): 7530–7535.

53 Vuong, T.V., Foumani, M., MacCormick, B. et al. (2016). Direct comparison of gluco-oligosaccharide oxidase variants and glucose oxidase: substrate range and H_2O_2 stability. *Sci. Rep.* 6 (1): 1–9.

54 Kanji, S., Buffie, J., Hutton, B. et al. (2005). Reliability of point-of-care testing for glucose measurement in critically ill adults. *Crit. Care Med.* 33 (12): 2778–2785.

55 Durand, F., Limoges, B., Mano, N. et al. (2011). Effect of substrate inhibition and cooperativity on the electrochemical responses of glucose dehydrogenase. Kinetic characterization of Wild and mutant types. *JACS* 133 (32): 12801–12809.

56 Okumura, N., Abo, T., Tsujimura, S., and Kano, K. (2006). Electron transfer kinetics between PQQ-dependent soluble glucose dehydrogenase and mediators. *Electrochemistry* 74 (8): 639–641.

57 Prévoteau, A. and Mano, N. (2013). How the reduction of O_2 on enzymes and/or redox mediators affects the calibration curve of "wired" glucose oxidase and glucose dehydrogenase biosensors. *Electrochim. Acta* 112: 318–326.

58 Lisdat, F. (2020). PQQ-GDH – structure, function and application in bioelectrochemistry. *Bioelectrochemistry* 134: 107496.

59 Ye, L., Hämmerle, M., Olsthoorn, A.J.J. et al. (1993). High current density "wired" quinoprotein glucose dehydrogenase electrode. *Anal. Chem.* 65 (3): 238–241.

60 Immanuel, S., Sivasubramanian, R., Gul, R., and Dar, M.A. (2020). Recent progress and perspectives on electrochemical regeneration of reduced nicotinamide adenine dinucleotide (NADH). *Chem. Asian J.* 15 (24): 4256–4270.

61 Yehezkeli, O., Tel-Vered, R., Raichlin, S., and Willner, I. (2011). Nano-engineered flavin-dependent glucose dehydrogenase/gold nanoparticle-modified electrodes for glucose sensing and biofuel cell applications. *ACS Nano* 5 (3): 2385–2391.

62 Filipiak, M.S., Vetter, D., Thodkar, K. et al. (2020). Electron transfer from FAD-dependent glucose dehydrogenase to single-sheet graphene electrodes. *Electrochim. Acta* 330: 134998.

63 Milton, R.D., Lim, K., Hickey, D.P., and Minteer, S.D. (2015). Employing FAD-dependent glucose dehydrogenase within a glucose/oxygen enzymatic fuel cell operating in human serum. *Bioelectrochemistry* 106: 56–63.

64 Cohen, R., Bitton, R.E., Herzallh, N.S. et al. (2021). Utilization of FAD-glucose dehydrogenase from T. Emersonii for Amperometric biosensing and biofuel cell devices. *Anal. Chem.* 93 (33): 11585–11591.

65 Ferri, S., Kojima, K., and Sode, K. (2011). Review of glucose oxidases and glucose dehydrogenases: a bird's eye view of glucose sensing enzymes. *J. Diabetes Sci. Technol.* 5 (5): 1068–1076.

66 Yamaoka, H., Yamashita, Y., Ferri, S., and Sode, K. (2008). Site directed mutagenesis studies of FAD-dependent glucose dehydrogenase catalytic subunit of Burkholderia cepacia. *Biotechnol. Lett* 30 (11): 1967–1972.

67 Sode, K., Igarashi, S., Morimoto, A., and Yoshida, H. (2002). Construction of engineered water-soluble PQQ glucose dehydrogenase with improved substrate specificity. *Biocatal. Biotransform.* 20 (6): 405–412.

68 Ito, K., Okuda-Shimazaki, J., Mori, K. et al. (2019). Designer fungus FAD glucose dehydrogenase capable of direct electron transfer. *Biosens. Bioelectron.* 123: 114–123.

69 Algov, I., Grushka, J., Zarivach, R., and Alfonta, L. (2017). Highly efficient Flavin-adenine dinucleotide glucose dehydrogenase fused to a minimal cytochrome C domain. *JACS* 139 (48): 17217–17220.

70 Ozawa, K., Iwasa, H., Sasaki, N. et al. (2017). Identification and characterization of thermostable glucose dehydrogenases from thermophilic filamentous fungi. *Appl. Microbiol. Biotechnol.* 101 (1): 173–183.
71 Dubey, M.K., Zehra, A., Aamir, M. et al. (2017). Improvement strategies, cost effective production, and potential applications of fungal glucose oxidase (GOD): current updates. *Front. Microbiol.* 8 (JUN): 266034.
72 Kovačević, G., Blažić, M., Draganić, B. et al. (2014). Cloning, heterologous expression, purification and characterization of M12 mutant of Aspergillus niger glucose oxidase in yeast Pichia pastoris KM71H. *Mol. Biotechnol.* 56 (4): 305–311.
73 Belyad, F., Karkhanei, A.A., and Raheb, J. (2018). Expression, characterization and one step purification of heterologous glucose oxidase gene from Aspergillus niger ATCC 9029 in Pichia pastoris. *EuPA Open Proteomics* 19: 1–5.
74 Prieto-Simón, B., Armatas, G.S., Pomonis, P.J. et al. (2004). Metal-dispersed Xerogel-based composite films for the development of interference free oxidase-based biosensors. *Chem. Mater.* 16 (6): 1026–1034.
75 Bharathi, S. and Nogami, M. (2001). A glucose biosensor based on electrodeposited biocomposites of gold nanoparticles and glucose oxidase enzyme. *Analyst* 126 (11): 1919–1922.
76 de Lara González, G.L., Kahlert, H., and Scholz, F. (2007). Catalytic reduction of hydrogen peroxide at metal hexacyanoferrate composite electrodes and applications in enzymatic analysis. *Electrochim. Acta* 52 (5): 1968–1974.
77 Dodevska, T., Horozova, E., and Dimcheva, N. (2006). Electrocatalytic reduction of hydrogen peroxide on modified graphite electrodes: application to the development of glucose biosensors. *Anal. Bioanal.Chem.* 386 (5): 1413–1418.
78 Yao, Y.L. and Shiu, K.K. (2008). A mediator-free Bienzyme Amperometric biosensor based on horseradish peroxidase and glucose oxidase immobilized on carbon nanotube modified electrode. *Electroanalysis* 20 (19): 2090–2095.
79 Zhu, L., Yang, R., Zhai, J., and Tian, C. (2007). Bienzymatic glucose biosensor based on co-immobilization of peroxidase and glucose oxidase on a carbon nanotubes electrode. *Biosens. Bioelectron.* 23 (4): 528–535.
80 Åsberg, P. and Inganäs, O. (2003). Hydrogels of a conducting conjugated polymer as 3-D enzyme electrode. *Biosens. Bioelectron.* 19 (3): 199–207.
81 Kulys, J., Wang, L., Hansen, H.E. et al. (1995). Methylene-green-mediated carbon paste glucose biosensor. *Electroanalysis* 7 (1): 92–94.
82 Kulys, J.J. and Čénas, N.K. (1983). Oxidation of glucose oxidase from Penicillium vitale by one- and two-electron acceptors. *Biochim. Biophys. Acta, Protein Struct. Mol. Enzymol.* 744 (1): 57–63.
83 Kulys, J., Palaima, A., and Urbelis, G. (1998). Employing heterocyclic Dihydropolyazines for Amperometric glucose sensing. *Anal. Lett.* 31 (4): 569–584.
84 Kulys, J., Razumas, V., Kazlauskaite, J. et al. (1994). Electrocatalytical oxidation of glucose on graphite using covalently immobilized glucose oxidase and 1,6-dithia- and 1,6-dioxapyrenes as mediators. *Electroanalysis* 6 (9): 740–745.
85 Ikeda, T., Hamada, H., Miki, K., and Senda, M. (1985). Glucose oxidase-immobilized benzoquinone-carbon paste electrode as a glucose senso. *Agric. Biol. Chem.* 49 (2): 541–543.
86 Jaffari, S.A. and Turner, A.P.F. (1997). Novel hexacyanoferrate(III) modified graphite disc electrodes and their application in enzyme electrodes—part I. *Biosens. Bioelectron.* 12 (1): 1–9.

87 Cecchet, F., Marcaccio, M., Margotti, M. et al. (2006). Redox mediation at 11-mercaptoundecanoic acid self-assembled monolayers on gold. *J. Phys. Chem. B* 110 (5): 2241–2248.

88 Ghica, M.E. and Brett, C.M.A. (2005). A glucose biosensor using methyl viologen redox mediator on carbon film electrodes. *Anal. Chim. Acta* 532 (2): 145–151.

89 Flexer, V., Ielmini, M.V., Calvo, E.J., and Bartlett, P.N. (2008). Extracting kinetic parameters for homogeneous [Os(bpy)2ClPyCOOH]+ mediated enzyme reactions from cyclic voltammetry and simulations. *Bioelectrochemistry* 74 (1): 201–209.

90 Mtemeri, L. and Hickey, D.P. (2023). Model-driven design of redox mediators: quantifying the impact of quinone structure on bioelectrocatalytic activity with glucose oxidase. *J. Phys. Chem. B* 127 (36): 7685–7693.

91 Chaubey, A. and Malhotra, B.D.D. (2002). Mediated biosensors. *Biosens. Bioelectron.* 17 (6–7): 441–456.

92 Kano, K. and Ikeda, T. (2000). Fundamentals and practices of mediated bioelectrocatalysis. *Anal. Sci.* 16 (10): 1013–1021.

93 Heller, A. and Feldman, B. (2008). Electrochemical glucose sensors and their applications in diabetes management. *Chem. Rev.* 108 (7): 2482–2505.

94 Pisoschi, A.M., Pop, A., Serban, A.I., and Fafaneata, C. (2014). Electrochemical methods for ascorbic acid determination. *Electrochim. Acta* 121: 443–460.

95 Dryhurst, G. (1972). Electrochemical oxidation of uric acid and xanthine at the pyrolytic graphite electrode. *J. Electrochem. Soc.* 119 (12): 1659.

96 Nematollahi, D., Shayani-Jam, H., Alimoradi, M., and Niroomand, S. (2009). Electrochemical oxidation of acetaminophen in aqueous solutions: kinetic evaluation of hydrolysis, hydroxylation and dimerization processes. *Electrochim. Acta* 54 (28): 7407–7415.

97 Nieh, C.-H.H., Tsujimura, S., Shirai, O., and Kano, K. (2013). Electrostatic and steric interaction between redox polymers and some flavoenzymes in mediated bioelectrocatalysis. *J. Electroanal. Chem.* 689: 26–30.

98 Gregg, B.A. and Heller, A. (1991). Redox polymer films containing enzymes. 1. A redox-conducting epoxy cement: synthesis, characterization, and electrocatalytic oxidation of hydroquinone. *J. Phys. Chem.* 95 (15): 5970–5975.

99 Joshi, P.P., Merchant, S.A., Wang, Y., and Schmidtke, D.W. (2005). Amperometric biosensors based on redox polymer–carbon nanotube–enzyme composites. *Anal. Chem.* 77 (10): 3183–3188.

100 Aoki, A. and Heller, A. (1993). Electron diffusion coefficients in hydrogels formed of cross-linked redox polymers. *J. Phys. Chem.* 97 (42): 11014–11019.

101 Degani, Y. and Heller, A. (1989). Electrical communication between redox centers of glucose oxidase and electrodes via electrostatically and covalently bound redox polymers. *JACS* 111 (6): 2357–2358.

102 Merchant, S.A., Meredith, M.T., Tran, T.O. et al. (2010). Effect of mediator spacing on electrochemical and enzymatic response of ferrocene redox polymers. *J. Phys. Chem. C* 114 (26): 11627–11634.

103 Gregg, B.A. and Heller, A. (1991). Redox polymer films containing enzymes. 2. Glucose oxidase containing enzyme electrodes. *J. Phys. Chem.* 95 (15): 5976–5980.

104 Meredith, M.T., Hickey, D.P., Redemann, J.P. et al. (2013). Effects of ferrocene methylation on ferrocene-modified linear poly(ethylenimine) bioanodes. *Electrochim. Acta* 92: 226–235.

105 Meredith, M.T., Kao, D.-Y., Hickey, D. et al. (2011). High current density ferrocene-modified linear poly(ethylenimine) bioanodes and their use in biofuel cells. *J. Electrochem. Soc.* 158 (2): B166.

106 Marcus, R.A. (1956). On the theory of oxidation-reduction reactions involving electron transfer. I. *J. Chem. Phys.* 24 (5): 966–978.

107 Marcus, R.A. (1965). On the theory of electron-transfer reactions. VI. Unified treatment for homogeneous and electrode reactions. *J. Chem. Phys.* 43 (2): 679–701.

108 Marcus, R.A. and Sutin, N. (1985). Electron transfers in chemistry and biology. *Biochim. Biophys. Acta, Rev. Biomembr.* 811 (3): 265–322.

109 Milton, R.D. and Minteer, S.D. (2017). Direct enzymatic bioelectrocatalysis: differentiating between myth and reality. *J. R. Soc. Interface* 14 (131): 20170253.

110 Feldberg, S.W. and Sutin, N. (2006). Distance dependence of heterogeneous electron transfer through the nonadiabatic and adiabatic regimes. *Chem. Phys.* 324 (1): 216–225.

111 Jeykumari, D.R.S. and Narayanan, S.S. (2008). Fabrication of bienzyme nanobiocomposite electrode using functionalized carbon nanotubes for biosensing applications. *Biosens. Bioelectron.* 23 (11): 1686–1693.

112 Harreither, W., Nicholls, P., Sygmund, C. et al. (2012). Investigation of the pH-dependent electron transfer mechanism of ascomycetous class II cellobiose dehydrogenases on electrodes. *Langmuir* 28 (16): 6714–6723.

113 Ryffel, G.U., Farruggio, A.P., Chavez, C.L. et al. (2012). Recombinantly produced cellobiose dehydrogenase from Corynascus thermophilus for glucose biosensors and biofuel cells. *Biotechnol. J.* 7 (11): 1359–1366.

114 Jayakumar, K., Reichhart, T.M.B., Schulz, C. et al. (2022). An oxygen insensitive amperometric glucose biosensor based on an engineered cellobiose dehydrogenase: direct versus mediated electron transfer responses. *ChemElectroChem* 9 (13): e202200418.

115 Bollella, P., Gorton, L., Ludwig, R., and Antiochia, R. (2017). A third generation glucose biosensor based on cellobiose dehydrogenase immobilized on a glassy carbon electrode decorated with electrodeposited gold nanoparticles: characterization and application in human saliva. *Sensors* 17 (8): 1912.

116 Kubihal, S., Goyal, A., Gupta, Y., and Khadgawat, R. (2021). Glucose measurement in body fluids: a ready reckoner for clinicians. *Diabetes Metab. Syndr. Clin. Res. Rev.* 15 (1): 45–53.

117 Murray-Bachmann, R., Ming Leung, T., Myers, A.K. et al. (2021). Reliability of continuous glucose monitoring system in the inpatient setting. *J. Clin. Transl. Endocrinol.* 25: 100262.

118 Taormina, C.R., Baca, J.T., Asher, S.A. et al. (2007). Analysis of tear glucose concentration with electrospray ionization mass spectrometry. *J. Am. Soc. Mass Spectrom.* 18 (2): 332–336.

119 Daum, K.M. and Hill, R.M. (1982). Human tear glucose. *Invest. Ophthalmol. Visual Sci.* 22 (4): 509–514.

120 Bruen, D., Delaney, C., Florea, L., and Diamond, D. (2017). Glucose sensing for diabetes monitoring: recent developments. *Sensors* 17 (8): 1866.

121 Gupta, S., Nayak, M., Sunitha, J. et al. (2017). Correlation of salivary glucose level with blood glucose level in diabetes mellitus. *J. Oral Maxillofac. Pathol.* 21 (3): 334–339.

122 Corrie, S.R., Coffey, J.W., Islam, J. et al. (2015). Blood, sweat, and tears: developing clinically relevant protein biosensors for integrated body fluid analysis. *Analyst* 140 (13): 4350–4364.

123 Saraoğlu, H.M. and Koçan, M. (2010). A study on non-invasive detection of blood glucose concentration from human palm perspiration by using artificial neural networks. *Expert Syst.* 27 (3): 156–165.

124 Wiorek, A., Parrilla, M., Cuartero, M., and Crespo, G.A. (2020). Epidermal patch with glucose biosensor: PH and temperature correction toward more accurate sweat analysis during sport practice. *Anal. Chem.* 92 (14): 10153–10161.

125 Pullano, S.A., Greco, M., Bianco, M.G. et al. (2022). Glucose biosensors in clinical practice: principles, limits and perspectives of currently used devices. *Theranostics* 12 (2): 493–511.

126 Hayford, J.T., Weydert, J.A., and Thompson, R.G. (1983). Validity of urine glucose measurements for estimating plasma glucose concentration. *Diabetes Care* 6 (1): 40–44.

127 Fine, J. (1965). Glucose content of normal urine. *Br. Med. J.* 1: 1209–1214.

128 Kondepati, V.R. and Heise, H.M. (2007). Recent progress in analytical instrumentation for glycemic control in diabetic and critically ill patients. *Anal. Bioanal.Chem.* 388 (3): 545–563.

129 Clarke, W.L., Cox, D., Gonder-Frederick, L.A. et al. (1987). Evaluating clinical accuracy of systems for self-monitoring of blood glucose. *Diabetes Care* 10 (5): 622–628.

130 Parkes, J.L., Slatin, S.L., Pardo, S., and Ginsberg, B.H. (2000). A new consensus error grid to evaluate the clinical significance of inaccuracies in the measurement of blood glucose. *Diabetes Care* 23 (8): 1143–1148.

131 Marija, K., Bernhard, K.F., and Beatrice, L.K. (2021). Blood-gas vs. central-laboratory analyzers: interchangeability and reference intervals for sodium, potassium, glucose, lactate and hemoglobin. *Heliyon* 7 (11): e08302.

132 Falk, J., Friesen, K.J., Okunnu, A., and Bugden, S. (2017). Patterns, policy and appropriateness: A 12-year utilization review of blood glucose test strip use in insulin users. *Can. J. Diabetes* 41 (4): 385–391.

133 Wring, S.A. and Hart, J.P. (1992). Chemically modified, screen-printed carbon electrodes. *Analyst* 117: 1281–1286.

134 Wang, J. (2008). Electrochemical glucose biosensors. *Chem. Rev.* 108 (2): 814–825.

135 Tang, Z., Du, X., Louie, R.F., and Kost, G.J. (2000). Effects of pH on glucose measurements with handheld glucose meters and a portable glucose analyzer for point-of-care testing. *Arch. Pathol. Lab. Med.* 124 (4): 577–582.

136 Pereira, A.J., Corrêa, T.D., De Almeida, F.P. et al. (2015). Inaccuracy of venous point-of-care glucose measurements in critically ill patients: a cross-sectional study. *PLoS One* 10 (6): e0129568.

137 Garg, S.K., Potts, R.O., Ackerman, N.R. et al. (1999). Correlation of fingerstick blood glucose measurements with GlucoWatch biographer glucose results in young subjects with type 1 diabetes. *Diabetes Care* 22 (10): 1708–1714.

138 Rao, G., Glikfeld, P., and Guy, R.H. (1993). Reverse iontophoresis: development of a noninvasive approach for glucose monitoring. *Pharm. Res.* 10 (12): 1751–1755.

139 Heikenfeld, J., Jajack, A., Rogers, J. et al. (2018). Wearable sensors: modalities, challenges, and prospects. *Lab Chip* 18 (2): 217–248.

140 Zebda, A., Cosnier, S., Alcaraz, J.-P.P. et al. (2013). Single glucose biofuel cells implanted in rats power electronic devices. *Sci. Rep.* 3: 1516.

141 Cinquin, P., Gondran, C., Giroud, F. et al. (2010). A glucose BioFuel cell implanted in rats. *PLoS One* 5 (5): e10476.

142 Cappon, G., Vettoretti, M., Sparacino, G., and Facchinetti, A. (2019). Continuous glucose monitoring sensors for diabetes management: a review of technologies and applications. *Diabetes Metab. J.* 43 (4): 383–397.

143 Jina, A., Tierney, M.J., Tamada, J.A. et al. (2014). Design, development, and evaluation of a novel microneedle array-based continuous glucose monitor. *J. Diabetes Sci. Technol.* 8 (3): 483–487.

144 Samant, P.P., Niedzwiecki, M.M., Raviele, N. et al. (2020). Sampling interstitial fluid from human skin using a microneedle patch. *Sci. Transl. Med.* 12 (571): 285.

145 Strambini, L.M., Longo, A., Scarano, S. et al. (2015). Self-powered microneedle-based biosensors for pain-free high-accuracy measurement of glycaemia in interstitial fluid. *Biosens. Bioelectron.* 66: 162–168.

146 Valdés-Ramírez, G., Li, Y.C., Kim, J. et al. (2014). Microneedle-based self-powered glucose sensor. *Electrochem. Commun.* 47: 58–62.

147 Liu, Y., Yu, Q., Luo, X. et al. (2021). Continuous monitoring of diabetes with an integrated microneedle biosensing device through 3D printing. *Microsyst. Nanoeng.* 7 (1): 1–12.

148 Chinnadayyala, S.R., Park, J., Satti, A.T. et al. (2021). Minimally invasive and continuous glucose monitoring sensor based on non-enzymatic porous platinum black-coated gold microneedles. *Electrochim. Acta* 369: 137691.

149 Dervisevic, M., Alba, M., Yan, L. et al. (2022). Transdermal electrochemical monitoring of glucose via high-density silicon microneedle Array patch. *Adv. Funct. Mater.* 32 (3): 2009850.

5

Ex Situ Biosensors

Jacquelyn E. McBride and Michelle Rasmussen

Lebanon Valley College, Department of Chemistry, Annville, PA, USA

5.1 Introduction

The use of bioelectrocatalysis for sensing applications has increased rapidly since the first reported sensor by Clark and Lyons in 1962. Electrocatalytic biosensors take advantage of the high sensitivity and specificity, fast response time, and lower cost of biological catalysts as the recognition element. This response is then transduced to a measurable electrochemical signal, either current, power, or potential generated from the presence of the analyte. While a large portion of research in this field is devoted to *in situ* or point-of-care biosensors for medical applications, biosensors have also become common for monitoring other analytes of interest. This chapter looks at the history and current progress in biosensors for organic carbon load, toxic compounds, and explosives.

5.1.1 Background

There are two main categories of bioelectrocatalytic sensors. The first category is amperometric sensors, in which the analytical signal is generated by a redox reaction at the working electrode. This type of sensor, as illustrated in Figure 5.1a, utilizes a three-electrode configuration with a reference electrode and a counter electrode for monitoring and/or controlling the electrochemical potential and current, respectively. An appropriate potential is applied for the desired reaction, and the measured current is proportional to the analyte. This category of sensor requires a potentiostat and an external power source.

The second category is sometimes referred to as self-powered biosensors. A self-powered sensor, as shown in Figure 5.1b, operates in a two-electrode configuration and is essentially a biofuel cell where the power and current output is proportional to the analyte. While this name was first coined in 2001, this type of sensor was developed earlier using microbial fuel cells. The analyte may be involved in a reaction at the anode, the cathode, or possibly both. This type of sensor requires no additional instrumentation and no external power source [1].

Within both categories, sensors can be further classified by how they respond to analyte concentrations. The more desirable response leads to a "turn-on" sensor where the analytical signal increases with increasing analyte concentration. The alternative is a "turn-off" sensor where the analytical signal decreases with increasing analyte concentration. A major drawback to turn off sensors is that they have more limited linear ranges. There is some finite signal with no analyte and

Bioelectrocatalysis: From Electron Transfer Processes to Emerging Technological Applications, First Edition.
Edited by Shelley D. Minteer and Matteo Grattieri.

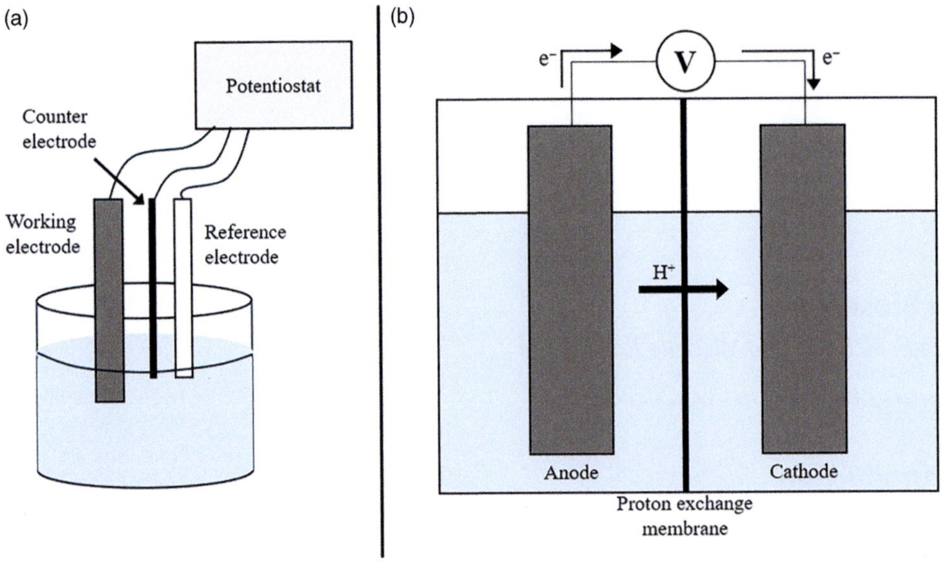

Figure 5.1 Schematic of biosensor configurations. (a) An amperometric biosensor where the reaction of interest occurs at the working electrode and (b) a self-powered biosensor where the reaction of interest may occur at either the anode or cathode depending on the design of the sensor.

some maximum concentration where the signal drops to zero, and turn off sensors are, therefore, limited to samples between those two concentrations. However, because bioelectrocatalytic sensors use biological catalysts, by considering inhibition of the catalyst, one can expand the pool of possible analytes to compounds that may not participate in a redox reaction but affect the catalytic activity.

5.2 Organic Carbon Load

Box 5.1 Indicator of Water Quality

One indicator of water quality is the organic carbon content. The type of carbon measured can vary. For example, total organic carbon (TOC) is all the carbon in organic compounds in a sample. Dissolved organic carbon (DOC) is any organic carbon that can pass through a 0.45-μm filter (i.e. not particulate). Chemical oxygen demand (COD) is a measure of the amount of oxygen needed to oxidize the organic carbon in a sample. For all three (TOC, DOC, and COD), the carbon is oxidized to carbon dioxide in the analysis by either combustion or a chemical oxidant. Biochemical oxygen demand (BOD) is another measure of carbon content that is determined by measuring the amount of dissolved oxygen (DO) consumed by aerobic microorganisms while decomposing organic matter in water. As the amount of organic matter increases, the DO consumption increases as well. Microbial-based electrodes, which make use of microorganisms as the catalyst, are an obvious choice for developing a sensor for BOD determination. For a more thorough examination of electrochemical BOD biosensors, refer to recent reviews published by Arlyapov et al. [2] and Zhou et al. [3].

Water pollution is largely caused by untreated wastewater effluent and improper land use practices due to population growth and increased urbanization. This leads to an increase in erosion and, therefore, sediment loadings of local waterways. The range of BOD values varies globally due to water usage and pollution levels. For example, a pristine river has a BOD of 1 mg L^{-1} or less, while a highly polluted river may have a value of 20 mg L^{-1} or higher. Efficiently treated sewage will have a value of 20 mg L^{-1} or less, while untreated sewage ranges from 200 to 600 mg L^{-1} depending on location [4].

The standard method for determining BOD, called the BOD_5, involves a five-day incubation period with the DO measured before and after incubation. It requires skill to perform the required procedures to get a reproducible measurement, and even then, the precision is affected by the type of microorganism as well as quantity and quality.

Microbial-based electrodes, where the biological catalyst consists of microorganisms, can measure the BOD in a faster, straightforward, and simpler process than the conventional method (Box 5.1). Biosensors developed for BOD measurement use a microbial film where organic substrates are oxidized and the amount of organic substrate can be quantified by the current generated.

5.2.1 BOD Determination by Oxygen Monitoring

The first BOD biosensor involved immobilizing bacteria onto a Clark-type oxygen electrode and measuring the current decrease as oxygen was consumed [5]. The group improved on this sensor by switching to a yeast biofilm and introduced a flow system for continuous monitoring [6]. Similar to the BOD_5 method, the DO content is measured before and after addition of the sample, but these early biosensors reduced the time needed for BOD measurements from five days to under one hour. Figure 5.2 shows a schematic for a BOD sensor based on DO content.

While this first BOD biosensor was a significant improvement over the traditional method, it still suffered from several limitations. Specifically, the use of a biofilm leads to low reproducibility and stability of measurements over time. More recent research has been focused on improving these areas [8, 9]. A significant challenge when working with biofilms is the lack of stability over time. The concentration of cellular reagents and byproducts is constantly changing, which affects the analytical signal and may also lead to inactivation of the biofilm. The incorporation of flow systems that allow for reagents to be added and byproducts to be removed can improve the lifetime and stability.

Another variable that has significant implications for long-term stability is the choice of microorganism. Typically, yeasts, such as *Trichosporon cutaneum*, *Arxula adeninivorans*, and *Saccharomyces cerevisiae*, or bacterial strains, such as *Escherichia coli*, *Bacillus subtilis*, and *Pseudomonas putida*, are used for BOD biosensors. Yeast biofilms tend to be more stable and able to oxidize a large range of organic substrates, while a single-strain bacterial biofilm will have fewer substrates. Unlike most sensing applications, where specificity can be extremely important, for a BOD biosensor, a larger range of substrates is preferable so that an accurate measure of organic carbon is obtained. To achieve this, a co-culture of strains or even activated sludge from wastewater treatment can be used as the biological catalyst. However, this type of biosensor will have lower reproducibility, as the microorganism composition will continually change over time. A single-strain sensor will have greater stability and therefore improved precision but will provide a less accurate measure of BOD.

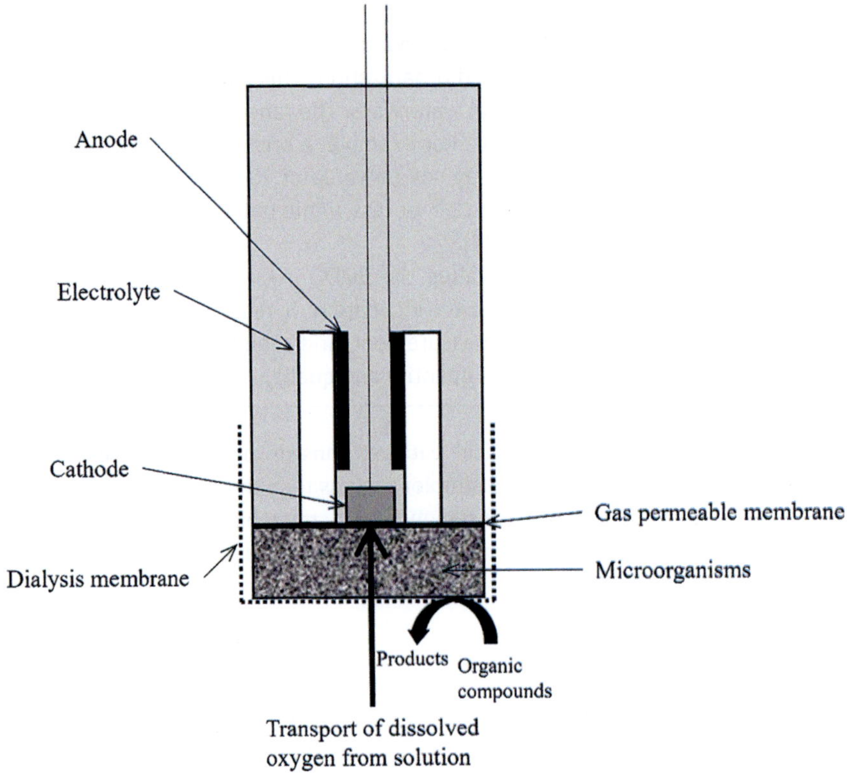

Figure 5.2 Schematic of BOD biosensor based on oxygen monitoring. *Source:* Adapted from Liu et al. [7]. Copyright 2002 Elsevier.

Te et al. recently reported a bioreactor-based BOD sensor. Their system involved a packed-bed bioreactor containing activated sludge. The sample is aerated, pumped into the bioreactor, and then to the oxygen sensor to measure the DO after respiration. A bypass was included for measuring the initial DO of the sample as well. The reported biosensor was able to determine BOD of real samples with no statistical difference to values obtained by the conventional BOD_5 method [10].

5.2.2 Mediator-Based BOD Sensors

One complication for BOD determination involving oxygen sensors is that the measurement depends on the initial DO content. This requires all samples to be aerated prior to introduction to the sensor, which increases the time required for each measurement. Incorporating an electron transport mediator can make it possible to develop a sensor that no longer depends on DO concentrations. The first example of this was reported in 2000 by Pasco et al., who used *E. coli* and ferricyanide in an air-free environment. The BOD was proportional to the ferrocyanide formed after incubation with the sample [11]. However, while removing the need to aerate samples, this sensor required all solutions to be deaerated, which still leads to longer testing times. Liu et al. investigated whether the presence of oxygen had a significant impact on the measurement of this type of sensor, showing that because the concentration of DO in water is so much lower than that of the mediator (ferricyanide), oxygen has a negligible effect on BOD determination. When examining the electron acceptors (oxygen and ferricyanide), they found that the oxygen consumption was largely due to normal respiration of the microorganisms, not from the oxidation of the organic carbon [12].

To improve mediated electron transfer, more recent research has been focused on producing reagent-free sensors where either a redox-active [13] or conductive polymer is used [14]. The polymer can immobilize the microorganisms while also providing electron transfer sites. This type of sensor typically has improved stability and simpler sample preparation. However, there is a trade-off when using exogenous electron acceptors; generally, mediators with appropriate redox potentials cannot move into or through cell membranes, while mediators that can easily penetrate cell membranes have potentials that are too low to be efficient.

5.2.3 Microbial Fuel Cell BOD Sensors

Exoelectrogenic bacteria, capable of directly transferring electrons to an electrode with no need for added mediators, offer a significant improvement on oxygen-based or mediator-based BOD sensors. Commonly used bacteria in this category are *Geobacter* and *Shewanella*. As discussed in Chapter 3, these bacteria can perform direct electron transfer using either outer membrane cytochrome *c* or nanowires. This type of sensor operates in a two-electrode configuration with the microbial reactions occurring at the anode. The current generated in the microbial fuel cell (MFC) is directly proportional to the organic matter in the solution and, therefore, gives a direct measure of the BOD [15–17].

Recently, Yan et al. reported a thorough investigation of the current production during BOD testing with an MFC sensor [18]. To obtain the most accurate measurement of BOD, the organic material should be completely oxidized, and the total charge produced during that time is calculated by integrating the current. However, this can cause long testing times, especially in carbon-rich samples where the slow decrease in current, the tail stage, can last for hours. Previous methods have stopped the testing early but that leaves a significant amount of organic material undigested. Yan et al. were able to propose a model to simulate the current in the tail stage, which allowed them to shorten testing times by 96%. Measurements are performed as normal but are stopped at the beginning of the tail stage. The charge is calculated from the recorded current and added to the tail stage current as predicted by their model.

5.3 Toxic Compounds

The ability to detect toxic compounds has importance in many areas, including the environment, agriculture, and food and drink industries. Common methods for detecting toxic compounds typically include chromatography (gas chromatography or high-performance liquid chromatography) along with mass spectrometry. These methods require expensive, bulky instrumentation and skilled technicians. In contrast, electrochemical biosensors are small and straightforward to use. They have many other advantages as well, including high sensitivity and specificity, low cost, and relatively simple fabrication and use, which make them an excellent choice for this application.

Toxic compounds found in the environment fall into a range of categories. Most of these compounds come from anthropogenic sources, including agriculture and manufacturing. Pesticides, other toxic organic compounds, and heavy metals are all found in the environment at nonnatural levels. Research focused on developing electrochemical biosensors for these toxins has been increasing for the last 20 years.

5.3.1 Pesticides

Pesticides from agriculture are one of the most common contaminants in local waterways and soil, which can lead to accumulation in food crops. Organophosphates are the primary component of many herbicides and insecticides. Numerous biosensors have been reported for the detection of organophosphates [19, 20]. The most common catalyst is the enzyme acetyl cholinesterase (AChE), which is irreversibly inhibited by organophosphate and non-organophosphate pesticides [21–25]. However, sensors that use this enzyme cannot distinguish between different organophosphates.

Zhang et al. reported an electrochemical bi-enzyme biosensor that can discriminate between organophosphate and non-organophosphate pesticides [26]. Their system used a layer-by-layer approach with carbon nanotubes coated with oppositely charged biopolymers (polyethyleneimine and DNA) and enzymes. Figure 5.3 shows the assembly of the electrode.

Organophosphate hydrolase was used for direct detection of organophosphate pesticides, while acetylcholinesterase was used for inhibition-based detection of both categories of pesticides. The sensor was first tested in a solution containing a known amount of acetylthiocholine (ACTh), then exposed to the sample containing the pesticide mixture. After exposure, the signal was measured again in the ACTh. The inhibition percentage of AChE was calculated based on the change in current before and after pesticide exposure, and this percentage was proportional to the total pesticide content in the sample. The reported system was tested with a mixture of paraoxon (an organophosphate) and carbaryl (a non-organophosphate) and was able to generate discriminative signals between the two classes of pesticides.

More recently, Arduini et al. used a paper-based electrochemical sensor to distinguish between three categories of pesticides: organophosphates, phenoxy acids, and triazines [27]. The sensor operated based on the degree of inhibition of three enzymes: butyrylcholinesterase, alkaline phosphatase, and tyrosinase. Each enzyme was inhibited to a different degree by the three model pesticides that were tested. The sensor was able to detect paraoxon, 2,4-dichlorophenoxyacetic acid, and atrazine at ppb levels with relative standard errors of <2% in doped river water samples.

5.3.2 Phenolic Compounds

Phenolic compounds are another category of organic toxins found in the environment from industrial waste. They are produced as byproducts or degradation products from a wide variety of industries, including the manufacture of pharmaceuticals, polymers, dyes, pesticides, and more. They can have significant impacts on the health of living organisms because they can easily pass through cell membranes and affect metabolic processes, including respiration and photosynthesis. They are also known to be mutagenic and genotoxic.

Most commonly, the enzyme tyrosinase is chosen as the catalyst for bioelectrochemical detection of phenolic compounds, including catechol, bisphenol A, and phenol [28–30]. The natural substrates for multicopper oxidases like tyrosinase are phenolic compounds, allowing for their direct detection by reduction of the product of the enzyme reaction. More recent research has focused on immobilization strategies to improve the sensitivity of tyrosinase-based sensors [31, 32].

Laccase is another multicopper oxidase that is regularly used as the catalyst for direct detection of phenol with bioelectrochemical sensors [33–35]. Laccase-based sensors have been used to detect catechol and pyrocatechol. Salamanca-Neto et al. used chemometric methods, specifically design of experiments (DOEs), to optimize a laccase electrode containing Pt nanoparticles and botryosphaeran, an exopolysaccharide produced by *Botryosphaeria rhodina* [36]. DOE was used to determine the optimal ratio of the three components for the electrode. After fabricating electrodes with the optimized mixture, the biosensor was used to measure 5-O-caffeoylquinic acid, a

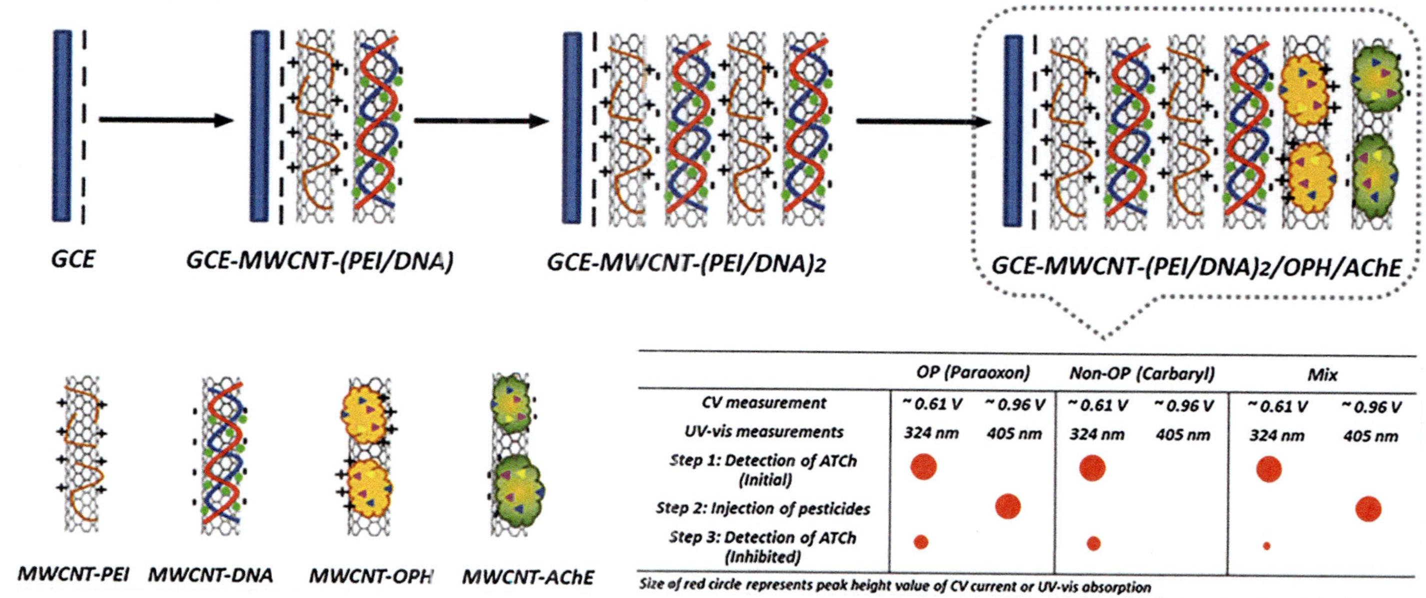

Figure 5.3 Schematic of layer-by-layer assembly of bi-enzyme sensor for pesticides. *Source:* Reproduced with permission from Zhang et al. [26]. Copyright 2015 Elsevier.

chlorogenic acid. Chlorogenic acids are phenolic compounds found in plants. The samples in this report were a variety of coffees from various sources; however, application of this sensor could be expanded to environmental samples as well.

In a recent work, a bioelectrochemical system using photosynthetic bacteria, *Rhodobacter capsulatus*, was reported [37]. While this system was not evaluated as a biosensor, it did show increased current production at increasing concentrations of dinitrophenol so it has a promise as a method for detecting phenolic compounds. The authors indicated that improving the stability of the biofilm will be critical for future implementation as a sensor.

5.3.3 Heavy Metals

Heavy metal contamination is a significant problem that can affect the environment and human health. Electrochemical biosensors have been developed for detecting various heavy metals using a variety of catalysts including urease [38, 39], *Chlorella vulgaris* [40], and *Enterobacter cloacae* [41]. These sensors were able to detect heavy metals like Cd^{2+}, Co^{2+}, Cr^{3+}, Cu^{2+}, Hg^{2+}, and Pb^{2+}. However, these sensors all produce a "turn-off" signal in response to analyte where the catalyst is inhibited or deactivated by the presence of the heavy metal. Therefore, distinguishing between heavy metals is not possible.

Grunzina et al. reported a system that used three bacterial strains (*Pseudomonas* sp., *Bacillus cereus*, and *E. coli*) that are inhibited to different degrees by the presence of Zn^{2+}, Co^{2+}, and Cu^{2+} [42]. This biosensor has the potential to distinguish between the three heavy metals, but further work needs to be done to show that this is possible. De Benedetto et al. used a glucose oxidase electrode to measure heavy metal ions by inhibition [43]. Their initial biosensor was able to detect copper and mercury ions. However, it had the same challenge as most inhibition-based sensors, which is a lack of specificity. Both copper and mercury ions inhibit glucose oxidase, resulting in a decrease in current. The group combined their sensor with chemometric methods and was able to detect Al^{3+}, Bi^{3+}, Cd^{2+}, Co^{2+}, Cr^{3+}, and Zn^{2+} [44].

5.3.4 Explosives

Another important application of bioelectrochemical sensors is for explosives sensing. Typical methods for detecting explosives involve large, complex instrumentation or living organisms (dogs, elephants, bees) [45]. Electrochemical biosensors can be fabricated to be small, require minimal sample preparation, and minimal training for use. When combined with high sensitivity and high specificity, electrochemical biosensors can be a good alternative to the conventional methods. However, the conventional methods have very low detection limits, and often, the target analytes are in the gas phase. Both of these challenges have limited the development of bioelectrochemical explosive sensors.

Gwenin et al. reported a novel nitroreductase that can be immobilized directly onto gold surfaces through a cysteine tag on the N-terminus with no need for any surface preparation or tethering compound [46]. Figure 5.4 shows the immobilization scheme for the nitroreductase-based biosensor.

Their amperometric biosensor used a ferrocene compound as an electron mediator in place of NADH, which is the natural electron donor. Regenerating NADH electrochemically is challenging because it results in an isomer that cannot be used by the enzyme. By using ferrocene dicarboxylic acid, this challenge can be removed. The reported sensor showed a linear change in current when exposed to increasing concentrations of 2,4-dinitroethylbenzene as a model explosive compound.

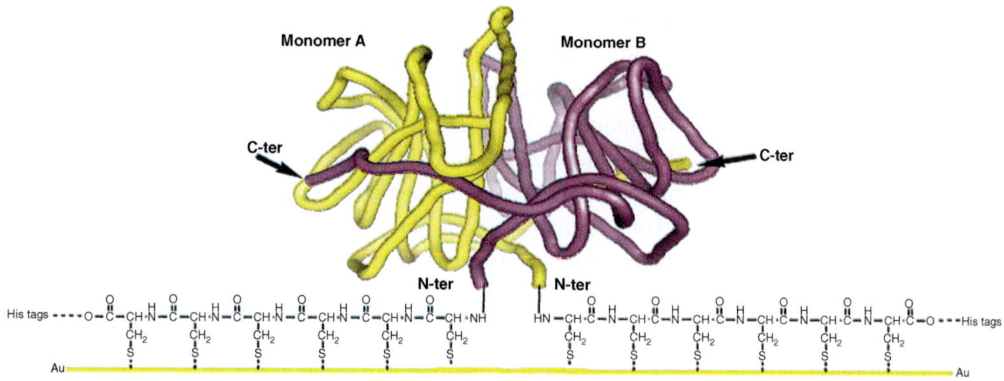

Figure 5.4 Schematic representation of nitroreductase immobilization on Au surface by cysteine tags on the N-terminus. *Source:* Reproduced with permission from Gwenin et al. [46]. Copyright 2007 Elsevier.

A photosystem II-based biosensor, reported by Bhalla et al., was able to detect picric acid and trinitrotoluene (TNT) [47]. Photosystem II (PSII) is the protein complex where water splitting and oxygen evolution occur in photosynthesis. Strong electron acceptors can inhibit electron transfer from PSII by blocking the active site for the natural acceptor, plastoquinone. Explosive compounds with nitro groups are good electron acceptors and, therefore, can inhibit PSII. In the reported biosensor, in the presence of TNT or picric acid, there is a decrease in photocurrent generated by PSII due to inhibition. This decrease is proportional to the concentration of either analyte.

5.4 Emerging Trends for Enhanced Performances

Synthetic biology and genetic engineering are likely to play a large role in future work on *ex situ* biosensors. It is becoming increasingly common to adjust enzymatic activity and/or substrate specificity using techniques like site-directed mutagenesis. Engineering of bacteria is another growing area for biosensors. Engineered microorganisms have been produced to address several limitations of this type of sensor, including to improve or add electrical contact with the electrode surface and to expand the window of possible analytes. For a more thorough discussion of this topic, refer to the recent review published by Gao et al. [48].

Another challenge faced by microbial-based biosensors is the time delay between exposure to analyte and generation of the signal. This delay can often be as long as thirty minutes, which severely limits the speed of analysis. Ajo-Franklin et al. reported an engineered *E. coli* with a synthetic electron transport chain that was able to reduce this delay time to two minutes [49]. The biosensor was developed to detect thiosulfate. The truly impactful part of the work is that the synthetic metabolic pathway is modular, meaning that the authors could easily remove one module or add another. Expanding this methodology to other analyte targets could lead to significant improvements to biosensors for a wide range of applications.

5.5 Conclusions and Future Outlook

Biosensor technologies have been developing rapidly for the past few decades. Electrochemical biosensors have several advantages that make them promising options for certain applications.

Depending on the choice of catalyst, they can be low cost, have high sensitivity, require minimal sample preparation, and involve simple analysis procedures. However, they still face many challenges to be practically useful. One major challenge is the required limit of detection for many environmental samples. For many compounds, the concentration found in the environment, while high enough to have potential health implications, is quite low. Additionally, environmental samples are complex, requiring a sensor or analytical method that can account for interferences of the sample matrix. Finally, as illustrated in this chapter, many of the reported bioelectrocatalytic-based sensors suffer from a lack of specificity. They can be used to measure the amount of a particular category of compound, organophosphates as an example, but it is generally not possible to distinguish between individual types of compound within that category. Future efforts should focus on expanding the target analytes, increasing the selectivity and specificity, and decreasing the detection limits of these sensors.

References

1 Grattieri, M. and Minteer, S.D. (2018). Self-powered biosensors. *ACS Sens.* 3: 44–53. https://doi.org/10.1021/acssensors.7b00818.

2 Arlyapov, V.A., Plekhanova, Y.V., Kamanina, O.A. et al. (2022). Microbial biosensors for rapid determination of biochemical oxygen demand: approaches, tendencies and development prospects. *Biosensors* 12: https://doi.org/10.3390/bios12100842.

3 Zhou, Y., Zheng, S., and Qin, W. (2024). Electrochemical biochemical oxygen demand biosensors and their applications in aquatic environmental monitoring. *Sens. Bio-Sens. Res.* 44: 100642. https://doi.org/10.1016/j.sbsr.2024.100642.

4 Verma, N. and Singh, A.K. (2012). Development of biological oxygen demand biosensor for monitoring the fermentation industry effluent. *ISRN Biotechnol.* 2013: 236062. https://doi.org/10.5402/2013/236062.

5 Karube, I., Matsunaga, T., Mitsuda, S., and Suzuki, S. (1977). Microbial electrode BOD sensors. *Biotechnol. Bioeng.* 19: 1535–1547. https://doi.org/10.1002/bit.260191010.

6 Hikuma, M., Suzuki, H., Yasuda, Y. et al. (1979). Amperometric estimation of BOD by using living immobilized yeasts. *Eur. J. Appl. Microbiol. Biotechnol.* 8: 289–297. https://doi.org/10.1007/BF00508793.

7 Liu, J. and Mattiasson, B. (2002). Microbial BOD sensors for wastewater analysis. *Water Res.* 36: 3786–3802. https://doi.org/10.1016/S0043-1354(02)00101-X.

8 Arlyapov, V.A., Yudina, N.Y., Machulin, A.V. et al. (2021). A biosensor based microorganisms immobilized in layer-by-layer films for the determination of biochemical oxygen demand. *Appl. Biochem. Microbiol.* 57: 133–141. https://doi.org/10.1134/S0003683821010038.

9 Kamanina, O.A., Lantsova, E.A., Rybochkin, P.V. et al. (2023). "3-in-1" Hybrid biocatalysts: association of yeast cells immobilized in a Sol–Gel matrix for determining sewage pollution. *ACS Appl. Mater. Interfaces* 15: 47779–47789. https://doi.org/10.1021/acsami.3c09897.

10 Le, B.-N.T., Nguyen, A.L., Ngo, T.T. et al. (2024). Application of biosensor for determination of biochemical oxygen demand and heavy metal toxicity in urban canal water. *Int. J. Environ. Sci. Technol.* 21: 7307–7320. https://doi.org/10.1007/s13762-024-05472-4.

11 Pasco, N., Baronian, K., Jeffries, C., and Hay, J. (2000). Biochemical mediator demand – a novel rapid alternative for measuring biochemical oxygen demand. *Appl. Microbiol. Biotechnol.* 53: 613–618. https://doi.org/10.1007/s002530051666.

12 Liu, L., Shang, L., Liu, C. et al. (2010). A new mediator method for BOD measurement under non-deaerated condition. *Talanta* 81: 1170–1175. https://doi.org/10.1016/j.talanta.2010.01.062.

13 Liu, L., Zhang, S., Xing, L. et al. (2012). A co-immobilized mediator and microorganism mediated method combined pretreatment by TiO_2 nanotubes used for BOD measurement. *Talanta* 93: 314–319. https://doi.org/10.1016/j.talanta.2012.02.039.

14 Zhao, C., Wang, G., Sun, M. et al. (2021). Bacterial cellulose immobilized *S. cerevisiae* as microbial sensor for rapid BOD detection. *Fibers Polym.* 22: 1208–1217. https://doi.org/10.1007/s12221-021-0650-5.

15 Alferov, S.V., Arlyapov, V.A., Alferov, V.A., and Reshetilov, A.N. (2018). Biofuel cell based on bacteria of the genus gluconobacter as a sensor for express analysis of biochemical oxygen demand. *Appl. Biochem. Microbiol.* 54: 689–694. https://doi.org/10.1134/S0003683818060029.

16 Moon, H., Chang, I.S., Kang, K.H. et al. (2004). Improving the dynamic response of a mediator-less microbial fuel cell as a biochemical oxygen demand (BOD) sensor. *Biotechnol. Lett.* 26: 1717–1721. https://doi.org/10.1007/s10529-004-3743-5.

17 Dang, N.P., Petrich, C., Pasztor, D. et al. (n.d.). *Shewanella baltica* in a microbial fuel cell for sensing of biological oxygen demand (BOD) of wastewater. *Anal. Lett.* 1–13. https://doi.org/10.1080/00032719.2024.2341087.

18 Yan, X., Wang, Z., Liao, C. et al. (2024). Understanding the tail current behavior of electroactive biofilms realizes the rapid measurement of biochemical oxygen demand. *Environ. Sci. Technol.* 58: 2881–2890. https://doi.org/10.1021/acs.est.3c07760.

19 Mostafa, G.A.E. (2010). Electrochemical biosensors for the detection of pesticides. *Open Electrochem. J.* 2: 22–42.

20 Songa, E.A., Arotiba, O.A., Owino, J.H.O. et al. (2009). Electrochemical detection of glyphosate herbicide using horseradish peroxidase immobilized on sulfonated polymer matrix. *Bioelectrochemistry* 75: 117–123. https://doi.org/10.1016/j.bioelechem.2009.02.007.

21 Andreescu, S., Noguer, T., Magearu, V., and Marty, J.-L. (2002). Screen-printed electrode based on AChE for the detection of pesticides in presence of organic solvents. *Talanta* 57: 169–176. https://doi.org/10.1016/S0039-9140(02)00017-6.

22 Thakkar, J.B., Gupta, S., and Prabha, C.R. (2019). Acetylcholine esterase enzyme doped multiwalled carbon nanotubes for the detection of organophosphorus pesticide using cyclic voltammetry. *Int. J. Biol. Macromol.* 137: 895–903. https://doi.org/10.1016/j.ijbiomac.2019.06.162.

23 Zhang, P., Sun, T., Rong, S. et al. (2019). A sensitive amperometric AChE-biosensor for organophosphate pesticides detection based on conjugated polymer and Ag-rGO-NH2 nanocomposite. *Bioelectrochemistry* 127: 163–170. https://doi.org/10.1016/j.bioelechem.2019.02.003.

24 Chen, H., Zuo, X., Su, S. et al. (2008). An electrochemical sensor for pesticide assays based on carbon nanotube-enhanced acetycholinesterase activity. *Analyst* 133: 1182–1186. https://doi.org/10.1039/B805334K.

25 Nagabooshanam, S., Roy, S., Mathur, A. et al. (2019). Electrochemical micro analytical device interfaced with portable potentiostat for rapid detection of chlorpyrifos using acetylcholinesterase conjugated metal organic framework using Internet of things. *Sci. Rep.* 9: 19862. https://doi.org/10.1038/s41598-019-56510-y.

26 Zhang, Y., Arugula, M.A., Wales, M. et al. (2015). A novel layer-by-layer assembled multi-enzyme/CNT biosensor for discriminative detection between organophosphorus and non-organophosphrus pesticides. *Spec. Issue Biosens.* 2014 (67): 287–295. https://doi.org/10.1016/j.bios.2014.08.036.

27 Arduini, F., Cinti, S., Caratelli, V. et al. (2019). Origami multiple paper-based electrochemical biosensors for pesticide detection. *Biosens. Bioelectron.* 126: 346–354. https://doi.org/10.1016/j.bios.2018.10.014.

28 Yildiz, H.B., Castillo, J., Guschin, D.A. et al. (2007). Phenol biosensor based on electrochemically controlled integration of tyrosinase in a redox polymer. *Microchim. Acta* 159: 27–34. https://doi.org/10.1007/s00604-007-0768-1.

29 Zehani, N., Fortgang, P., Saddek Lachgar, M. et al. (2015). Highly sensitive electrochemical biosensor for bisphenol A detection based on a diazonium-functionalized boron-doped diamond electrode modified with a multi-walled carbon nanotube-tyrosinase hybrid film. *Biosens. Bioelectron.* 74: 830–835. https://doi.org/10.1016/j.bios.2015.07.051.

30 Wee, Y., Park, S., Kwon, Y.H. et al. (2019). Tyrosinase-immobilized CNT based biosensor for highly-sensitive detection of phenolic compounds. *Biosens. Bioelectron.* 132: 279–285. https://doi.org/10.1016/j.bios.2019.03.008.

31 Verma, S., Thakur, D., Pandey, C.M., and Kumar, D. (2023). Recent prospects of carbonaceous nanomaterials-based laccase biosensor for electrochemical detection of phenolic compounds. *Biosensors* 13: https://doi.org/10.3390/bios13030305.

32 Forzato, C., Vida, V., and Berti, F. (2020). Biosensors and sensing systems for rapid analysis of phenolic compounds from plants: a comprehensive review. *Biosensors* 10: https://doi.org/10.3390/bios10090105.

33 Nazari, M., Kashanian, S., and Rafipour, R. (2015). Laccase immobilization on the electrode surface to design a biosensor for the detection of phenolic compound such as catechol. *Spectrochim. Acta, Part A* 145: 130–138. https://doi.org/10.1016/j.saa.2015.01.126.

34 Salvo-Comino, C., Garcia-Hernandez, C., Garcia-Cabezon, C., and Rodriguez-Mendez, M.L. (2020). Promoting laccase sensing activity for catechol detection using LBL assemblies of chitosan/ionic liquid/phthalocyanine as immobilization surfaces. *Bioelectrochemistry* 132: 107407. https://doi.org/10.1016/j.bioelechem.2019.107407.

35 Boubezari, I., Bessueille, F., Bonhomme, A. et al. (2020). Laccase-based biosensor encapsulated in a galactomannan-chitosan composite for the evaluation of phenolic compounds. *Biosensors* 10: https://doi.org/10.3390/bios10060070.

36 Salamanca-Neto, C.A.R., Marcheafave, G.G., Scremin, J. et al. (2020). Chemometric-assisted construction of a biosensing device to measure chlorogenic acid content in brewed coffee beverages to discriminate quality. *Food Chem.* 315: 126306. https://doi.org/10.1016/j.foodchem.2020.126306.

37 de Moura Torquato, L.D., Matteucci, R.M., Stufano, P. et al. (2023). Photobioelectrocatalysis of Intact Photosynthetic Bacteria Exposed to Dinitrophenol. *ChemElectroChem* 10: e202300013. https://doi.org/10.1002/celc.202300013.

38 Pal, P., Bhattacharyay, D., Mukhopadhyay, A., and Sarkar, P. (2009). The detection of mercury, cadium, and arsenic by the deactivation of urease on rhodinized carbon. *Environ. Eng. Sci.* 26: 25–32. https://doi.org/10.1089/ees.2007.0148.

39 Ilangovan, R., Daniel, D., Krastanov, A. et al. (2006). Enzyme based biosensor for heavy metal ions determination. *Biotechnol. Biotechnol. Equip.* 20: 184–189. https://doi.org/10.1080/13102818.2006.10817330.

40 D. Claude, G. Houssemeddine, B. Andriy, C. Jean-Marc, Whole cell algal biosensors for urban waters monitoring, in: 2007. https://api.semanticscholar.org/CorpusID:54519063.

41 Rasmussen, M. and Minteer, S.D. (2015). Long-term arsenic monitoring with an *Enterobacter cloacae* microbial fuel cell. *Bioelectrochemistry* Amst. Neth. 106: 207–212. https://doi.org/10.1016/j.bioelechem.2015.03.009.

42 Gruzina, T.G., Zadorozhnyaya, A.M., Gutnik, G.A. et al. (2007). A bacterial multisensor for determination of the contents of heavy metals in water. *J. Water Chem. Technol.* 29: 50–53. https://doi.org/10.3103/S1063455X07010080.

43 Malitesta, C. and Guascito, M.R. (2005). Heavy metal determination by biosensors based on enzyme immobilised by electropolymerisation. *Biosens. Bioelectron.* Sel. Pap. Eighth World Congr. Biosens. Part I 20: 1643–1647. https://doi.org/10.1016/j.bios.2004.08.003.

44 De Benedetto, G.E., Di Masi, S., Pennetta, A., and Malitesta, C. (2019). Response surface methodology for the optimisation of electrochemical biosensors for heavy metals detection. *Biosensors* 9: https://doi.org/10.3390/bios9010026.

45 Smith, R.G., D'Souza, N., and Nicklin, S. (2008). A review of biosensors and biologically-inspired systems for explosives detection. *Analyst* 133: 571–584. https://doi.org/10.1039/B717933M.

46 Gwenin, C.D., Kalaji, M., Williams, P.A., and Jones, R.M. (2007). The orientationally controlled assembly of genetically modified enzymes in an amperometric biosensor. *Biosens. Bioelectron.* Chem Biosensing Transistors Mater. Syst. 22: 2869–2875. https://doi.org/10.1016/j.bios.2006.12.012.

47 Bhalla, V., Zhao, X., and Zazubovich, V. (2011). Detection of explosive compounds using photosystem II-based biosensor. *J. Electroanal. Chem.* 657: 84–90. https://doi.org/10.1016/j.jelechem.2011.03.026.

48 Zhu, J., Wang, B., Zhang, Y. et al. (2023). Living electrochemical biosensing: Engineered electroactive bacteria for biosensor development and the emerging trends. *Biosens. Bioelectron.* 237: 115480. https://doi.org/10.1016/j.bios.2023.115480.

49 Atkinson, J.T., Su, L., Zhang, X. et al. (2022). Real-time bioelectronic sensing of environmental contaminants. *Nature* 611: 548–553. https://doi.org/10.1038/s41586-022-05356-y.

6

Biofuel Cells and Biosolar Cells

Matteo Grattieri[1,2] *and Shelley D. Minteer*[3]

[1]*Department of Chemistry, Università degli Studi di Bari Aldo Moro, Bari, Italy*
[2]*Istituto per i Processi Chimico Fisici, Consiglio Nazionale delle Ricerche, Bari, Italy*
[3]*Kummer Institute Center for Resource Sustainability, Missouri University of Science and Technology, Rolla, MO, USA*

6.1 Introduction

A fuel cell is a device that converts chemical energy into electrical energy, having the oxidation of a fuel (i.e. an electron donor) taking place at the anode with the simultaneous reduction of an electron acceptor (i.e. oxygen) at the cathode [1]. Commonly, a membrane is utilized to separate the anodic and cathodic chambers, where the two half-reactions take place, allowing the flow of electrons in an external circuit for power production. Equations (6.1) and (6.2) describe the hydrogen oxidation reaction (anodic process) and the oxygen reduction reaction (cathodic process) taking place in a classic proton-exchange membrane (PEM) fuel cell, respectively.

$$2H_2 \rightarrow 4H^+ + 4e^- \quad E = 0.00\ \text{V vs. SHE} \tag{6.1}$$

$$O_2 + 4H^+ + 4e^- \rightarrow 2H_2O \quad E = 1.23\ \text{V vs. SHE} \tag{6.2}$$

Fuel cells are nowadays an established technology that commonly employ conductive metal nanoparticles as electrocatalysts and operate in a broad range of temperatures (i.e. from 45 to 150 °C) [2, 3]. While the employed catalysts have several advantages, such as high stability and high catalytic activity, they also suffer from various issues, including the need for rare precious metal catalysts or the passivation of the metals, which require specific operation conditions such as the use of high-purity fuels.

Conversely from metal-based electrocatalysts, the use of biological materials offers an appealing alternative that has attracted the interest of researchers for more than a century [4–6]. Among the unique features of bio-electrocatalysts is their renewable nature, as they can either be extracted from a wide range of organisms or use intact organisms that can be directly employed as catalysts. Fuel cells utilizing biological materials as catalysts for at least one of the electrode processes are defined as "biofuel cells." Figure 6.1 reports a scheme of such a system for a case where enzymes are utilized at both the anode and the cathode electrodes.

From this initial classification, further subclasses of biofuel cells can be defined. By detailing the type of biological catalyst employed, we can distinguish between enzymatic biofuel cells (Figure 6.2A) and microbial biofuel cells (Figure 6.2B), where the former utilize isolated enzymes as catalysts coupled to the abiotic electrodes, and the latter utilize intact, metabolically active microbial cells. The terminology has commonly removed the prefix "bio," resulting in these two

Bioelectrocatalysis: From Electron Transfer Processes to Emerging Technological Applications, First Edition.
Edited by Shelley D. Minteer and Matteo Grattieri.

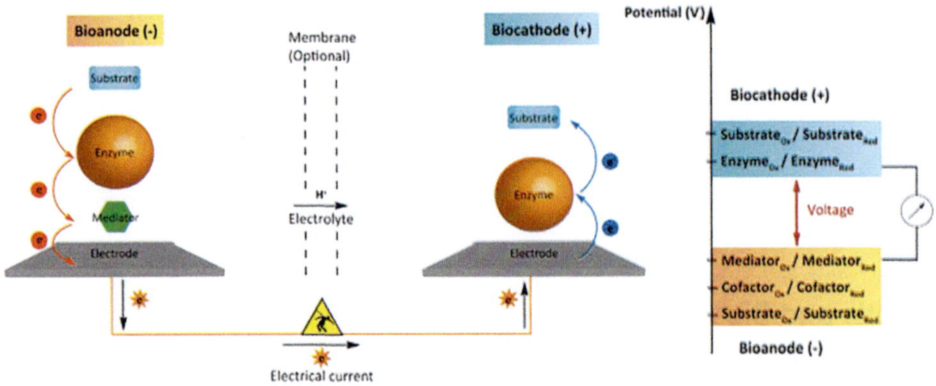

Figure 6.1 Scheme of an enzymatic biofuel cell with both the anodic and cathodic processes catalyzed by enzymes. A redox mediator is employed at the bioanode, while direct electron transfer is obtained at the cathode. *Source:* Reproduced with permission from Rasmussen et al. [7]. Copyright 2016 Elsevier B.V.

categories of biofuel cells being defined as "enzymatic fuel cells" (EFCs) and "microbial fuel cells" (MFCs) [11, 12]. When photosynthetic entities [13–18] or intact organisms [19–22] are utilized as biocatalysts and light triggers the photosynthetic process, a photo biofuel cell can be obtained. Among the light-dependent systems, a specific subclass can be defined when water is utilized as the electron donor. In such a particular case, the obtained device can be considered as a biohybrid alternative to conventional photovoltaic devices, and the biofuel cell is commonly defined as a "biosolar cell" or "biophotovoltaic" (Figure 6.2C) [10, 23–26].

The following sections of this chapter will describe EFCs, MFCs, and biosolar cells, along with the main challenges for their potential applications and the current research directions.

6.2 Enzymatic Fuel Cells

6.2.1 Historical Overview: From Early Studies to Current Days

The concept of EFC was first invented by Yahiro et al. in 1964, who reported a biofuel cell composed of a glucose oxidase bioanode and a Pt-based cathode operating at neutral pH [27]. The anodic and cathodic half reactions involved (at pH 7) were:

$$2C_6H_{12}O_6 \rightarrow 2C_6H_{10}O_6 + 4H^+ + 4e^- \quad E' = -0.36\ \text{V vs. SHE} \tag{6.3}$$

$$O_2 + 4H^+ + 4e^- \rightarrow 2H_2O \quad E' = +0.81\ \text{V vs. SHE} \tag{6.4}$$

However, compared to the maximum theoretical potential of 1.17 V for an electrochemical cell performing these half-reactions, the EFC reported in their seminal work achieved a relatively limited difference in potential, in the range of 175–350 mV. The result can be explained considering the various overpotentials at play in the system, including the use of a redox mediator for the anodic process, and the overpotential for the oxygen reduction reaction (ORR). Despite the limited performance of that pioneering EFC, the work of Yahiro et al. set the basis for the possibility of utilizing an oxidoreductase enzyme to catalyze the oxidation of a fuel, transferring the obtained electrons to an electrode surface utilizing small redox mediators. A few years later, in 1978, an important improvement for EFC was reported by Berezin et al., who demonstrated the possibility of accomplishing direct electron transfer from the active redox sites of an enzyme, specifically laccases, to an

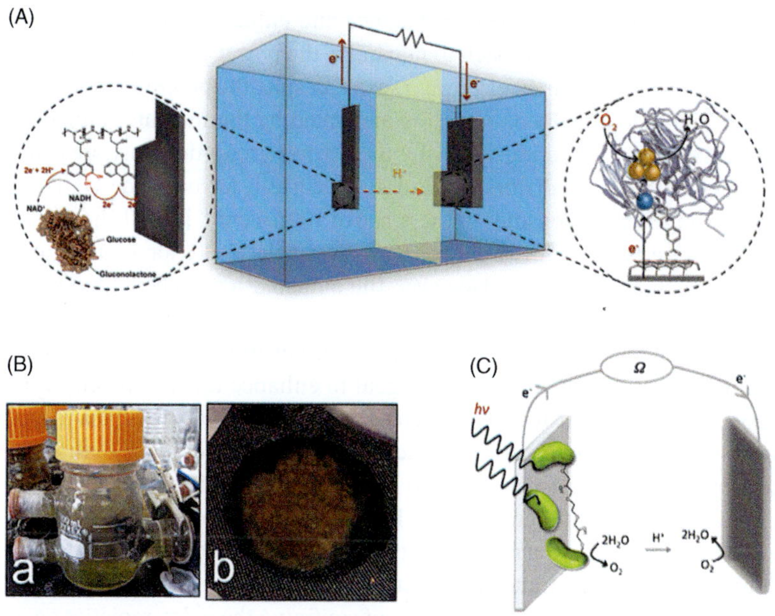

Figure 6.2 (A) Glucose/O_2 enzymatic fuel cell with a NAD-dependent glucose dehydrogenase-based bioanode and laccase-based biocathode. *Source:* Reproduced from Ref. [8] with permission from the Royal Society of Chemistry. (B) Single chamber-microbial fuel cell with bacterial cells colonizing both the anode and cathode electrode (a); biofilm growing on the air-breathing cathode (b). *Source:* Reproduced with permission from Cristiani et al. [9]. Copyright 2013 Elsevier B.V. (C) Scheme of a biosolar cell with an intact chloroplast-based photoanode. *Source:* Adapted from Ref. [10] with permission from the Royal Society of Chemistry.

electrode [28]. The work not only reported the first enzyme-based cathode for the oxygen reduction reaction, but also showed the critical possibility of avoiding the potential loss due to the difference in potential between the enzymatic redox center and the redox mediator employed for electron transfer.

This overview on the origins of EFC underlines the advantages that enzyme-based electrodes have over other type of biocatalysts for developing fuel cells: (i) high catalytic activity with no need of rare metal-based catalysts, (ii) facilitated electron transfer to the electrode due to the limited distance between their active redox sites and the electrodes (from few angstrom to ~70 Å) [29], and (iii) high specificity due to substrate recognition features in the enzyme structure that allow selectively oxidizing a compound, potentially eliminating the need to remove interfering substrates and allowing for a simpler fuel cell setup. Specifically, the latter feature allows for the bioanode and biocathode to operate in a single membrane-less electrochemical cell, which opens the development of a miniaturized system [30, 31]. In contrast, different limitations affect enzymatic biofuel cells. The selectivity of enzymes for a specific substrate often results in their inability to completely oxidize the substrate, leading to a decrease in energy density. Furthermore, EFCs are also characterized by lower current densities compared to classical fuel cells and high cost due to enzyme isolation and purification. An additional aspect to consider is that the enzymes used as catalysts generally exhibit limited stability compared to abiotic catalysts [32].

The issue of lower current densities has led to several studies focused on maximizing the electron transfer process. On the one hand, with the aim of improving direct electron transfer from enzymes to electrodes, approaches such as site-selective orientation of enzymes on electrodes or the use of conductive hydrogels can minimize the donor–acceptor distance and enhance current densities [33, 34]. On the other hand, a widely utilized approach to maximize the power production of EFC is the use of mediated electron transfer over direct electron transfer, despite the potential loss issue resulting from the addition of the redox mediator, which is motivated by the mediator allowing for higher electron transfer rates [35]. Pioneering works by the Turner group demonstrated the potential of utilizing organometallic redox mediators in solution, specifically ferrocene derivatives, for enzymatic electrodes, achieving current densities in the range of 100 μA cm^{-2} [36]. It should be noted that many of these works were aided in developing enzyme-based electrodes for applications not necessarily limited to EFC such as biosensors. However, the obtained results benefited both research fields.

Following the use of redox mediators as diffusible molecules present in the electrolyte, more complex electrode architectures were investigated with the goal to enhance further the electron transfer process and the resulting current densities. For this purpose, the possibility to bind a redox mediator and the enzyme using flexible chains was reported by Schuhmann et al. [37]. Such a system consisted of glucose oxidase-based anodes obtained by covalently binding the enzyme to ferrocene derivatives via spacer chains of different lengths. The work highlighted the importance of a sufficient length (i.e. >10 bonds) to allow the redox moieties penetrating the enzyme to interact with the redox centers for accomplishing the transfer of electrons.

A different approach gained particular attention, consisting of confining the redox moieties on a polymer backbone rather than directly on the enzyme, thereby obtaining flexible redox polymers that can be used to modify the electrode surface and entrap the enzymes [38–41]. In the 2000s, the use of redox polymers has since expanded significantly, becoming an extremely popular approach in EFC works [42, 43]. Transition metal complexes were utilized in the early studies, with the use of organic redox pendants that became popular in more recent studies [44]. Figure 6.3 illustrates various redox polymers employed in EFC, ranging from the Os-based redox polymers used in pioneering studies to a recent system featuring naphthoquinone pendants. Interestingly, the use of a naphthoquinone linear polyethyleneimine polymer by Milton et al. allowed for the combination of both a low onset potential for the anodic process and a high catalytic current density in an EFC for glucose oxidation catalyzed by glucose dehydrogenase at the anode [45]. The glucose/O_2 EFC achieved an OCP of about 0.86 V, with a current and power density of 5.4 ± 0.5 mA cm^{-2} and 2.3 ± 0.2 mW cm^{-2}, respectively. It is thus clear that the development of artificial redox mediation approaches has greatly improved the performance of EFC since the pioneering reports.

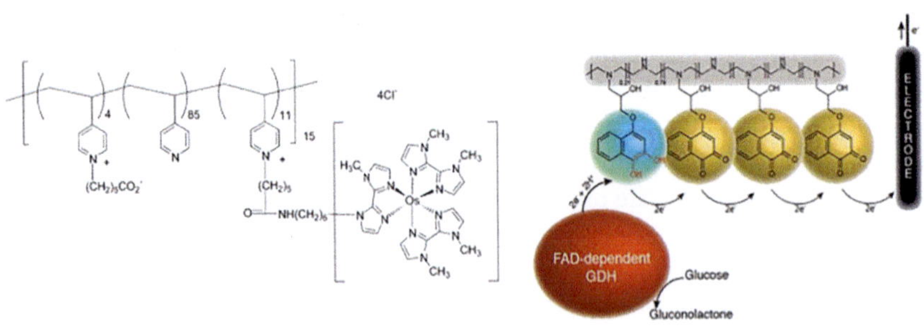

Figure 6.3 Examples of different redox polymers employed in enzymatic fuel cells. Left, poly(4-vinylpyridine)-[Os(N,N′-dialkylated-2,2′-biimidazole)$_3$]$^{2+/3+}$ adapted with permission from Mao et al. [40]. Copyright 2003 American Chemical Society. Right, naphthoquinone linear polyethyleneimine. *Source:* Adapted from Milton et al. [45] with permission from the Royal Society of Chemistry.

6.2.2 Deep Oxidation of Fuel

The EFCs described so far in this chapter were composed of one or a maximum of two enzymes. As previously introduced, depending on the specific fuel utilized in the system, achieving the complete oxidation of the fuel (i.e. obtaining CO_2) is often not possible. Under such conditions, the energy density of the EFC is significantly decreased, thereby affecting its performance. To tackle this issue, an approach that gained particular attention is the use of multiple enzymes at one electrode to obtain an "enzyme cascade" for the oxidation of one fuel [46]. This concept was pioneered in the work of Arechederra and Minteer [47], who showed how the use of three enzymes greatly improved the energy density and power output of an EFC. In the work, glycerol was utilized as a model fuel accounting for a theoretical maximum energy density of 6260 W h L^{-1}, as shown in Figure 6.4. If an EFC with only a pyrroloquinoline quinone (PQQ)-dependent alcohol dehydrogenase bioanode is utilized, an enzyme capable of undergoing direct electron transfer with the electrode, the oxidation of the substrate would lead to tartronaldehyde, accounting for an energy recovery of only 32%. The combination of this enzyme with a second enzyme, PQQ-dependent aldehyde dehydrogenase, results in five sequential oxidations, yielding mesoxalic acid with a 70% energy recovery. It was demonstrated that oxalate oxidase exhibits activity toward mesoxalic acid, due to the structural similarities between its natural substrate, oxalic acid, and mesoxalic acid. By immobilizing oxalate oxidase on Toray paper electrodes with tetrabutylammonium-modified Nafion, it was possible to accomplish three more sequential oxidations, for a 100% energy recovery in the EFC. It is interesting to note that the biological pathway for glycerol metabolism in living cells involves glycolysis and the Krebs cycle, for a total of 16 enzymes. Conversely, the three-enzyme cascade obtained in the EFC allowed complete glycerol oxidation by rationally designing it to include an enzyme that could oxidize both mesoxalic acid and oxalic acid.

From this pioneering work, several enzyme cascades have been reported in the literature, with complicated architectures, up to 15 enzymes, that allowed deep oxidation of various substrates such as glucose [48–50], methanol, and ethanol [51]. With a 15-enzyme cascade, Zhu et al. achieved a coulombic efficiency of ~95% from the oxidation of a sugar mixture composed of glucose, fructose, and sucrose, achieving a maximum power density of 1.08 mW cm^{-2} [52].

It is essential to emphasize that, while the use of an enzyme cascade offers significant advantages for deep fuel oxidation, it also introduces increased complexity in the EFC that must be designed [53]. Specifically, since the various enzymes employed in the cascade may have different optimal pH levels for their operation, the overall performance may be limited by an enzyme being less active compared to the others. Furthermore, an enzyme having a slower kinetic for substrate oxidation would influence the energy generation. To tackle the last aspect, the amount of each specific enzyme utilized in the cascade should be carefully defined, taking into consideration the different enzyme kinetics. Another possible mitigation is the use of hybrid catalytic cascades, where enzymes are coupled to abiotic catalysts for deep fuel oxidation. Following such an approach, Franco et al. reported the complete oxidation of ethanol and lactate with a catalytic system composed of pyrene-TEMPO (2,2,6,6-tetramethylpiperidinyl-N-oxyl) and oxalate decarboxylase enzyme [54, 55]. The hybrid system for ethanol oxidation retained $74 \pm 9\%$ of the initial power density after 30 days, passing from 390 ± 30 to 290 ± 20 μW cm^{-2}.

6.2.3 Substrate Channeling

To further improve the performance of EFC with enzyme cascades, an interesting approach is the use of substrate channeling. Such an approach is inspired by nature, which has evolved

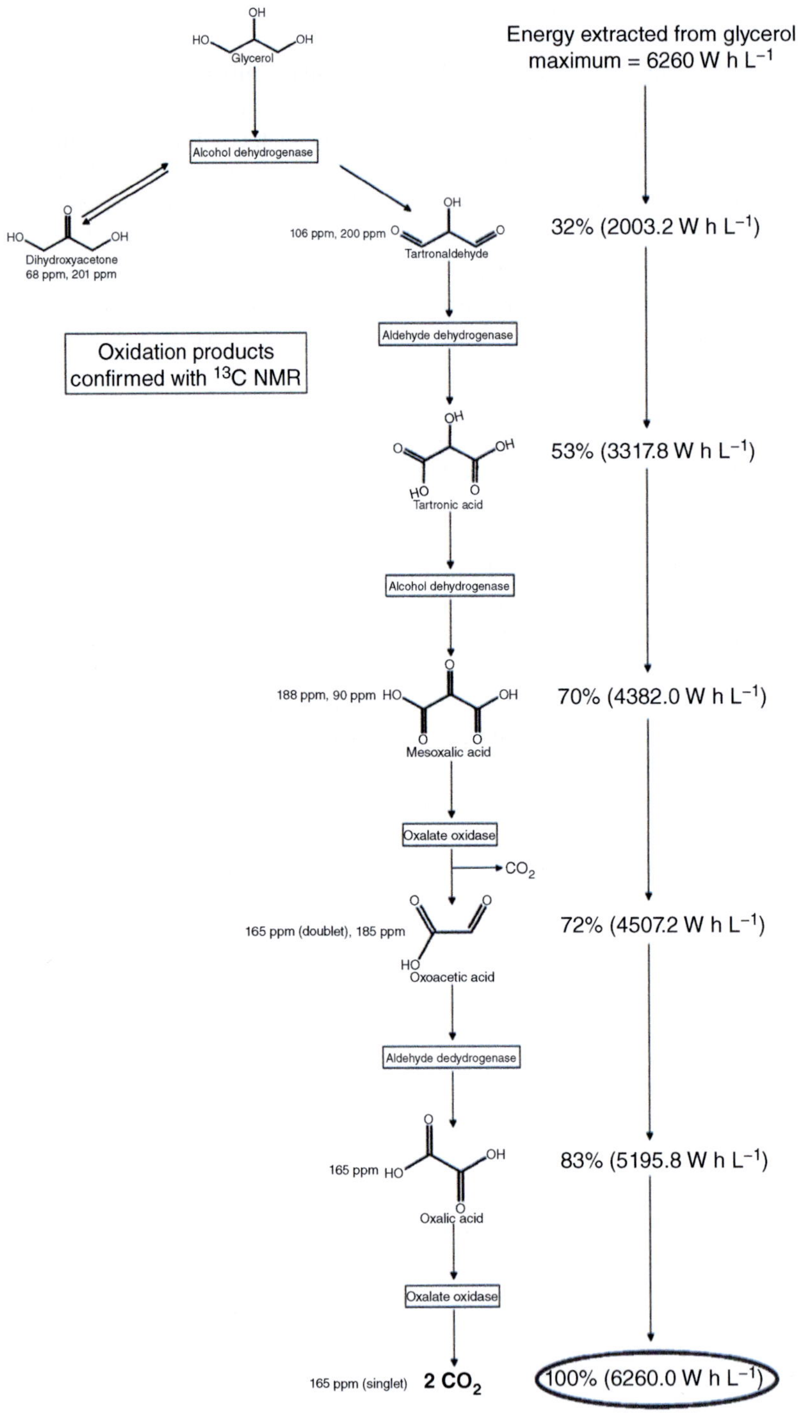

Figure 6.4 Three-enzyme cascade allowing the complete oxidation of glycerol. *Source:* Reproduced from Arechederra and Minteer [47]. Copyright 2009 Wiley-VCH Verlag GmbH & Co.

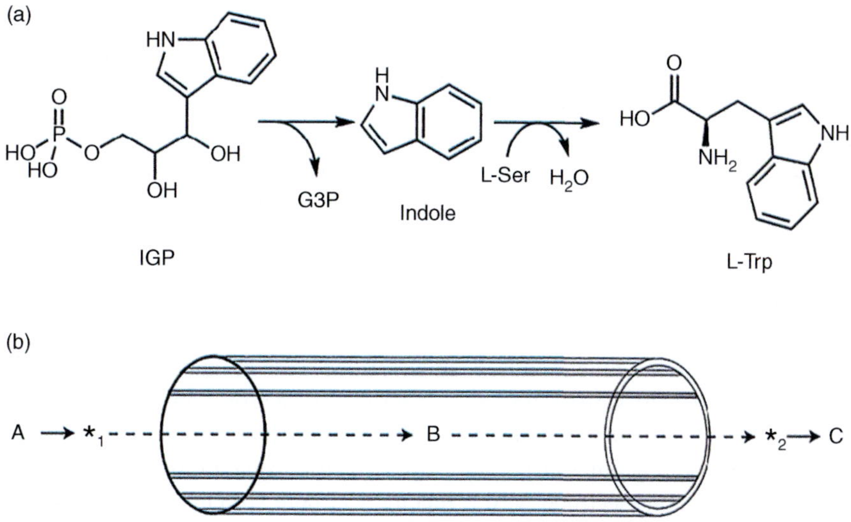

Figure 6.5 (a) Sequence of reactions involved in the tryptophan synthase substrate channeling. (b) Scheme of the substrate channeling process. *Source:* Reproduced with permission from Wheeldon et al. [57]. Copyright 2016 Macmillan Publishers Limited.

specific structures where different enzymes catalyze sequential reactions with a stoichiometric arrangement of the active sites, as well as of the chemical and physical mechanisms controlling the mass transport of reactants and reaction intermediates. As a result, substrate channeling is achieved when the direct transfer of a reactant from one active site to another is obtained, and the intermediates between the various enzymatic steps present in the cascade are not in equilibrium with the bulk solution [56]. A classic example of this natural process is the one observed in the intramolecular tunnel in the bifunctional enzyme tryptophan synthase. The enzyme catalyzes a two-step reaction cascade, as shown in Figure 6.5a, where indole-3-glycerol-phosphate is first converted into indole and glyceraldehyde-3-phosphate at one active site. Following, indole is condensed with serine to produce tryptophan and water at the second active site. A scheme of the intramolecular tunnel that allows substrate channels to take place is shown in Figure 6.5b.

Reaching substrate channeling in an EFC results in increased efficiencies and yields of the system. An example of EFC with an enzyme cascade accomplishing substrate channeling has been reported by Xia et al. [58], who demonstrated an enzyme cascade-based bioanode utilizing alcohol dehydrogenase and aldehyde dehydrogenase for the oxidation of methanol. Substrate channeling was achieved through the use of zinc domains, which served as anchors to immobilize the enzymes on DNA-modified multiwalled carbon nanotubes at a desired distance. Additionally, the zinc domains enabled the direct capture of the two enzymes from cell lysate, achieving a protein purification-free approach.

When designing an artificial enzyme cascade-based electrode with substrate channeling, some critical aspects should be taken into considerations [57]: (i) when the distance between two active sites is greater than 1 nm, having proximity between the sites is not sufficient to ensure channeling; (ii) substrate challenging can be achieved also for higher distances (i.e. up to 10 nm), provided that some conditions are present such as electrostatic guidance, compartmentalization, and bounded surface diffusion; and (iii) it is preferable to have clusters with high concentration of active sites, so that the reaction intermediates will be rapidly consumed avoiding their diffusion in the bulk.

6.2.4 Nanostructured Electrodes for High-Performance EFC

The use of nanomaterials to obtain nanostructured electrodes is another promising approach to developing EFCs with improved performance. The enhanced current density obtained when employing such electrodes with enzymes can be attributed to the high enzyme loading enabled by the advanced materials, compared to flat electrodes [59]. Carbon nanomaterials have been extensively explored for this purpose. Ciaccafava et al. reported an EFC with a bioanode based on a hyperthermophilic hydrogenase on carboxylic-functionalized carbon nanotubes coated pyrolytic graphite electrode and a biocathode based on bilirubin oxidase from *Myrothecium verrucaria* [60]. The EFC achieved a power density of 300 μW cm^{-2}, while the use of the enzymes on unmodified pyrolytic graphite electrode allowed to reach only 1 μW cm^{-2}. Modification of glassy carbon electrodes with p-phenylenediamine-functionalized single-walled carbon nanotubes allowed Tasca et al. to achieve current densities of 500 μA cm^{-2} under direct electron transfer conditions with cellobiose dehydrogenase-based bioanodes for lactose oxidation [61]. Remarkably, Tsujimura et al. achieved a current density of about 100 mA cm^{-2} from glucose oxidation by developing a mesoporous MgO-templated carbon electrode with a hierarchical pore structure coated with a biocatalytic hydrogel composed of a conductive redox polymer (poly(1-vinyl-imidazole) complex of $[Os(bpy)_2Cl]$) and deglycosylated flavin adenine dinucleotide-dependent glucose dehydrogenase from *Aspergillus terreus* [62]. The obtained current density was 30 times higher than the maximum value achieved on a flat glassy carbon electrode.

Despite the considerable advantage of current and power density resulting from the use of carbon nanomaterials, an important aspect to consider is the potential toxicity of carbon nanomaterials [63, 64], which might prevent their use to avoid release in the environment [65].

As an alternative to carbon nanomaterials, metallic nanomaterials have been investigated for the development of EFC. Particular attention has been paid to the use of gold nanoparticles (Au-NPs), which have shown little to no toxicity [66], although a clear understanding of the changes in physicochemical properties of Au-NPs resulting from their interaction with biological materials requires further studies [67]. Wang et al. reported a sugar/oxygen EFC with a bioanode based on cellobiose dehydrogenase from *Corynascus thermophilus* and a cathode based on bilirubin oxidase from *M. verrucaria*, both immobilized on gold electrodes modified with 20-nm Au-NPs [68]. The device achieved a maximum power density of 15 μW cm^{-2}, compared to 3 μW cm^{-2} for the EFC having the electrodes prepared without Au-NPs. More recently, it was demonstrated that modifying carbon substrates, such as graphdiyne, with small Au nanoparticles (diameter 5 nm) facilitated electron transfer and enzyme loading, resulting in current densities of 20 μA cm^{-2} from glucose bioanodes with glucose-6-phosphate dehydrogenase, approximately 3.5 times higher than those of electrodes without Au nanoparticles [69].

6.3 Microbial Fuel Cells

The possibility of coupling microbial cells to electrodes for current generation in MFC was first reported in 1911, with the seminal study of Michael Cresse Potter [4]. After that pioneering study, few other reports on the coupling of intact microorganisms with electrodes followed before MFCs gained attention in the 1960s [5, 6], when NASA was interested in studying this technology for electricity generation in space missions utilizing organic waste as fuel [70]. Later on, Bennetto and collaborators drew new attention to MFC by demonstrating the possibility of utilizing different diffusible redox mediators to enhance the current generation performance of these devices [71, 72].

Specifically, the capability of small molecules, such as thionine, to enhance electron transfer (EET) with different bacterial species has been reported, including *Proteus vulgaris* and *Escherichia coli*. They also showed that the performance of the MFCs could be enhanced by chemically immobilizing bacterial cells on the electrode surface [73]. Building upon these early studies, the interest in MFC research and electroactive bacterial cells soared in the 2000s [74, 75], with various studies reporting on the possibility to achieve power densities in the range of few $\mu W\ cm^{-2}$ (few $mW\ m^{-2}$) [76–79]. Various studies began to investigate how to optimize MFCs and their operational conditions to enhance power production performance, including optimal pH, electrolyte, and dissolved oxygen concentrations [80]. Early studies on MFC employed a two-chamber configuration, where a salt bridge, or more commonly a PEM, separated the anolyte and catholyte, as commonly done in fuel cell technology. These configurations required the use of a sacrificial electron acceptor at the cathode, such as ferricyanide ($K_3[Fe(CN)_6]$), which has a low reduction overpotential and allows the use of low-cost cathodes (i.e. carbon-based), resulting in minimal potential loss [81]. However, the need for a continuous supply of $K_3[Fe(CN)_6]$ hinders the practical implementation of this configuration. A common alternative electron acceptor is oxygen, but its use in a two-chamber setup also carries some challenges, such as the need for continuous gas bubbling to avoid diffusion limitations on the performance of the device.

To improve the performance of the MFCs, an interesting approach was proposed with a couple of microbial anodes, performing the oxidation of organic matter, and enzymatic cathodes, performing oxygen reduction with high efficiency. Such devices are defined as hybrid-MFC, and were reported by Schaetzle et al., with a laccas-based cathode utilizing 2,2-azinobis(3-ethylbenzothiazoline-6-sulfonic acid) diammonium salt as redox mediator [82].

Another approach to deal with the requirement of gas bubbling limiting the implementation of two-chamber configurations is the use of air-breathing cathodes, where O_2 from the air is allowed to diffuse through the cathode, which also became popular for MFC [83–85]. In this type of configuration, the anode and cathode electrodes are exposed to the same solution, with the bacterial cells that can colonize both electrode surfaces, and the power densities achieved are improved to about $50\ \mu W\ cm^{-2}$ ($0.5\ W\ m^{-2}$). A scheme of two-chamber and air-breathing (single-chamber) MFCs is reported in Figure 6.6. The possibility to employ an air-breathing MFC configuration comes from the oxygen consumption in the proximity of the cathode, where O_2 takes part in the reduction reaction at the three-point interface and the eventual oxygen remaining is further consumed by the bacteria colonizing the electrode, ensuring anaerobic conditions at the anode [9]. Further optimization of the air-breathing electrodes has focused on avoiding rare metal catalysts while retaining

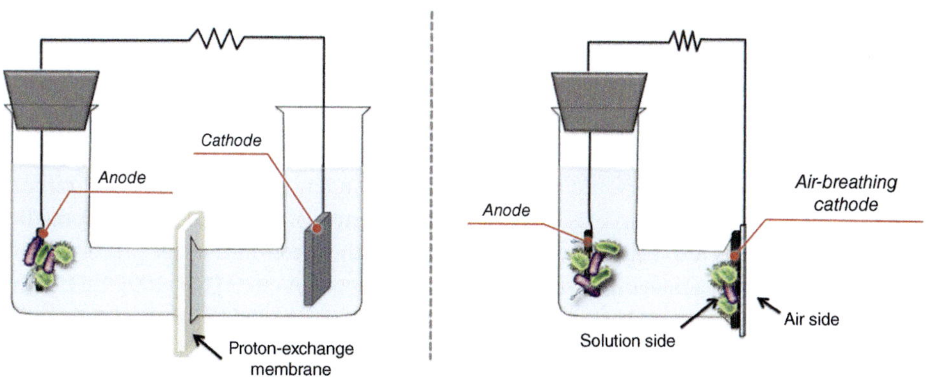

Figure 6.6 Scheme of two-chamber (left) and air-breathing MFCs.

good catalytic properties for the oxygen reduction reaction [86]. Hybrid-MFCs have also been developed with an air-breathing configuration, as reported by Higgins et al., who, by employing a direct-electron transfer laccase cathode, achieved a maximum power density of 26 W m^{-3} [87].

6.3.1 Complex Substrates and Microbial Species for Power Production

Among the reasons for the high interest in the use of MFC for power generation is that the use of intact microorganisms allows dealing with one of the issues previously discussed for EFC: incomplete fuel oxidation. Utilizing intact, metabolically active microorganisms allows retaining their complete enzymatic machinery, thus enabling multiple enzymatic reactions, as in EFC with enzymatic cascades, while providing the self-regenerating features unique to living organisms, where inactive enzymes are continuously replaced. This capability has resulted in a broad variety of complex substrates being utilized as electron donors for the microbial communities in MFC [88, 89]. Starting from the use of specific carbohydrates, reports on MFCs are now available with a broad variety of high-molecular-weight compounds being fed to the microorganisms colonizing the anode electrode for current generation [90]. It is worth mentioning that power can be obtained from a broad range of monosaccharides in lignocellulosic biomass, achieving remarkable power densities in the range of 1–2 W m^{-2} [91]. Furthermore, many food and food processing wastewater, as well as industrial wastewater containing highly toxic compounds such as the petrochemical and pharmaceutical industries, allowed power production, although achieving lower densities, in the range of 50–200 mW m^{-2} [92, 93]. Interestingly, MFCs have also been successfully applied with substrates characterized by high salinity (>0.6 M NaCl), where bacterial species coping with such extreme conditions allowed current generation [94]. Specifically, the use of bacterial species isolated from hypersaline environments (total dissolved salts >35 g L^{-1}) enabled power generation in MFCs operating with electrolytes containing more than 60 g L^{-1} of total dissolved salts [95–98]. These operating conditions are of interest for developing devices generating current while treating wastewaters with high salt content such as leather industry wastewater and oilfield wastewater [99]. The MFCs operating with hypersaline solutions reached lower power production compared to systems operating at lower salinities but could still achieve values in the range of 20–100 mW m^{-2}, depending on the specific salinity and the type of cathode electrode utilized (Pt-based or Pt-free) [100–103]. When a two-chamber MFC setup was utilized, with ferricyanide supply in the catholyte and neutral red utilized as a redox mediator in the anolyte, a remarkable power density of more than 0.5 W m^{-2} was achieved with a total dissolved salt content higher than 150 g L^{-1} due to the presence of an halophile archaea (*Haloferax volcanii*) as biocatalyst [104]. This result allows us to discuss, in addition to the diversity of substrates and electrolytes that can be utilized as electron donors in MFCs, the broad variety of microorganisms that can serve as biocatalysts "wired" to electrodes. To date, more than 90 different bacterial species have been reported as electroactive [105]. Even more interestingly, the possibility that electroactivity is an adaptation available to all existing microorganisms has been recently formulated [106]. Such a possibility has been proposed considering that the available studies on MFC are no longer limited to a few microorganisms but rather cover archaea, bacteria, and even eukaryotes. In fact, the effectiveness of both diffusible redox mediators and redox polymers has also been reported for mediating the electron transfer process from intact yeast cells to electrodes, achieving current densities of various μA cm^{-2} [107–109]. EET with microalgae has also been reported [110], and various studies have extended the biofuel cell landscape to power-generating devices employing aquatic and terrestrial higher plants, achieving power densities greater than 150 mW m^{-2} [111–115]. Interestingly, recent studies have also shown the possibility of obtaining biofuel cells that employ mammalian cells directly, such as fibroblasts [116].

These results open the doors to new research directions toward the implementation of new types of intact biocatalysts into biofuel cells, further expanding the potential applications of such devices.

6.3.2 Current Approaches for Improving Power Density Production

Considerable efforts have been directed in recent years toward overcoming the hindered extracellular electron transfer process that limits the current generation and power production of MFCs. Most of the recent approaches focus on modifying the membrane of bacterial cells to provide artificial electron transfer pathways or in developing advanced abiotic/biotic interfaces. Among the various approaches, the possibility of utilizing carbon dots to tune the bacterial membrane has been recently reported. Specifically, Guo et al. synthesized ultrasmall carbon dots containing nitrogen using a hydrothermal process with *Hydrocotyle vulgaris* as a carbon source [117]. Carbon dots, with an average diameter of 1.3 nm, were absorbed by *Shewanella oneidensis* MR-1 cells, and an MFC prepared with the modified bacterial cells on carbon cloth anodes allowed a 13-fold higher power density compared to an MFC with unmodified bacteria (332 and 24 mW m^{-2}, respectively).

As previously mentioned for EFCs, the use of nanomaterials has also been intensively explored for developing microbial electrodes with abiotic/biotic interfaces that facilitate the colonization of microorganisms and enable high current densities [118]. Specifically, the combination of carbon nanotubes and core–shell nanostructures of Fe and Co has recently enabled achieving a power density of 3 W m^{-2} in mixed culture MFCs with 100-mm-thick biofilms [119]. By comparison, the power density of MFCs operating under the same conditions but with bare carbon felt electrodes achieved a maximum value of 2 W m^{-2}. Recently, Cao et al. reported that the introduction of transmembrane and outer-membrane nanoparticles in *Shewanella*-based MFCs has allowed for the remarkable power density of 6.6 W m^{-2} [120]. Such performance was achieved by developing reduced graphene oxide-silver nanoparticle scaffolds that released silver ions responsible for producing Ag-modified *Shewanella* cells with enhanced electron transfer capabilities. The coulombic efficiency of the obtained devices passed 80%.

By combining gold nanoparticles and polyaniline modification of carbon felt electrodes, Thapa et al. developed an MFC based on *E. coli* cells, which normally exhibit little extracellular electron transfer capabilities, that could achieve a power density of 1.4 W m^{-2} [121]. As a drawback, the cathode of the MFC was based on 0.05 M $K_3[Fe(CN)_6]$, but the effectiveness of Au-NPs in supporting the electron transfer process is promising for future studies.

From a synthetic biology perspective, significant advancements have been performed [122], and the complete *S. oneidensis* Mtr pathway, comprising a total of 4 heterologously expressed cytochromes with 28 heme cofactors distributed across the periplasm and the inner and outer cellular membranes, was recently expressed in *E. coli* cells, allowing for a threefold increase in current densities [123].

In addition to the modification of the electrode with nanomaterials, engineering of the MFC setup has been recently reported by Zhang et al. to achieve record-high power density [124]. The engineered configuration consisted of a flow-MFC where the bacterial cells are not immobilized in a biofilm on the electrode surface but rather continuously flow in the electrolyte, as shown in Figure 6.7a. Diffusible redox mediators were added in solution, mediating the extracellular electron transfer process from *S. oneidensis* cells to porous carbon felt electrodes. Specifically, anthraquinone-2,7-disulfonic acid disodium was used in the anolyte at a concentration of 25 mM, resulting in a remarkable power output of about 17 mW cm^{-2} (170 W m^{-2}). While the reported flow MFC utilized a standard substrate for the study (lactate) and a potassium ferricyanide solution

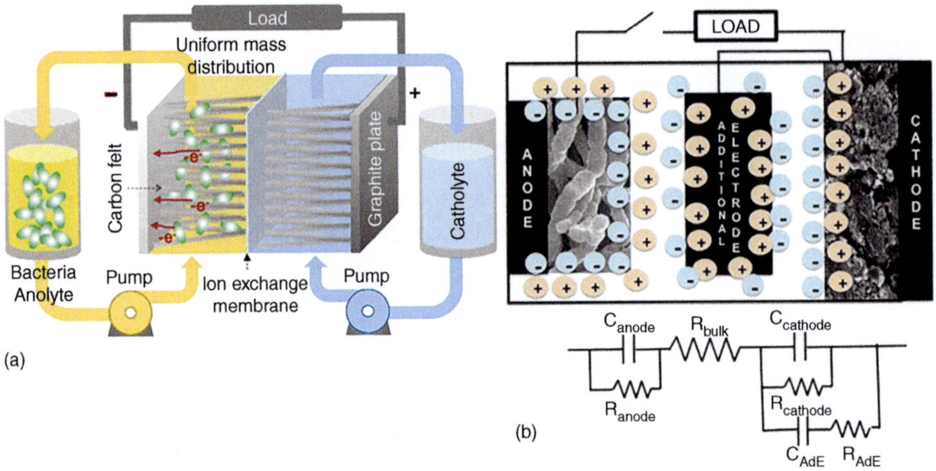

Figure 6.7 (a) Scheme of the flow-MFC where bacterial cells continuously flow in the anolyte and through a porous carbon felt electrode. Favins and quinone-derivatives are added to the anolyte to mediate the extracellular electron transfer process. For the cathodic process, aqueous solution of 0.1 M $K_3Fe(CN)_6$ and 0.2 M NaCl was utilized in the catholyte. *Source:* Adapted with permission from Zhang et al. [124]. Copyright 2024 The Authors. (b) Scheme of self-powered supercapacitive MFC with additional electrode configuration and electrical equivalent circuit for the device. *Source:* Adapted with permission from Santoro et al. [125]. Copyright 2015 Elsevier.

with sodium chloride in the catholyte, the study represents a promising approach for the future development of MFC for power generation applications.

Another interesting operating condition that has been reported to allow high power production in MFC is their use as an intermittent power production system, rather than continuous operation [126]. Santoro et al. showed how the coupling of a carbon brush anode, colonized with a mixed consortium of bacteria, with a cathode based on bilirubin oxidase, short-circuited to an additional carbon brush electrode, achieved 19 mW (84 W m^{-2}) during pulsed discharges of 10 ms [125]. Such an MFC was classified as a self-powered supercapacitor, where the metabolic activity of the bacterial cells "recharge" the device during the resting phase by transferring electrons to the anode (Figure 6.7b). In a recent work, Poli et al. investigated different configurations for the coupling of the MFC electrodes with internal or external supercapacitors, providing interesting insights into the implementation of such systems for powering practical devices [127].

Looking toward future developments, an interesting approach recently reported for the rapid preparation of bioelectrodes for MFC is the use of additive manufacturing (3D-printing) with bioinks comprising the bacterial culture, redox mediators, and the entrapping matrix. Yang et al. utilized an ink made of sodium alginate and carbon black modified with riboflavin to obtain *Shewanella* 3D-printed bioanodes, achieving power densities in the range of 250 mW m^{-2} [128].

6.4 Biosolar Cells

The possibility to directly convert sunlight into electrical energy has attracted the scientific community for more than 100 years [129]. Within the various devices developed at this purpose, the use of biological catalysts to obtain biofuel cells utilizing water as electron donor, which are commonly defined as "biosolar cells," can be dated back to the 1960s and the first report of a biosolar cells employing intact microbial cells as photoelectrocatalysts was reported by Tanaka et al. in 1985 [130]. In their pioneering study, the photosynthetic bacterium *Anabaena variabilis* M2 was utilized with the diffusible redox mediator 2-hydroxy-1,4-naphthoquinone at the anode. The device was completed by a ferricyanide-based cathode, resulting in the generation of current densities in the range of 1.25 μA cm^{-2} under illumination. From this seminal work, significant interest of the scientific community has been posed on developing biosolar cells, and several advancements have been made both employing isolated photosynthetic biological entities and intact organisms (Box 6.1) coupled to electrode surfaces, both for the development of biophotoanodes, biophotocathodes, and complete biosolar cells [131, 132].

Box 6.1 The Molecular Machinery of Photosynthesis

Photosynthesis is the biological process that allows converting solar energy into chemical energy in prokaryotes (i.e. green and purple bacteria, heliobacteria, and cyanobacteria) and eukaryotic organisms (i.e. algae and plants). The enzymatic machinery to accomplish such a task has evolved over billions of years, with the appearance of photosynthetic anoxygenic bacteria (not evolving oxygen) dating back to 3.7 billion years ago. Purple and green bacteria are both classified as anoxygenic bacteria. They evolved giving symbiotic interaction and resulting in photosynthetic apparatuses composed of two photosystems (Scheme 6.1 center), photosystem II (PSII) resembling the reaction center (RC) of purple bacteria (Scheme 6.1 left) and photosystem I (PSI) resembling the RC of green bacteria (Scheme 6.1 right) [133].

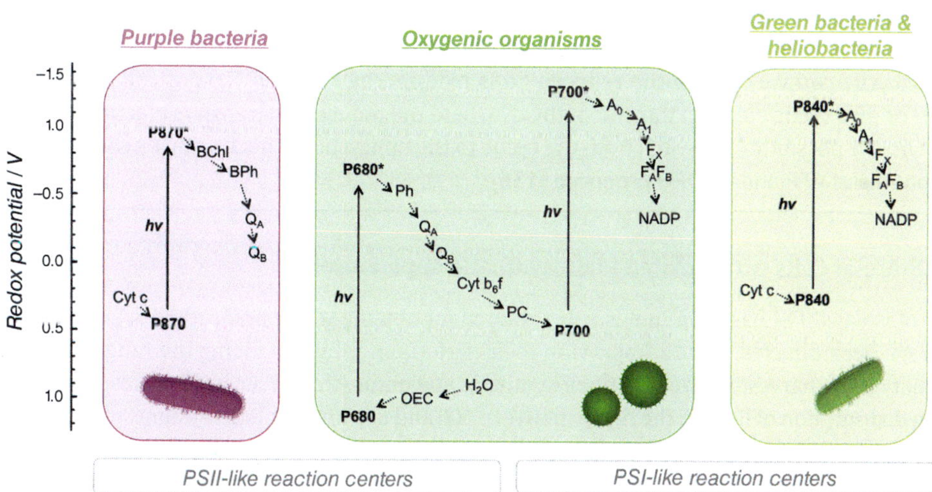

Scheme 6.1 Photosynthetic apparatuses with their redox intermediates and corresponding redox potentials for purple bacteria (left), oxygenic organisms (cyanobacteria, algae, and plants, center), and green bacteria and heliobacteria (right). Continuous arrows indicate photoexcitation events. Dashed lines indicate electron transfer events. Cyt: cytochrome; *hv*: photon energy; BChl: bacteriochlorophyll; BPh: bacteriopheophytin; Q_A: primary quinone acceptor; Q_B: secondary quinone acceptor; OEC: oxygen-evolving center; Ph: pheophytin; Cytb_6f: cytochrome b6f complex; PC: Plastocyanin; A_0 and A_1: quinones; Fx: ferredoxin; NADP: nicotinamide adenine dinucleotide phosphate; PSI: photosystem I; PSII: photosystem II.

(Continued)

Box 6.1 (Continued)

In purple bacteria, the RC, a dimer of bacteriochlorophyll (BChl) *a* molecules (P870) commonly referred to as "special pair," absorbs photons in the near-infrared region of the spectrum at a wavelength depending on the bacterial species, resulting in the promotion of the dimer in its excited state (P870*), 1.5 V above ground level. The photoexcited electron is then transferred through a series of consecutive steps, involving a monomeric BChl, a bacteriopheophytin (BPh), a primary quinone acceptor (Q_A), and a secondary quinone acceptor (Q_B). A second photoexcitation event allows obtaining a double-reduced doubly protonated Q_B (QH_2), taking up two protons from the cytoplasm. QH_2 is then released from the RC to reach the binding pocket of cytochrome bc_1. The Q_B binding site is replenished with a quinone from the lipid-soluble quinone pool (QP). Cytochrome bc_1 utilizes the electrons from QH_2 to re-reduce the electron donor of the RC, while protons are translocated across the photosynthetic membrane, generating a transmembrane proton gradient that drives the synthesis of adenosine triphosphate (ATP) [134]. In green bacteria, the RC is again composed of a dimer absorbing photon at 840 nm (P840), generating a photoexcited state 1.2 V above ground level. The electron is then shuttled through a series of molecules as previously described for purple bacteria, finally reaching ferredoxin (Fx), a protein containing a single [2Fe-2S] center. For oxygenic organisms, PSII and PSI work in pairs in a mechanism called the Z-scheme. In PSII, water is oxidized at the oxygen-evolving center (OEC) to obtain O_2 [135]. From the OEC, electrons are transferred to the chlorophyll dimer (P680) that undergoes photoxidation, covering an energy gap of about 2 V. Similarly to the previous cases, the photoexcited electron is passed through a series of electron acceptors until reducing plastocyanin that transfers the electron to the PSI. Here, at the special pair P700, a second excitation event takes place, and the electrons can then follow (i) a noncyclic pathway where the reduced Fd is oxidized by Fx-$NADP^+$ reductase to obtain NADPH or (ii) a cyclic electron transfer pathway where the reduced Fd transfers electrons back to the QP. The possibility to switch from the cyclic to the noncyclic pathways allows controlling the amounts of ATP and NADPH produced [136].

6.4.1 Biosolar Cells with Isolated Photosynthetic Apparatuses

The use of isolated PSI has been intensively explored for obtaining biophotocathodes [25]. Recent efforts have been directed toward improving their performance by enhancing the range of wavelength that can be harvested from the photosynthetic apparatus (the "green gap" corresponding to the limited absorption of light in the region between 500 and 650 nm for all the systems utilizing the green pigment chlorophyll) [137], developing advanced electrode architectures allowing improved current densities [138, 139], and obtaining metal-free carbon-based electrodes for low cost, sustainable systems [140]. It should be noted that isolated PSI has also been utilized for developing biophotoanodes. In a work by Gizzie et al., a solid-state biosolar cell exhibiting prolonged stability was reported, where PSI was immobilized on a TiO_2 anode by electropolymerizing polyaniline at the electrode surface [23]. The obtained biosolar cell, completed by a silver cathode, could operate for three repeated alternations of light-on/off cycles with a 12-h duration.

The photosynthetic apparatus of anoxygenic phototrophs, the RC based on BChls, has also been investigated for the development of biophotoelectrodes [141, 142]. Friebe et al. reported a mesoporous electrode composed of an inverse-opal mesoporous indium tin oxide electrode with RC-light harvesting 1 complexes deposited with cytochrome c operating for up to 33 days under alternated light/dark conditions [143]. While the current density generated by the biophotocathode

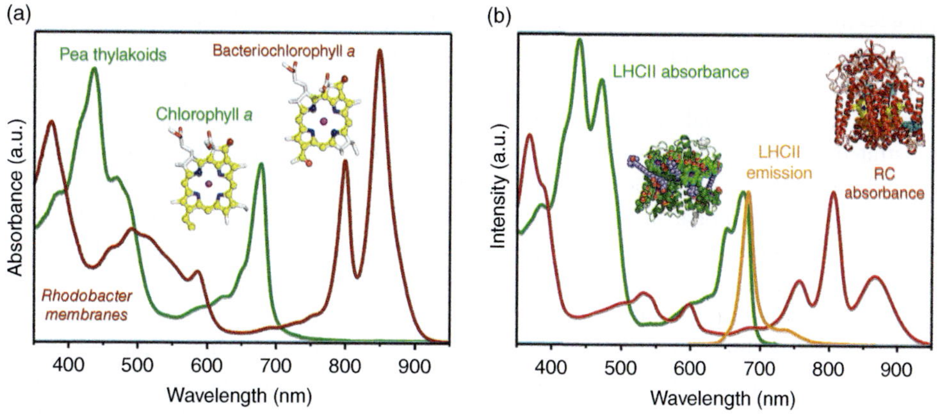

Figure 6.8 Absorption spectra of thylakoid membranes from oxygenic phototrophs and from anoxygenic phototrophs. *Source:* Adapted with permission from Liu et al. 2020 [144]. Copyright The Authors http://creativecommons.org/licenses/by/4.0/.

decreased by approximately 90% in 48 h, from 320 μA cm^{-2} to about 30 μA cm^{-2}, the addition of 20-μM cytochrome c to the electrolyte allowed for partial recovery of the photocurrent density, which then decayed to 5 μA cm^{-2} after 10 days of operation. In a recent study, recombinant plant light harvesting complexes were covalently locked with the RC from the purple bacterium *Rhodobacter sphaeroides* to obtain chimeras performing polychromatic solar energy harvesting [144]. It is interesting to note how the absorbance spectra of plant and bacterial carotenoids are complementary, as shown in Figure 6.8. Specifically, the emission of LHCII partially overlaps with the absorption of the RC at 750 nm.

Isolated PSII has been explored as a biophotocatalyst for the development of water splitting anodes. Sokol et al. reported the possibility of efficiently wiring PSII to ITO electrodes modified with an Os-based redox polymer, achieving high current densities in the range of 230 μA cm^{-2} [145]. The further addition of a diffusible redox mediator (2,6-dichloro-1,4-benzoquinone) enabled an improvement in current densities up to 375 mA cm^{-2}. However, one of the critical challenges in the use of isolated PSII for the development of biosolar cells is its short stability, which has limited the use of PSII-based photoelectrodes in the hour-time scale [146]. Nature has addressed this limitation by developing a dedicated PSII repair cycle that enables the degradation of damaged components and the reassembly of active PSII in chloroplasts or cyanobacteria [147, 148]. Accordingly, thylakoid membranes have gained interest for the development of biosolar cells, as they allow for maintaining photosynthetic protein complexes in their native environment. Pankratov et al. reported a biosolar cell based on isolated thylakoid membranes extracted from *Spinacia oleracea* and CNTs-modified electrodes with an osmium-based redox polymer, achieving ~100 μA cm^{-2} [149].

6.4.2 Biosolar Cells with Intact Organisms and Organelles

Targeting the possibility of developing self-repairing biocatalysts, using intact chloroplasts can be further beneficial to the stability of the biosolar cells compared to the use of isolated thylakoid membranes, since the chloroplasts have self-repair mechanisms against photodamage. A naphthoquinone-based redox polymer, combined with a diffusible mediator (2,6-dichlorobenzoquinone), was reported by Hasan et al. to significantly improve the photocurrent density generated from chloroplast-based photoanodes, achieving ~30 μA cm^{-2} [150].

Leveraging on the capability of photosynthetic bacteria to replicate efficiently, Darus et al. developed a biosolar cell with the biophotoanode composed of a mat-building marine photosynthetic consortium dominated by cyanobacteria and polymerized Os-complexes, achieving day/night operation for multiple days and stable current densities of $32 \pm 3\,\mu A\,cm^{-2}$ [151]. Zhang et al. also demonstrated the remarkable stability of intact cyanobacteria-based photoanodes forming biofilms on indium tin oxide electrodes compared to PSII-based counterparts [152]. Specifically, the biophotoanodes with cyanobacteria biofilms resulted in about two to three times lower current density generation (about $0.2\,\mu A\,cm^{-2}$), but greatly passed the stability of isolated PSII-based biophotoanodes with increasing photo-responses even after 5 days of operation.

With the aim of better understanding and improving the biophotocurrent generation process in intact-organism-based biosolar cells, various studies have recently combined experimental and computational approaches [153]. The combination of electrochemical experimental evidence and quantum mechanical calculations, specifically density functional theory, allowed clarifying the rate-determining step in purple-bacteria-based biophotoanodes operating with diffusible quinone-based mediators [154]. The experimentally measured current density obtained employing the various mediators was fit with the computationally derived electrochemical properties of the mediators, unveiling the one-electron, proton-decoupled electron transfer taking place in the lipophilic membrane of purple bacteria as the rate-determining step for photocurrent generation. For the case of micro-algae-based biophotoelectrodes, a modeling approach for electrochemical and fluorescence experimental data obtained with *Chlamydomonas reinhardtii*-biophotoanodes operating in the presence of the redox mediator 2,6-dichloro-1,4-benzoquinone concluded that the slow outflow of the diffusible redox mediator outside the lipophilic membranes of the micro-alga limits current densities [155]. These studies contribute to our mechanistic understanding of the extracellular electron transfer process between intact photosynthetic organisms and electrodes and provide critical insights to develop improved biohybrid systems.

In addition to diffusible redox mediators or redox polymers, recent studies have focused on modifying photosynthetic bacteria with inorganic nanoparticles to enhance photocurrent generation and power densities in biosolar cells. Liu and Choi showed the possibility to intracellularly synthesize gold nanoparticles (Au-NPs) in *Synechocystis* sp. PCC 6803 [156]. The biohybrid cyanobacteria obtained presented membrane-bound Au-NPs and their use in a biosolar cell allowed a 33-fold increase in power densities compared to a device operating with regular *Synechocystis* sp. PCC 6803, achieving about $19\,\mu W\,cm^{-2}$ ($190\,mW\,m^{-2}$). To deepen the understanding of how to control and maximize the interaction of Au-NPs with the cell membrane of photosynthetic bacteria Lasala et al. synthesized Au-NPs with different surface functionalization and prepared biohybrids of purple bacteria [157]. It was demonstrated that surface functionalization with cysteine in its zwitterionic form enabled the uniform distribution of NPs in the bacterial membrane, resulting in increased current densities and a decreased onset potential for photocurrent generation, down to −0.16 V vs. Ag|AgCl.

6.4.3 Engineered Biosolar Cell Setups for Improved Power Production

In addition to the efforts to facilitate the harvesting of photo-induced electrons from photosynthetic entities to electrode surfaces previously discussed, recent studies have been directed toward engineering devices (i.e. electrode morphology and operational conditions) for high power production. By using structured indium-tin-oxide electrodes with porosities on the micrometer length scale coupled to cyanobacteria *Nostoc punctiforme* and *Synechocystis* sp. PCC6803 current generation could be increased by two orders of magnitude compared to nonporous electrodes, achieving $12\,mA\,m^{-2}$ (vs. $0.04\,mA\,m^{-2}$ with nonporous electrodes) [158].

The use of a microfluidic device with separation of the charging (cyanobacteria reducing a diffusible redox mediator) and power delivery processes (the discharge of the reduced mediator at the anode electrode) allowed achieving a power density of 540 mW m^{-2} with mutant cyanobacterial cells [159]. Specifically, *Synechocystis* mutant cells deficient in the main photosynthetic and respiratory electron sinks (the terminal oxidases and the flavodiiron complexes Flv1/3 and Flv2/4) reduced the redox mediator $K_3[Fe(CN)_6]$ in the charging unit for a period of 3 h. Following this, the fluid containing the reduced charge carrier was injected into the power production unit, where control of the flow rate allowed for achieving the reported performance. Interestingly, such operating conditions would enable charging the device during the day and producing power only when needed.

The operation of biosolar cells as self-powered supercapacitors is another interesting approach to developing devices with improved power performance. Pankratova et al. reported the first supercapacitive biosolar cell operating with thylakoid membranes in contact with Os-based redox polymers at the biophotoanode, which was short-circuited to a charge storage component composed of a gold-disc electrode modified with Au-NPs [26]. The biosolar cell was completed by a bilirubin oxidase-based cathode further short-circuited with another charge storage component (Figure 6.9). Under pulse-operation, the device achieved a maximum power density of 56 μW cm^{-2} (560 mW m^{-2}) with 3-second discharges at 280 μA cm^{-2}. The performance of the supercapacitive

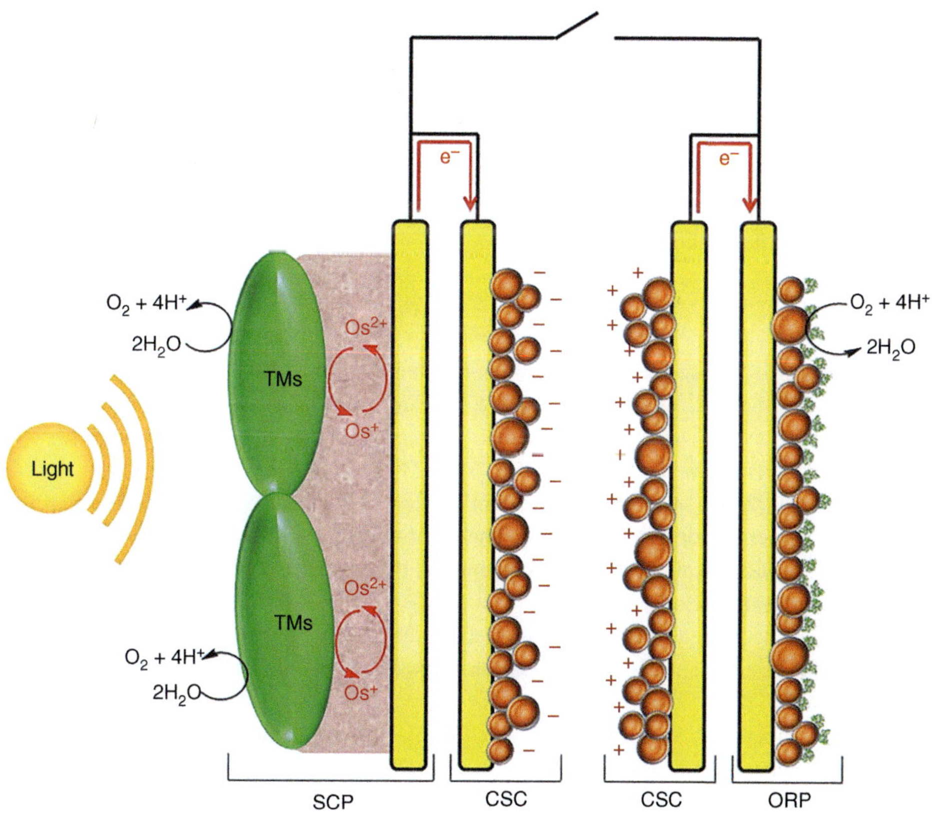

Figure 6.9 Scheme of the supercapacitive biosolar cell with thylakoid membranes-based biophotoanode (SCP: solar converting part), an Os-based redox polymer, a charge storage component (CSC), and a bilirubin oxidase-based biocathode (ORP: oxygen reducing part). *Source:* Reproduced with permission from Pankratova et al. [26]. Copyright 2017 WILEY-VCH Verlag GmbH & Co.

biosolar cell was further improved by employing functionalized carbon nanotubes in a direct electron transfer thylakoid membranes-biosolar cell, achieving $1.5\,W\,m^{-2}$ with discharges of 1 second at $420\,\mu A\,cm^{-2}$ [160].

6.5 Conclusions: Future Perspective and Emerging Approaches

The field of biofuel cells and biosolar cells, both employing isolated catalysts or intact organisms, has experienced exciting advancements over the last decade, underlined by the improvements in electrode materials, the use of redox polymers, rationally designed nanomaterials, synthetic biology, and engineered device setups that have resulted in the achievement of remarkable power densities. It is interesting to note that recently, a cyanobacteria-based biosolar cell has allowed powering of a microprocessor for over half a year, showing the possibility to utilize ambient light as the only energy input to power practical devices [161]. Looking to the next decade of research, the knowledge gained has laid the foundation for developing high-performing biofuel and biosolar cells with improved stability, paving the way for their practical implementation. To achieve this goal, it will be critical for the community of scientists and engineers working on bioelectrochemical systems to continue the collaborative environment that has allowed this field to thrive in the past years.

Acknowledgments

M. Grattieri would like to acknowledge the funding from Fondazione CON IL SUD, Grant "Brains to South 2018," project number 2018-PDR-00914. S.D. Minteer would like to thank the National Science Foundation, grant number (2406605).

References

1 Hamann, C.H., Hamnett, A., and Vielstich, W. (2007). *Electrochemistry*. Weinheim: Wiley-VCH.

2 Varcoe, J.R., Atanassov, P., Dekel, D.R. et al. (2014). Anion-exchange membranes in electrochemical energy systems. *Energy Environ. Sci.* 7: 3135–3191.

3 Li, J., Sougrati, M.T., Zitolo, A. et al. (2020). Identification of durable and non-durable FeNx sites in Fe–N–C materials for proton exchange membrane fuel cells. *Nat. Catal.* 4: 10–19.

4 Potter, M.C. (1911). Electrical effects accompanying the decomposition of organic compounds. *Proc. R. Soc. Lond.* 260–276.

5 Cohen, B. (1931). The bacterial culture as an electrical half-cell. *J. Bacteriol.* 21: 18–19.

6 Davis, J.B. and Yarbrough, H.F.J. (1962). Preliminary experiments on a microbial fuel cell. *Science* 137: 615–616.

7 Rasmussen, M., Abdellaoui, S., and Minteer, S.D. (2016). Enzymatic biofuel cells: 30 years of critical advancements. *Biosens. Bioelectron.* 76: 91–102.

8 Abdellaoui, S., Milton, R.D., Quah, T., and Minteer, S.D. (2016). NAD-dependent dehydrogenase bioelectrocatalysis: the ability of a naphthoquinone redox polymer to regenerate NAD. *Chem. Commun.* 52: 1147–1150.

9 Cristiani, P., Carvalho, M.L., Guerrini, E. et al. (2013). Cathodic and anodic biofilms in single chamber microbial fuel cells. *Bioelectrochemistry* 92: 6–13.

10 Grattieri, M., Chen, H., and Minteer, S.D. (2020). Chloroplast biosolar cell and self-powered herbicide monitoring. *Chem. Commun.* 56: 13161–13164.

11 Cooney, M.J., Svoboda, V., Lau, C. et al. (2008). Enzyme catalysed biofuel cells. *Energy Environ. Sci.* 1: 320.

12 Santoro, C., Arbizzani, C., Erable, B., and Ieropoulos, I. (2017). Microbial fuel cells: From fundamentals to applications. A review. *J. Power Sources* 356: 225–244.

13 Meunier, C.F., Yang, X.-Y., Rooke, J.C., and Su, B.-L. (2011). Biofuel cells based on the immobilization of photosynthetically active bioentities. *ChemCatChem* 3: 476–488.

14 Yehezkeli, O., Tel-Vered, R., Wasserman, J. et al. (2012). Integrated photosystem II-based photo-bioelectrochemical cells. *Nat. Commun.* 3: 742.

15 Kato, M., Cardona, T., Rutherford, A.W., and Reisner, E. (2012). Photoelectrochemical water oxidation with photosystem II integrated in a mesoporous indium–tin oxide electrode. *J. Am. Chem. Soc.* 134: 8332–8335.

16 Hasan, K., Dilgin, Y., Emek, S.C. et al. (2014). Photoelectrochemical communication between thylakoid membranes and gold electrodes through different quinone derivatives. *ChemElectroChem* 1: 131–139.

17 Rasmussen, M. and Minteer, S.D. (2014). Investigating the mechanism of thylakoid direct electron transfer for photocurrent generation. *Electrochim. Acta* 126: 68–73.

18 S. Morlock, S.K. Subramanian, A. Zouni, F. Lisdat, Scalable Three-Dimensional Photobioelectrodes Made of Reduced Graphene Oxide Combined with Photosystem I, *ACS Appl. Mater. Interfaces*, 13 (2021) 11237–11246.

19 K. Hasan, H. Bekir Yildiz, E. Sperling, P.O. Conghaile, M.A. Packer, D. Leech, C. Hagerhall, L. Gorton, Photo-electrochemical communication between cyanobacteria (Leptolyngbia sp.) and osmium redox polymer modified electrodes, *Phys. Chem. Chem. Phys.*, 16 (2014) 24676–24680.

20 Rasmussen, M. and Minteer, S.D. (2014). Photobioelectrochemistry: solar energy conversion and biofuel production with photosynthetic catalysts. *J. Electrochem. Soc.* 161: H647–H655.

21 Hasan, K., Grippo, V., Sperling, E. et al. (2017). Evaluation of photocurrent generation from different photosynthetic organisms. *ChemElectroChem* 4: 412–417.

22 Grattieri, M. (2020). Purple bacteria photo-bioelectrochemistry: enthralling challenges and opportunities. *Photochem. Photobiol. Sci.* 19: 424–435.

23 Gizzie, E.A., Niczgoda, J.S., Robinson, M.T. et al. (2015). Photosystem I-polyaniline/TiO_2 solid-state solar cells: simple devices for biohybrid solar energy conversion. *Energy Environ. Sci.* 8: 3572–3576.

24 Kirchhofer, N.D., Rasmussen, M.A., Dahlquist, F.W. et al. (2015). The photobioelectrochemical activity of thylakoid bioanodes is increased via photocurrent generation and improved contacts by membrane-intercalating conjugated oligoelectrolytes. *Energy Environ. Sci.* 8: 2698–2706.

25 Robinson, M.T., Gizzie, E.A., Mwambutsa, F. et al. (2017). Mediated approaches to Photosystem I-based biophotovoltaics. *Curr. Opin. Electrochem.* 5: 211–217.

26 Pankratova, G., Pankratov, D., Hasan, K. et al. (2017). Supercapacitive photo-bioanodes and biosolar cells: a novel approach for solar energy harnessing. *Adv. Energy Mater.* 7: 1602285.

27 Yahiro, A.T., Lee, S.M., and Kimble, D.O. (1964). Bioelectrochemistry: I. Enzyme utilizing bio-fuel cell studies. *Biochim. Biophys. Acta, Spec. Sect. Biophys. Subj.* 88: 375–383.

28 Berezin, I.V., Bogdanovskaia, V.A., Varfolomeev, S.D. et al. (1978). Bioelectrocatalysis-equilibrium oxygen potential in presence of laccase. *Dokl. Akad. Nauk SSSR* 240: 615–618.

29 Reginald, S.S., Lee, H., Fazil, N. et al. (2022). Control of carbon monoxide dehydrogenase orientation by site-specific immobilization enables direct electrical contact between enzyme cofactor and solid surface. *Commun. Biol.* 5.

30 Chen, T., Barton, S.C., Binyamin, G. et al. (2001). A miniature biofuel cell. *J. Am. Chem. Soc.* 123: 8630–8631.

31 Heller, A. (2004). Miniature biofuel cells. *Phys. Chem. Chem. Phys.* 6: 209–216.

32 Wang, L., Wu, X., Su, B.S.Q.W. et al. (2021). Enzymatic biofuel cell: opportunities and intrinsic challenges in futuristic applications. *Adv. Energy Sustainability Res.* 2.

33 Yates, N.D.J., Fascione, M.A., and Parkin, A. (2018). Methodologies for "wiring" redox proteins/enzymes to electrode surfaces. *Chem. Eur. J.* 24: 12164–12182.

34 Hickey, D.P., Lim, K., Cai, R. et al. (2018). Pyrene hydrogel for promoting direct bioelectrochemistry: ATP-independent electroenzymatic reduction of N_2. *Chem. Sci.* 9: 5172–5177.

35 Chen, H., Simoska, O., Lim, K. et al. (2020). Fundamentals, applications, and future directions of bioelectrocatalysis. *Chem. Rev.* 120: 12903–12993.

36 Cass, A.E., Davis, G., Francis, G.D. et al. (1984). Ferrocene-mediated enzyme electrode for amperometric determination of glucose. *Anal. Chem.* 56: 667–671.

37 Schuhmann, W., Ohara, T.J., Schmidt, H.L., and Heller, A. (1991). Electron transfer between glucose oxidase and electrodes via redox mediators bound with flexible chains to the enzyme surface. *J. Am. Chem. Soc.* 113: 1394–1397.

38 Dicks, J.M., Hattori, S., Karube, I. et al. (1989). Ferrocene modified polypyrrole with immobilised glucose oxidase and its application in amperometric glucose microbiosensors. *Ann. Biol. Clin. (Paris)* 47: 607–619.

39 Hodak, J., Etchenique, R., Calvo, E.J. et al. (1997). Layer-by-layer self-assembly of glucose oxidase with a poly(allylamine)ferrocene redox mediator. *Langmuir* 13: 2708–2716.

40 Mao, F., Mano, N., and Heller, A. (2003). Long tethers binding redox centers to polymer backbones enhance electron transport in enzyme "wiring" hydrogels. *J. Am. Chem. Soc.* 125: 4951–4957.

41 Mano, N., Mao, F., and Heller, A. (2003). Characteristics of a miniature compartment-less glucose-O_2 biofuel cell and its operation in a living plant. *J. Am. Chem. Soc.* 125: 6588–6594.

42 Scodeller, P., Carballo, R., Szamocki, R. et al. (2010). Layer-by-layer self-assembled osmium polymer-mediated laccase oxygen cathodes for biofuel cells: the role of hydrogen peroxide. *J. Am. Chem. Soc.* 132: 11132–11140.

43 Yuan, M. and Minteer, S.D. (2019). Redox polymers in electrochemical systems: From methods of mediation to energy storage. *Curr. Opin. Electrochem.* 15: 1–6.

44 Ruff, A. (2017). Redox polymers in bioelectrochemistry: common playgrounds and novel concepts. *Curr. Opin. Electrochem.* 5: 66–73.

45 Milton, R.D., Hickey, D.P., Abdellaoui, S. et al. (2015). Rational design of quinones for high power density biofuel cells. *Chem. Sci.* 6: 4867–4875.

46 Macazo, F.C. and Minteer, S.D. (2017). Enzyme cascades in biofuel cells. *Curr. Opin. Electrochem.* 5: 114–120.

47 Arechederra, R.L. and Minteer, S.D. (2009). complete oxidation of glycerol in an enzymatic biofuel cell. *Fuel Cells* 9: 63–69.

48 Tasca, F., Gorton, L., Kujawa, M. et al. (2010). Increasing the coulombic efficiency of glucose biofuel cell anodes by combination of redox enzymes. *Biosens. Bioelectron.* 25: 1710–1716.

49 Xu, S. and Minteer, S.D. (2011). Enzymatic biofuel cell for oxidation of glucose to CO_2. *ACS Catal.* 2: 91–94.

50 Zhu, Z., Sun, F., Zhang, X., and Zhang, Y.H.P. (2012). Deep oxidation of glucose in enzymatic fuel cells through a synthetic enzymatic pathway containing a cascade of two thermostable dehydrogenases. *Biosens. Bioelectron.* 36: 110–115.

51 Lau, C., Moehlenbrock, M.J., Arechederra, R.L. et al. (2015). Paper based biofuel cells: Incorporating enzymatic cascades for ethanol and methanol oxidation. *Int. J. Hydrog. Energy* 40: 14661–14666.

52 Zhu, Z., Ma, C., and Percival Zhang, Y.H. (2018). Co-utilization of mixed sugars in an enzymatic fuel cell based on an in vitro enzymatic pathway. *Electrochim. Acta* 263: 184–191.

53 Xiao, X., Xia, H.-Q., Wu, R. et al. (2019). Tackling the challenges of enzymatic (bio)fuel cells. *Chem. Rev.* 119: 9509–9558.

54 Franco, J.H., Klunder, K.J., Lee, J. et al. (2020). Enhanced electrochemical oxidation of ethanol using a hybrid catalyst cascade architecture containing pyrene-TEMPO, oxalate decarboxylase and carboxylated multi-walled carbon nanotube. *Biosens. Bioelectron.* 154: 112077.

55 Franco, J.H., Grattieri, M., de Andrade, A.R., and Minteer, S.D. (2021). Unveiling complete lactate oxidation through a hybrid catalytic cascade. *Electrochim. Acta* 376: 138044.

56 Miles, E.W., Rhee, S., and Davies, D.R. (1999). The molecular basis of substrate channeling. *J. Biol. Chem.* 274: 12193–12196.

57 Wheeldon, I., Minteer, S.D., Banta, S. et al. (2016). Substrate channelling as an approach to cascade reactions. *Nat. Chem.* 8: 299–309.

58 Xia, L., Van Nguyen, K., Holade, Y. et al. (2017). Improving the performance of methanol biofuel cells utilizing an enzyme cascade bioanode with DNA-bridged substrate channeling. *ACS Energy Lett.* 2: 1435–1438.

59 Zhao, C.-E., Gai, P., Song, R. et al. (2017). Nanostructured material-based biofuel cells: recent advances and future prospects. *Chem. Soc. Rev.* 46: 1545–1564.

60 Ciaccafava, A., De Poulpiquet, A., Techer, V. et al. (2012). An innovative powerful and mediatorless H_2/O_2 biofuel cell based on an outstanding bioanode. *Electrochem. Commun.* 23: 25–28.

61 Tasca, F., Harreither, W., Ludwig, R. et al. (2011). Cellobiose dehydrogenase aryl diazonium modified single walled carbon nanotubes: enhanced direct electron transfer through a positively charged surface. *Anal. Chem.* 83: 3042–3049.

62 Tsujimura, S., Murata, K., and Akatsuka, W. (2014). Exceptionally high glucose current on a hierarchically structured porous carbon electrode with "wired" flavin adenine dinucleotide-dependent glucose dehydrogenase. *J. Am. Chem. Soc.* 136: 14432–14437.

63 Magrez, A., Kasas, S., Salicio, V.R. et al. (2006). Cellular toxicity of carbon-based nanomaterials. *Nano Lett.* 6: 1121–1125.

64 Yuan, X., Zhang, X., Sun, L. et al. (2019). Cellular toxicity and immunological effects of carbon-based nanomaterials. *Part. Fibre Toxicol.* 16.

65 Peng, Z., Liu, X., Zhang, W. et al. (2020). Advances in the application, toxicity and degradation of carbon nanomaterials in environment: a review. *Environ. Int.* 134: 105298.

66 Carnovale, C., Bryant, G., Shukla, R., and Bansal, V. (2019). Identifying trends in gold nanoparticle toxicity and uptake: size, shape, capping ligand, and biological corona. *ACS Omega* 4: 242–256.

67 Sani, A., Cao, C., and Cui, D. (2021). Toxicity of gold nanoparticles (AuNPs): a review. *Biochem. Biophys. Rep.* 26: 100991.

68 Wang, X., Falk, M., Ortiz, R. et al. (2012). Mediatorless sugar/oxygen enzymatic fuel cells based on gold nanoparticle-modified electrodes. *Biosens. Bioelectron.* 31: 219–225.

69 Wang, Z., Li, Y., Shi, Z. et al. (2024). Implementation of π-π interaction in AuNPs@GDY to boost the bioelectrocatalysis in enzymatic biofuel cells. *Bioelectrochemistry* 158: 108712.

70 J.H. Canfield, B.H. Goldner, R. Lutwack, Research on applied bioelectrochemistry. First quarterly progress report, 14 Mar. - 30 Jun. 1963, in, MAGNA Corporation, Anaheim, California, 1963.

71 Delaney, G.M., Bennetto, P., Mason, J.R. et al. (1984). Electron-transfer coupling in microbial fuel cells. 2. performance of fuel cells containing selected microorganism-mediator-substrate combinations. *J. Chem. Technol. Biotechnol.* 34B: 13–27.

72 Bennetto, H.P., Stirling, J.L., Tanaka, K., and Vega, C.A. (1983). Anodic reactions in microbial fuel cells. *Biotechnol. Bioeng.* 25: 559–568.

73 Allen, M.R. and Bennetto, P. (1993). Microbial fuel-cells: electricity production from carbohydrates. *Appl. Biochem. Biotechnol.* 39 (40): 27–40.

74 Kim, B.H., Kim, H.J., Hyun, M.S., and Park, D.H. (1999). Direct electrode reaction of Fe(III)-reducing bacterium, Shewanella putrefaciens. *J. Microbiol. Biotechnol.* 9: 127–131.

75 Kim, B.H., Ikeda, T., Park, H.S. et al. (1999). Electrochemical activity of an Fe(III)-reducing bacterium, Shewanella putrefaciens IR-1, in the presence of alternative electron acceptors. *Biotechnol. Tech.* 13: 475–478.

76 Reimers, C.E., Tender, L.M., Fertig, S.J., and Wang, W. (2001). Harvesting energy from the marine sediment–water interface. *Environ. Sci. Technol.* 35: 192–195.

77 Tender, L.M., Reimers, C.E., Stecher, H.A. 3rd, et al. (2002). Harnessing microbially generated power on the seafloor. *Nat. Biotechnol.* 20: 821–825.

78 Kim, H.J., Park, H.S., Hyun, M.S. et al. (2002). A mediator-less microbial fuel cell using a metal reducing bacterium, Shewanella putrefaciens. *Enzym. Microb. Technol.* 30: 145–152.

79 Bond, D.R. and Lovley, D.R. (2003). Electricity production by Geobacter sulfurreducens attached to electrodes. *Appl. Environ. Microbiol.* 69: 1548–1555.

80 Gil, G.-C., Chang, I.-S., Kim, B.H. et al. (2003). Operational parameters affecting the performannce of a mediator-less microbial fuel cell. *Biosens. Bioelectron.* 18: 327–334.

81 Logan, B.E., Hamelers, B., Rozendal, R. et al. (2006). Microbial fuel cells: methodology & technology. *Environ. Sci. Technol.* 40: 5181–5192.

82 Schaetzle, O., Barrière, F., and Schroder, U. (2009). An improved microbial fuel cell with laccase as the oxygen reduction catalyst. *Energy Environ. Sci.* 2: 96–99.

83 Liu, H. and Logan, B.E. (2004). Electricity generation using an air-cathode single chamber microbial fuel cell in the presence and absence of a proton exchange membrane. *Environ. Sci. Technol.* 38: 4040–4046.

84 Liu, H., Cheng, S., and Logan, B.E. (2005). Production of electricity from acetate or butyrate using a single-chamber microbial fuel cell. *Environ. Sci. Technol.* 39: 658–662.

85 Cheng, S., Liu, H., and Logan, B.E. (2006). Increased performance of single-chamber microbial fuel cells using an improved cathode structure. *Electrochem. Commun.* 8: 489–494.

86 Santoro, C., Artyushkova, K., Babanova, S. et al. (2014). Parameters characterization and optimization of activated carbon (AC) cathodes for microbial fuel cell application. *Bioresour. Technol.* 163: 54–63.

87 Higgins, S.R., Lau, C., Atanassov, P. et al. (2011). Hybrid biofuel cell: microbial fuel cell with an enzymatic air-breathing cathode. *ACS Catal.* 1: 994–997.

88 Pant, D., Van Bogaert, G., Diels, L., and Vanbroekhoven, K. (2010). A review of the substrates used in microbial fuel cells (MFCs) for sustainable energy production. *Bioresour. Technol.* 101: 1533–1543.

89 Chae, K.J., Choi, M.J., Lee, J.W. et al. (2009). Effect of different substrates on the performance, bacterial diversity, and bacterial viability in microbial fuel cell. *Bioresour. Technol.* 100: 3518–3525.

90 Pandey, P., Shinde, V.N., Deopurkar, R.L. et al. (2016). Recent advances in the use of different substrates in microbial fuel cells toward wastewater treatment and simultaneous energy recovery. *Appl. Energy* 168: 706–723.

91 Catal, T., Li, K., Bermek, H., and Liu, H. (2008). Electricity production from twelve monosaccharides using microbial fuel cells. *J. Power Sources* 175: 196–200.

92 Velvizhi, G. and Venkata Mohan, S. (2012). Electrogenic activity and electron losses under increasing organic load of recalcitrant pharmaceutical wastewater. *Int. J. Hydrog. Energy* 37: 5969–5978.

93 Chandrasekhar, K. and Venkata Mohan, S. (2012). Bio-electrochemical remediation of real field petroleum sludge as an electron donor with simultaneous power generation facilitates biotransformation of PAH: effect of substrate concentration. *Bioresour. Technol.* 110: 517–525.

94 Grattieri, M. and Minteer, S.D. (2018). Microbial fuel cells in saline and hypersaline environments: advancements, challenges and future perspectives. *Bioelectrochemistry* 120: 127–137.

95 Miller, L.G. and Oremland, R.S. (2008). Electricity generation by anaerobic bacteria and anoxic sediments from hypersaline soda lakes. *Extremophiles* 12: 837–848.

96 Paul, V.G., Minteer, S.D., Treu, B.L., and Mormile, M.R. (2014). Ability of a haloalkaliphilic bacterium isolated from Soap Lake, Washington to generate electricity at pH 11.0 and 7% salinity. *Environ. Technol.* 35: 1003–1011.

97 Rousseau, R., Dominguez-Benetton, X., Délia, M.-L., and Bergel, A. (2013). Microbial bioanodes with high salinity tolerance for microbial fuel cells and microbial electrolysis cells. *Electrochem. Commun.* 33: 1–4.

98 Rousseau, R., Santaella, C., Bonnafous, A. et al. (2016). Halotolerant bioanodes: the applied potential modulates the electrochemical characteristics, the biofilm structure and the ratio of the two dominant genera. *Bioelectrochemistry* 112: 24–32.

99 Roustazadeh Sheikhyousefi, P., Nasr Esfahany, M., Colombo, A. et al. (2017). Investigation of different configurations of microbial fuel cells for the treatment of oilfield produced water. *Appl. Energy* 192: 457–465.

100 Monzon, O., Yang, Y., Yu, C. et al. (2015). Microbial fuel cells under extreme salinity: performance and microbial analysis. *Environ. Chem.* 12: 293–299.

101 Monzon, O., Yang, Y., Li, Q., and Alvarez, P.J.J. (2016). Quorum sensing autoinducers enhance biofilm formation and power production in a hypersaline microbial fuel cell. *Biochem. Eng. J.* 109: 222–227.

102 O. Monzon, Y. Yang, J. Kim, A. Heldenbrand, Q. Li, P.J.J. Alvarez, Microbial fuel cell fed by Barnett Shale produced water: Power production by hypersaline autochthonous bacteria and coupling to a desalination unit, *Biochem. Eng. J.*, 117 (2017) 87–91.

103 M. Grattieri, M. Suvira, K. Hasan, S.D. Minteer, Halotolerant Extremophile Bacteria from the Great Salt Lake for Recycling Pollutants in Microbial Fuel Cells, *J. Power Sources*, 356 (2017) 310–318.

104 Abrevaya, X.C., Sacco, N., Mauas, P.J., and Corton, E. (2011). Archaea-based microbial fuel cell operating at high ionic strength conditions. *Extremophiles* 15: 633–642.

105 Koch, C. and Harnisch, F. (2016). Is there a specific ecological niche for electroactive microorganisms? *ChemElectroChem* 3: 1282–1295.

106 Digel, L., Bonné, R., and Aiyer, K. (2024). Are all microbes electroactive? *Cell Rep. Phys. Sci.* 5: 102200.

107 Gunawardena, A., Fernando, S., and To, F. (2008). Performance of a Yeast-mediated biological fuel cell. *Int. J. Mol. Sci.* 9: 1893–1907.

108 Raghavulu, S.V., Goud, R.K., Sarma, P.N., and Mohan, S.V. (2011). Saccharomyces cerevisiae as anodic biocatalyst for power generation in biofuel cell: Influence of redox condition and substrate load. *Bioresour. Technol.* 102: 2751–2757.

109 Benjamin Ocheja, O., Wahid, E., Honorio Franco, J. et al. (2024). Polydopamine-immobilized yeast cells for portable electrochemical biosensors applied in environmental copper sensing. *Bioelectrochemistry* 157: 108658.

110 Shlosberg, Y., Krupnik, N., Tóth, T.N. et al. (2022). Bioelectricity generation from live marine photosynthetic macroalgae. *Biosens. Bioelectron.* 198: 113824.

111 Hubenova, Y. and Mitov, M. (2012). Conversion of solar energy into electricity by using duckweed in direct photosynthetic plant fuel cell. *Bioelectrochemistry* 87: 185–191.

112 Hubenova, Y. and Mitov, M. (2015). Enhanced metabolic and redox activity of vascular aquatic plant Lemna valdiviana under polarization in direct photosynthetic plant fuel cell. *Bioelectrochemistry* 106: 226–231.

113 Shlosberg, Y., Meirovich, M.M., Yehezkeli, O. et al. (2022). Production of photocurrent and hydrogen gas from intact plant leaves. *Biosens. Bioelectron.* 215: 114558.

114 Choque, S., Zuñiga, C., González, A. et al. (2023). Development of biophoto anodes using Ulvophyceae macroalgae. *J Mater Chem A* 11: 2661–2669.

115 Hubenova, E. and Hubenova, Y. (2025). Bacteria and eukaryotes as anodic biocatalysts in microbial fuel cells – a comprehensive review. *J. Power Sources* 629: 236002.

116 Shlosberg, Y., Tchaicheeyan, O., and Lesman, A. (2024). Bio-electricity production from fibroblasts and their cultivation medium. *Electrochim. Acta* 506: 145020.

117 Guo, X., Yang, C., Wu, J. et al. (2024). Ultra-small carbon dots boost bioelectricity generation by accelerating extracellular electron transfer. *J. Power Sources* 610: 234711.

118 Mouhib, M., Antonucci, A., Reggente, M. et al. (2019). Enhancing bioelectricity generation in microbial fuel cells and biophotovoltaics using nanomaterials. *Nano Res* 12: 2184–2199.

119 Wang, Y., Cheng, X., Liu, K. et al. (2022). 3D hierarchical Co8FeS8-FeCo2O4/N-CNTs@CF with an enhanced microorganisms–anode interface for improving microbial fuel cell performance. *ACS Appl. Mater. Interfaces* 14: 35809–35821.

120 Cao, B., Zhao, Z., Peng, L. et al. (2021). Silver nanoparticles boost charge-extraction efficiency in Shewanella microbial fuel cells. *Science* 373: 1336–1340.

121 Thapa, K., Liu, W., Zhang, Y. et al. (2024). Boosting the power performance of microbial fuel cells by using dual nanomaterial-modified carbon felt electrodes. *Energy Fuel* 38: 21412–21422.

122 Dong, F., Simoska, O., Gaffney, E., and Minteer, S.D. (2022). Applying synthetic biology strategies to bioelectrochemical systems. *Electrochem. Sci. Adv.* 2: e2100197.

123 Mouhib, M., Reggente, M., Li, L. et al. (2023). Extracellular electron transfer pathways to enhance the electroactivity of modified Escherichia coli. *Joule* 7: 2092–2106.

124 Zhang, L., Zhang, Y., Liu, Y. et al. (2024). High power density redox-mediated Shewanella microbial flow fuel cells. *Nat. Commun.* 15.

125 Santoro, C., Soavi, F., Serov, A. et al. (2016). Self-powered supercapacitive microbial fuel cell: the ultimate way of boosting and harvesting power. *Biosens. Bioelectron.* 78: 229–235.

126 Dewan, A., Beyenal, H., and Lewandowski, Z. (2009). Intermittent energy harvesting improves the performance of microbial fuel cells. *Environ. Sci. Technol.* 43: 4600–4605.

127 Poli, F., Santoro, C., and Soavi, F. (2023). Improving microbial fuel cells power output using internal and external optimized, tailored and totally green supercapacitor. *J. Power Sources* 564: 232780.

128 Yang, J., Xu, P., Li, H. et al. (2024). Enhancing extracellular electron transfer of a 3D-printed Shewanella bioanode with riboflavin-modified carbon black bioink. *ACS Appl. Bio Mater.* 7: 2734–2740.

129 Ciamician, G. (1912). The photochemistry of the future. *Science* 36: 385–394.

130 Tanaka, K., Tamamushi, R., and Ogawa, T. (1985). Bioelectrochemical fuel-cells operated by the cyanobacterium. *J. Chem. Technol. Biotechnol.* 35: 191–197.

131 Grattieri, M., Beaver, K., Gaffney, E.M. et al. (2020). Advancing the fundamental understanding and practical applications of photo-bioelectrocatalysis. *Chem. Commun.* 56: 8553–8568.

132 Morlock, S., Subramanian, S.K., Zouni, A., and Lisdat, F. (2022). Bio-inorganic hybrid structures for direct electron transfer to photosystem I in photobioelectrodes. *Biosens. Bioelectron.* 214: 114495.

133 Blankenship, R.E. (2010). Early evolution of photosynthesis. *Plant Physiol.* 154: 434–438.

134 Allen, J.P. and Williams, J.C. (1998). Photosynthetic reaction centers. *FEBS Lett.* 438: 5–9.

135 McEvoy, J.P. and Brudvig, G.W. (2006). Water-splitting chemistry of photosystem II. *Chem. Rev.* 106: 4455–4483.

136 Laughlin, T.G., Bayne, A.N., Trempe, J.-F. et al. (2019). Structure of the complex I-like molecule NDH ofoxygenic photosynthesis. *Nature* 566: 411–414.

137 Morlock, S., Subramanian, S.K., Zouni, A., and Lisdat, F. (2023). Closing the green gap of photosystem I with synthetic fluorophores for enhanced photocurrent generation in photobiocathodes. *Chem. Sci.* 14: 1696–1708.

138 Wolfe, K.D., Dervishogullari, D., Stachurski, C.D. et al. (2019). Photosystem I multilayers within porous indium tin oxide cathodes enhance mediated electron transfer. *ChemElectroChem* 7: 596–603.

139 Ciornii, D., Riedel, M., Stieger, K.R. et al. (2017). Bioelectronic circuit on a 3D electrode architecture: enzymatic catalysis interconnected with photosystem I. *J. Am. Chem. Soc.* 139: 16478–16481.

140 Stachurski, C.D., Williams, J.M., Tabaquin, P.M. et al. (2023). Utilizing Toray paper as a metal-free, high surface area electrode for photosystem i–driven mediated electron transfer. *Energ. Technol.* 11.

141 Buscemi, G., Vona, D., Ragni, R. et al. (2021). Polydopamine/ethylenediamine nanoparticles embedding a photosynthetic bacterial reaction center for efficient photocurrent generation. *Adv. Sustainable Syst.* 2000303.

142 Friebe, V.M. and Frese, R.N. (2017). Photosynthetic reaction center-based biophotovoltaics. *Curr. Opin. Electrochem.* 5: 126–134.

143 Friebe, V.M., Barszcz, A.J., Jones, M.R., and Frese, R.N. (2022). Sustaining electron transfer pathways extends biohybrid photoelectrode stability to years. *Angew. Chem. Int. Ed.* 61.

144 Liu, J., Friebe, V.M., Frese, R.N., and Jones, M.R. (2020). Polychromatic solar energy conversion in pigment-protein chimeras that unite the two kingdoms of (bacterio)chlorophyll-based photosynthesis. *Nat. Commun.* 11: 1542.

145 Sokol, K.P., Mersch, D., Hartmann, V. et al. (2016). Rational wiring of photosystem II to hierarchical indium tin oxide electrodes using redox polymers. *Energy Environ. Sci.* 9: 3698–3709.

146 Roach, T. and Krieger-Liszkay, A. (2014). Regulation of photosynthetic electron transport and photoinhibition. *Curr. Protein Pept. Sci.* 15: 351–362.

147 Nath, K., Jajoo, A., Poudyal, R.S. et al. (2013). Towards a critical understanding of the photosystem II repair mechanism and its regulation during stress conditions. *FEBS Lett.* 587: 3372–3381.

148 Theis, J. and Schroda, M. (2016). Revisiting the photosystem II repair cycle. *Plant Signal. Behav.* 11: e1218587.

149 Pankratov, D., Zhao, J., Nur, M.A. et al. (2019). The influence of surface composition of carbon nanotubes on the photobioelectrochemical activity of thylakoid bioanodes mediated by osmium-complex modified redox polymer. *Electrochim. Acta* 310: 20–25.

150 Hasan, K., Milton, R.D., Grattieri, M. et al. (2017). Photobioelectrocatalysis of intact chloroplasts for solar energy conversion. *ACS Catal.* 7: 2257–2265.

151 Darus, L., Sadakane, T., Ledezma, P. et al. (2016). Redox-polymers enable uninterrupted day/night photo-driven electricity generation in biophotovoltaic devices. *J. Electrochem. Soc.* 164: H3037–H3040.

152 Zhang, J.Z., Bombelli, P., Sokol, K.P. et al. (2018). Photoelectrochemistry of photosystem II in vitro vs in vivo. *J. Am. Chem. Soc.* 140: 6–9.

153 Gaffney, E.M., Grattieri, M., Rhodes, Z., and Minteer, S.D. (2020). Exploration of computational approaches for understanding microbial electrochemical systems: opportunities and future directions. *J. Electrochem. Soc.* 167: 065502.

154 Grattieri, M., Rhodes, Z., Hickey, D.P. et al. (2019). Understanding biophotocurrent generation in photosynthetic Purple bacteria. *ACS Catal.* 9: 867–873.

155 Beauzamy, L., Lemaître, F., and Derr, J. (2020). Underlying mechanisms in microbial solar cells: how modeling can help. *Sustainable Energy Fuels* 4: 6004–6010.

156 Liu, L. and Choi, S. (2021). Enhanced biophotoelectricity generation in cyanobacterial biophotovoltaics with intracellularly biosynthesized gold nanoparticles. *J. Power Sources* 506: 230251.

157 Lasala, P., Matteucci, R.M., Volpicella, S.R. et al. (2024). Deciphering the role of inorganic nanoparticles' surface functionalization on biohybrid microbial photoelectrodes. *ACS Appl. Mater. Interfaces* 16: 58598–58608.

158 Wenzel, T., Hartter, D., Bombelli, P. et al. (2018). Porous translucent electrodes enhance current generation from photosynthetic biofilms. *Nat. Commun.* 9: 1299.

159 Saar, K.L., Bombelli, P., Lea-Smith, D.J. et al. (2018). Enhancing power density of biophotovoltaics by decoupling storage and power delivery. *Nat. Energy* 3: 75–81.

160 Pankratov, D., Pankratova, G., Dyachkova, T.P. et al. (2017). Supercapacitive biosolar cell driven by direct electron transfer between photosynthetic membranes and CNT networks with enhanced performance. *ACS Energy Lett.* 2: 2635–2639.

161 Bombelli, P., Savanth, A., Scarampi, A. et al. (2022). Powering a microprocessor by photosynthesis. *Energy Environ. Sci.* 15: 2529–2536.

7

Microbial Electrochemical Technologies for Used Water Treatment

Ruggero Rossi

Department of Environmental Health and Engineering, Johns Hopkins University, Baltimore, MD, USA

7.1 Introduction

Wastewater treatment represents the most obvious utilization of bioelectrochemical systems, coupling the degradation of organic matter and contaminants in solution with producing useful electricity or valuable chemicals. The purpose of conventional wastewater treatment technologies is to remove pollutants from used water streams to prevent the contamination of receiving water bodies while protecting the environment and the health of its inhabitants [1]. The lack of proper wastewater handling and sanitation technologies directly contributes to the spread of deadly diseases, such as cholera, typhoid fever, and rotavirus, which are responsible for the death of more than 800 children under five years old every day across the globe. The two main reasons behind this improper handling of wastewater are the high cost associated with wastewater treatment, which could be untenable in some regions of the world, and the unavailability of a well-developed wastewater and energy infrastructure. In fact, the areas of the world where a large portion of the population has limited access to electricity are also typically characterized by limited access to safe water.

Currently, wastewater treatment and conveyance consume about 4% of the total U.S. energy demand, with an energy intensity ranging from 0.5 up to 2 kWh m^{-3}, with biological treatment comprising up to half of this figure [2, 3]. Several studies indicated that the energy content of typical domestic and industrial wastewater is larger than the energy needed to treat them; thus, recovering a small portion of this energy could contribute to making wastewater treatment energy-neutral. For example, it has been calculated that domestic, industrial, and animal wastewater together contains around 17 GW of potential power, and 15 GW is currently consumed for their treatment [4]. In a smaller-scale scenario, it was estimated in 2004 that in a conventional wastewater treatment plant in Toronto, there was approximately 9.3 times more energy in the wastewater compared to that consumed to treat it [5].

Reducing the energy demand for wastewater treatment with microbial electrochemical technologies could facilitate proper water management practices in areas where low energy availability prohibits the implementation of conventional wastewater treatment technologies. In addition, the implementation of microbial electrochemical technologies for water treatment can accelerate the development of a circular economy valorizing wastes for the production of energy and value-added products. Anaerobic digestion can contribute to this process by generating biogas from digested

Bioelectrocatalysis: From Electron Transfer Processes to Emerging Technological Applications, First Edition.
Edited by Shelley D. Minteer and Matteo Grattieri.

sludge, which can be transformed into electricity or further refined and injected in the natural gas pipeline. However, the high temperature that anaerobic digestion typically requires coupled with a relatively low rate of methane generation can make this process untenable in cold climates or for small reclamation facilities where sludge production is limited. According to the American Biogas Council, of the 14,000 wastewater treatment plants in the United States, only 1100 employ anaerobic digesters as of 2023.

Microbial electrochemical technologies rely on the activity of bacteria to convert the biodegradable organic matter contained in domestic and industrial wastewater into useful electricity or chemicals. Microbial fuel cells (MFCs) use exoelectrogenic bacteria on the electrode to oxidize the organic matter in wastewater. Early studies on MFCs leveraging microorganisms to produce electricity were conducted by Potter back in 1911, but it was not until 2004 that the first example of an MFC used to recover electricity from wastewater, while reducing the organic matter concentration to levels suitable for discharge was published in the scientific literature [6, 7]. Exoelectrogenic bacteria converted the organic matter contained in the wastewater to electrons, which were transferred to an external electron acceptor, the anode. The electrons produced at the anode were consumed at the cathode, where the oxygen reduction reaction (ORR) was carried out, reducing the oxygen contained in the air.

A maximum voltage of 1.1 V can be theoretically obtained in an MFC, assuming acetate is the primary metabolite in wastewater and oxygen is the final electrode acceptor at neutral pH, according to the following equations [8]:

Anode:

$$CH_3COO^- + 4H_2O \rightleftharpoons 2HCO_3^- + 9H^+ + 8e^- \; E = -0.296\,V \; \left(CH_3COO^-, 2HCO_3^- = 5\,mM\right)$$

Cathode:

$$O_2 + 4H^+ + 4e^- \rightleftharpoons 2H_2O \; E = +0.805\,V\,(pO_2 = 0.2\ \ atm)$$

Typically, a voltage much lower than 1.1 V is obtained in MFCs treating domestic and industrial wastewaters due to the losses associated with the microbial metabolism, mass-transfer limitations, and activation losses for the cathodic reaction. Nevertheless, the positive voltage allows to actively generate electrical current, fulfilling the aim of minimizing energy consumption during wastewater treatment while recovering energy from waste streams.

Compared to conventional chemical fuel cells fed hydrogen and oxygen, power production from MFCs is by far less predictable. Power production in MFCs is a function of the microbial community colonizing the anode and its ability to convert organic matter into electricity, and both these parameters are subject to a dynamic equilibrium as a function of the environmental conditions (temperature, wastewater strength, substrate content) in which the MFC operates. Oxidizers at the cathode other than oxygen have been investigated in the lab and can inflate maximum voltage and power production that can be produced in MFCs. Ferricyanide and manganese species were the most common electrode acceptors in the cathode when oxygen was not used. The use of these alternative oxidizers can be advantageous in that they can be coupled with obligate anaerobes on the anode, avoiding the presence of oxygen in the reactor, but their use outside the lab for wastewater treatment is not practical due to the high cost and the challenges associated with their regeneration and handling [9, 10].

Box 7.1 Chemical (COD) and biochemical (BOD) oxygen demand

The chemical oxygen demand (COD) of wastewater indicates the amount of oxygen (O_2) required to oxidize the organic matter to carbon dioxide (CO_2) and water (H_2O) and represents an easily accessible source of useful energy. In conventional wastewater treatment, the biodegradable portion (BOD) of the COD is transformed by bacteria into inorganic carbon and sludge, which is a semi-solid material with high COD and primarily contains biomass. The produced sludge can be digested in an anaerobic digester to produce natural gas, which is a mixture of methane, CO_2, and trace amount of other gases, which can be reused on-site for energy generation or can be further purified and injected into the methane pipeline. In small treatment plants, anaerobic digestion of the sludge does not represent an economical approach due to the minimal production of sludge during the treatment process. In these sites, anaerobic treatment of wastewater is typically preferred as it results in lower sludge productivity and a reduced cost for the plant for sludge disposal.

Exoelectrogenic bacteria need to utilize a portion of the energy obtained from the oxidation of the organic matter in solution to sustain the microbial metabolism. Nevertheless, the theoretical energy production of MFCs can be substantial, and is directly related to the concentration of biodegradable material in the wastewater and the efficiency of bacteria in transferring the obtained electrons to the anode (Box 7.1). Virtually any source of biodegradable organic matter can be used for power generation in MFCs, but not all the organic matter can be immediately transformed into electricity. For example, *Geobacter* spp., exoelectrogenic bacteria commonly found on the anode of MFCs, is unable to utilize glucose, sucrose, or similarly highly reduced hydrocarbons as a substrate [11]. *Geobacter* strains have been shown to be capable of oxidizing typical end products of fermentation such as lactate and formate, but with lower efficiency compared to acetate [12]. Therefore, the degradation of highly variable substrates typically present in domestic and industrial wastewaters in MFCs requires the development of syntrophic interactions with different microorganisms that break down more complex chemicals into simpler metabolites that can be easily converted by exoelectrogenic bacteria into electrical current [13]. This cooperative work of different microorganisms in MFCs allows using a broad range of fuels for renewable electricity generation from waste biomass (Box 7.2).

Box 7.2 Exoelectrogenic microorganisms powering MFCs

Exoelectrogenic bacteria are naturally present in the environment and domestic wastewater; therefore, MFCs treating real liquid samples do not typically need an inoculum. *Geobacter sulfurreducens*, *Geobacter metallireducens*, and *Geobacter anodireducens* are dissimilatory metal-reducing bacteria and some of the most common and studied exoelectrogenic bacteria found on the anode of MFCs [14]. A relevant early study investigated the impact of the inoculum sources on MFC performance, using three different wastewater samples from as many geographical locations to inoculate lab-scale MFCs. After an initial adaptation period of approximately one month, all the reactors produced stable performance and the initially

(Continued)

Box 7.2 (Continued)

variable microbial community converged toward a stable community with high similarity, comprising primarily (>50%) *Geobacter* spp. [15], indicating that the environmental conditions, rather than the inoculum source, typically control the microbial ecology in MFCs. Another model exoelectrogen as popular as *Geobacter* in the scientific community is *Shewanella oneidensis* MR-1 [16]. Despite the predominance of exoelectrogenic bacteria in MFCs, the ability of microorganisms to interact with external electrodes is extended to all three domains of life and includes Archaea and Eukarya, with obviously variable electricity production rate efficiency.

7.2 Integrating Microbial Electrochemical Technologies in Wastewater Treatment Plants

Conventional wastewater treatment technologies generally consist of preliminary processes (pumping, screening, and grit removal) followed by primary clarifiers to remove settleable particles and floatable oil and grease, biological treatment for pollutant abatement, secondary clarification, and finally disinfection. Primary clarifiers contribute to the removal of approximately one-third of the BOD and half of the total solids entering the treatment plant with minimal energy demand in a hydraulic retention time (HRT) of 1 to 3 hours. Secondary biological treatment converts the remaining BOD into bacterial biomass, sludge, which is separated from the treated water in secondary clarifiers. Secondary biological treatment represents the core of the wastewater treatment process and typically requires continuous aeration of the wastewater in order to provide sufficient oxygen to the bacterial biomass to rapidly oxidize and remove the biodegradable organic matter. Aeration for biological treatment can account for up to half of the operation costs in typical wastewater treatment plants, and up to 50% of the electricity used during treatment is consumed for aeration [3].

MFCs are biological reactors and can replace, or integrate, biological treatment in conventional wastewater treatment trains, providing the following advantages:

i) *Recovery of the chemical energy in the organic matter as electricity or useful products* (such as H_2 gas in microbial electrolysis cells [MECs]) *rather than sludge, which needs to be disposed.* This is important in that it helps create a circular economy in which the chemical energy in wastewater is transformed into value-added products. The amount of electricity generated in MFCs depends on the biodegradable organic content and the efficiency of the cells in transforming the organic content into electricity.
ii) *Modularity.* Microbial electrochemical technologies can be assembled in a modular fashion and potentially operate as stand-alone reactors, avoiding the need for integration in an existing wastewater infrastructure. A for-profit company currently active in the United States and commercializing microbial electrochemical technologies for wastewater treatment (Aquacycl) operates modular MFCs in shipping containers. This is important as it enables the utilization of MFCs for wastewater treatment in areas where used water conveyance has not been properly developed.
iii) *Lower sludge production compared to conventional treatment technologies.* The anodic oxidation of the organic pollutants in wastewater in an anaerobic process, with consequently lower sludge production (approximately one-fifth) compared to aerobic wastewater treatment [17]. Sludge

and solids management represent a large portion of the cost during wastewater treatment ($300–800/t) [18, 19], and using MFCs for wastewater treatment can reduce solid production, substantially reducing operating costs for their handling and disposal.

iv) *Avoid the energy costs associated with aeration when using air cathodes in MFCs.* In MFCs, the oxygen needed for the oxidation of organic matter is provided to the cathode and not directly to the microorganisms as in conventional wastewater treatment technology. This allows engineering reactor configurations in which passive air flow enables a stable oxygen concentration near the cathode, similar to the one in air. Passive aeration through the cathode reduces costs while potentially maintaining a high rate for substrate removal.

The vast majority of the literature on MFCs for wastewater treatment has been conducted on lab and bench-scale reactors with a maximum liquid volume of a few liters. In order to demonstrate their applicability for treatment of wastewater, which is produced globally at a rate of ~380 billion m^3 every year, the volume and treatment capacity of these reactors should be increased. There are now several reports available in the literature focusing on the scale-up of MFCs and their integration in wastewater treatment plants. Eighteen studies have been published on MFCs with a volume larger than 100 L, which is considered sufficiently large to enable wastewater treatment for a few persons based on the average wastewater production per capita (200 L day^{-1}) and the typical HRT for wastewater treatment (six hours) [1, 2]. Some of the performance parameters of these studies are reported in Table 7.1 (Box 7.3).

Box 7.3 Critical parameters in pilot MFC operation

MFC performance for wastewater treatment is reported as a function of the power density produced (normalized either by area or liquid volume), informing on the efficiency in generating electricity and as a function of the amount of pollutants removed from the wastewater in a certain amount of time (COD removal). Coulombic efficiency (CE) is another valuable parameter for MFCs treating wastewaters and measures the efficiency of the MFC in converting the organic matter into electricity. Power density (average or maximum) and CE are reported in most of the MFC studies, independently of their volume. For large-scale MFCs, other parameters have been introduced to provide a better quantification of their performance. The net energy recovery (NER) is calculated by normalizing the average power produced over time by either the flow rate (Wh m^{-3}) or by the flow rate multiplied by the COD removal (Wh kg_{COD}^{-1}), proving a clear indication of the performance as a function of the wastewater flow or pollutant removal efficiency [39, 40]. Other useful parameters for pilot-scale reactors that are sometimes reported include the organic loading rate, which is calculated by dividing the COD by the HRT, and the volumetric treatment rate, which is obtained by dividing the COD removed by the HRT [41]. Both these parameters can provide useful information on the efficiency of the MFCs in processing input wastewater.

7.3 Large-Scale MFCs for Wastewater Treatment – Material Selection

Although MFCs have been proposed as an alternative technology for wastewater treatment for more than a decade ago, they have only been recently considered for deployment due to challenges related to materials, performance, and sufficient experience with scale-up for conducting

Table 7.1 Summary of the performance and configuration of pilot-scale MFCs with a liquid volume larger than 100 L.

Wastewater type	Volume	Flow rate ($L\,h^{-1}$)	Configuration	Anode	Cathode	Separator[a)]	Catholyte	COD removal (%)	CE (%)[b)]	Power density (mWm^{-2})	Packing density ($m^2\,m^{-3}$)	Energy recovery (Wh kg_{COD}^{-1})	Study
Domestic	1600	200	Air cathode	Carbon felt	Carbon-based cathode	AEM	–	60 ± 8	3.6	8	1.0	2.6	[20]
Domestic	1000	91	Air cathode	Carbon brushes	Activated carbon on SS support	–	–	34 ± 7	1.6	81	7.2	25	[21]
Domestic	246	21	Air cathode	Carbon brushes	Carbon-based cathode	AEM	–	–	–	–	–	–	[22]
Domestic	250	2	Air cathode	Carbon brushes	Pt/C on carbon mesh	–	–	79 ± 7	5	45	8.0	206	[23]
Domestic	255	6	Air cathode	Carbon brushes	Activated carbon on SS support	–	–	41 ± 16	30	56	5.2	184	[24]
Swine	110	28	Air cathode	Carbon brushes	Carbon-based cathode	–	–	36 ± 15	7	95	6.3	110	[25]
Domestic	850	71	Air cathode	Carbon brushes	Activated carbon on SS support	–	–	49	9	61	24	88	[26]
Domestic	720	40	Air cathode	Goetite on carbon felt	Co-Zn-Fe-Sn-Cu catalyst	Clayware ceramic membrane	–	78	–	6	1.9	1.5	[27]
Domestic	226	25	Air cathode	Nonwoven graphite fabric	Carbon-based cathode	AEM	–	31	20	29	17.6	–	[28]

Domestic	1000	–	Submerged cathode	Reticulated vitreous carbon	Reticulated vitreous carbon	VANADion membrane	Groundwater	88 ± 8	5–15	44	6.8	224	[29]
Domestic	1500	200	Submerged cathode	Carbon brushes	Carbon brushes	porous polyurethane sponge	Aerated anolyte percolating	91	0.64	29	15.0	–	[30]
Domestic	1500	200	Submerged cathode	Carbon brushes	Carbon brushes	porous polyurethane sponge	Aerated anolyte percolating		–	34	15.0	–	[31]
Domestic + AD filtrate	800	33	Submerged cathode	Carbon brushes	Carbon brushes	–	Same as anolyte	19	–	11	1.0	–	[32]
Domestic	1000	56	Submerged cathode	Carbon brushes	Carbon brushes	–	Same as anolyte	90	–	8	0.3	–	[33]
Domestic	260	–	Submerged cathode	Carbon fabric	Carbon fabric	–	Same as anolyte	80	–	9	1.4	–	[34]
Urine	455	19	Submerged cathode	Carbon veil sheet	Activated carbon	–	Same as anolyte	69	–	10	134	25	[35]
Domestic	1000	333	Submerged cathode	Granular activated carbon	Granular activated carbon	CEM	Aerated anolyte effluent	70	41	110	–	880	[36, 37]
Domestic	192	11	Submerged cathode	Carbon brushes	N-doped activated carbon powder on carbon cloth	CEM	Anolyte effluent	77	–	–	–	127	[38]

a) AEM, anion exchange membrane; CEM, cation exchange membrane.
b) CE, Coulombic efficiency.

large-scale tests. Large reactors are needed to process the volume of wastewater produced daily by typical households, and scaling up MFCs for wastewater treatment requires a careful balance between maintaining a reasonable cost of the system while maximizing performance in terms of energy recovery and effluent water quality.

Materials for MFCs can have a large impact on the system cost. Typical elements include anode and cathode electrodes, a separator if used, and the whole reactor body. The initial MFC designs used precious metal catalysts such as platinum to facilitate the ORR at the cathode. Pt is a common electrocatalyst for ORR in fuel cells but its use in MFCs is untenable due to the excessive cost and the poor stability in complex media such as wastewater [42]. Non-precious metal catalysts similar to the ones developed for hydrogen fuel cells were previously proposed for replacing Pt, but cumbersome synthesis procedures and unknown stability in real environments can limit their utilization at scale [43]. Several studies indicated that carbon-based cathode materials represent the ideal choice for microbial electrochemical technologies, coupling low cost and large availability with acceptable catalytic activities toward ORR [44, 45]. Activated carbon has been successfully employed in large-scale air cathode (up to 0.7 m^2) MFCs for wastewater treatment [21, 24, 26, 39, 46]. Polymer binders added to the activated carbon allowed to create a structurally sound catalyst layer that can be applied onto a metallic support acting as a current collector. Additional polymer layers can be casted on top of the catalyst layer to facilitate oxygen diffusion to the activated carbon and avoid water leakage [47, 48]. Submerged cathodes, where the oxygen for the ORR is dissolved in the catholyte, typically make use of electrodes with high surface area, such as granular activated carbon or graphite fiber brushes. Air cathodes generally outperform submerged cathodes at any scale in terms of maximum power production but are more expensive to manufacture and more difficult to engineer, particularly at larger scale [49].

Anode electrodes are typically carbon-based materials, such as carbon felt, cloth, paper, foams, graphite sheets, plates, granules, or brushes [50]. The main advantage of using carbon-based materials as anodes in MFCs lies in their good biocompatibility with the exoelectrogenic biofilm. What differentiate the materials listed above is their conductivity, which is larger for graphitic carbon compared to amorphous carbon, and porosity, which is typically larger for amorphous carbon compared to graphitic carbon. Therefore, a trade-off exists between large surface area and high electrical conductivity. Electrodes with high surface area are advantageous in that they allow increasing the amount of exoelectrogenic bacteria on the electrode in the same volume of less porous electrodes, but the lower conductivity, which could not be relevant at lab scale due to the small distance between where the electrons are generated and consumed, can limit performance at scale [51].

A thick biofilm increases the resistance of the bioanode toward contaminants and oxygen, which can limit MFC performance by providing the biofilm an electron sink alternative to the electrode. A dense biofilm, counterintuitively, does not appear to be limited by the transport of substrate as the microorganisms closer to the electrode maintain their ability to transfer electrons from the active bacteria in contact with the solution [52, 53]. Graphite brushes and granular activated carbon were the preferred anodes in the vast majority of pilot-scale MFCs for wastewater treatment (Figure 7.1, Table 7.1), providing plenty of electroactive area for biofilm growth in the reactor. Previous studies investigated the treatment of carbon-based anodes for enhanced surface area through heat treatment or decoration of the electrode surface with additional catalysts [54, 55]. To date, only heat treatment has been used for large-scale systems [21].

Separators are not always necessary in large-scale MFCs. A separator can be used to divide the cell into anode and cathode chambers, avoiding comingle of anolyte and catholyte [56]. Separators have been justified in a few large-scale studies for minimizing oxygen intrusion in the anodic biofilm. This can be useful if a liquid catholyte is used in the cell and air is sparged in the catholyte

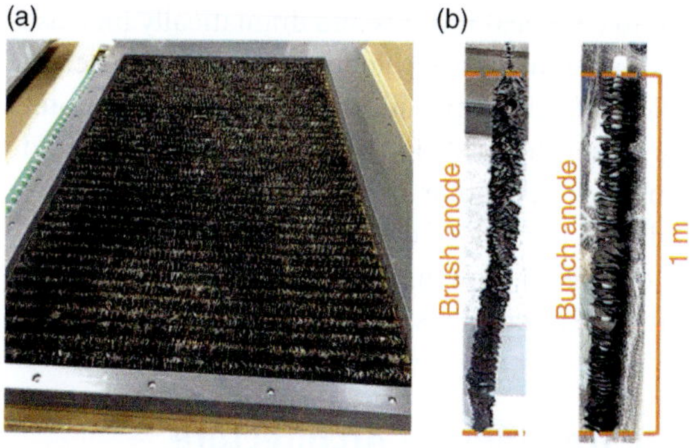

Figure 7.1 Photos of anodes in pilot-scale MFCs. (a) In the study by Rossi et al. [26], 40 carbon fiber brush anodes (2.5 cm diameter, 61 cm brush length) were used for each module, and 17 anode modules were used in the whole MFC. *Source:* Reproduced with permission from Rossi et al. [26] Copyright 2022 Elsevier. (b) In the study by Xie et al. [20], graphite fiber brushes were compared with bunch anodes (4 cm diameter, 100 cm brush length), manufactured by stacking carbon felt electrodes in a stainless-steel wire as a current collector. *Source:* Reprinted with permission from Xie et al. [20] Copyright 2024 Elsevier.

to maintain a large oxygen concentration for the ORR. Early studies focused on the use of ion exchange membranes for MFCs, resembling early work on fuel cells, but it was later shown that these expensive membranes provided limited advantages in terms of better performance while representing a large additional cost [56]. Nevertheless, a few pilot-scale reactors opted for the use of either anion exchange membranes [20, 22, 28] or ion-selective membranes for redox flow-batteries [29]. Alternative to ion exchange membranes, unexpensive cloth separators were used in scaled-up MFCs and placed in close contact with the cathode to limit biofilm growth, but the effectiveness of this approach remains to be thoroughly investigated at larger scale [57].

The stability of the materials used in MFCs and other microbial electrochemical systems is a major source of concern for implementation at scale. Performance stability has been investigated in several studies in terms of power production and wastewater treatment efficiency but the main underlying processes governing performance degradation remain to be completely unveiled. Anodes in MFCs have been operated for several years with no adverse effects on performance. Typical carbonaceous anodes do not corrode during MFC operation (even though attention must be used in the selection of metal wires and current collectors, which are susceptible to corrosion), and a stable biofilm operation and microbial ecology are ensured if the environmental conditions are maintained constant.

Performance decrease for cathodes and separator over time is more problematic compared to anodes. The development of biofilms on cathode limits the transport of ions and oxygen, resulting in mass-transfer limitation, which impair cathodic activity [58]. A decrease in cathode performance is not only caused by biofilm growth. The ORR enables the development of a local high pH near the cathode, which facilitates the precipitation of carbonate salts in the catalyst environment, blocking active sites and resulting in large decrease of the performance over time [58–60]. A maximum power decrease of 36% was recorded for a lab-scale MFC fed a synthetic media after six months of operation, and 53% of this decrease was due to the precipitation of solids in the cathode, while biofouling was responsible for 37% of power degradation [59]. In large-scale MFCs treating real waste streams, the performance decrease was accelerated by the low buffer capacity of the media compared to synthetic streams which lead to a higher pH near the cathode. For example, the power produced from an 85 L MFC fed domestic wastewater decreased by 34% after only one month of operation. Initial performance can be restored by removing the biofilm and soaking the cathode in an acid solution, but this can result in chemical hazards and dramatically increase costs [61]. Mitigation of fouling at the cathode while enabling the development of a biofilm rich in exoelectrogens at the anode remains a major challenge in the application of MFCs for wastewater treatment that has yet to be resolved. Membrane bioreactors (MBRs) are subject to fouling and performance is restored by dosing a cleaning solution at regular time intervals. It is unlikely that the same approach can be applied to MFC due to the risk of damaging the anodic biofilm. Separators can foul due to chemical precipitation or biofilm growth. Both these phenomena limit the transport of ions across the electrode, increasing the internal resistance of the MFC.

7.4 Large-Scale MFCs for Wastewater Treatment – Architecture

It is currently unclear what the best reactor configuration is for large-scale MFCs for wastewater treatment, and the lack of understanding of the impact of dimension on MFC performance limits our ability to develop bench-scale prototypes for predicting the performance at scale. The large body of literature on the scale-up of hydrogen fuel cells is not of aid due to the large performance difference in terms of power and current density (four to five orders of magnitude smaller for MFCs compared to hydrogen fuel cells) and the focus of MFCs on achieving not only good electrochemical performance but also high wastewater treatment efficiency. Primarily, two approaches were used for scaling up MFCs for wastewater treatment: (i) using several repeating units of small reactors connected in series or in parallel and (ii) using a large electrochemical cell comprising several anode–cathode modules immersed in a large plug-flow reactor. The modularity of the repeating-unit MFCs could be advantageous as individual cells can be added or removed as a function of the wastewater flow or composition, and bench-scale reactors can be used directly, avoiding expensive and time-consuming trials for scaling up the reactor and electrode size. However, hydraulic and electrical connection of several different MFCs can be challenging (Box 7.4). In contrast, increasing the electrode dimension from the current cm^2 up to m^2 or dm^2 has not been extensively attempted before and potential limitations in this route have yet to be identified.

Box 7.4 Voltage and current reversal in microbial electrochemical systems

The poor electrochemical performance of MFCs can be overcome by connecting more than one reactor either in parallel to increase current output or in series to increase voltage output. However, inherent constraints in MFCs result in the formation of shunt current and a phenomenon called voltage or current reversal. During voltage reversal, the voltage of one MFC in the series is reversed compared to its initial value, and the voltage of a neighboring MFC "feeds" the reversed cell, resulting in an overall decrease of the voltage output. In the reversed cell, what was the anode is now the cathode and vice versa, so that electrode corrosion and biofilm detachments start to occur. Voltage reversal can be overcome by connecting MFCs in parallel rather than in series, but current reversal can then occur, where small differences in the performance between cells change the overall polarity and electrons are consumed in the parallel circuit.

Large-scale MFCs can be classified based on their cathode configuration, employing liquid catholytes or air cathodes. In liquid catholytes, the cathode is submerged in solution, and the oxygen dissolved in water provides the oxidizer for the cathodic reaction. The low concentration of dissolved oxygen in water and wastewaters (<0.2 mM) reduces the efficiency of this configuration as air needs to be typically pumped in the catholyte to replenish the oxygen consumed at the cathode or a large catholyte flow rate is needed. Both these approaches can result in large energy consumption for pumping. In one study using a submerged cathode, the catholyte was aerated intermittently, and optimization of the amount of air pumped in the catholyte allowed to minimize energy consumption in the reactor. Nevertheless, maximum energy production from the MFC (1.40×10^{-3} kWh m^{-3}) was approximately one order of magnitude smaller than the energy consumed for aeration (2.34×10^{-2} kWh m^{-3}) [31].

In addition to the operational costs to oxygenate the catholyte, liquid catholyte configurations typically incur additional capital costs due to the use of separators between anode and cathode to minimize oxygen intrusion near the anode. In some MFCs with a liquid catholyte configuration, it was reported that a thick biofilm developed on the cathode. Despite the general understanding that a cathodic biofilm could limit oxygen transport to the electrode and therefore decrease performance, it was reported an increase in cathodic activity with the biofilm compared to its absence [62]. This phenomenon was ascribed to the fact that the biofilm on the cathode catalyzed the reoxidation of metals in the cathode environment or could directly accept electrons from the electrode.

An air-cathode configuration that has been tested in pilot-scale reactors (from 80 to 850 L) from different groups was based on several small air cathodes soldered to a stainless-steel frame [21, 24, 26]. This design was advantageous in that the stainless-steel frame provided structural integrity to the cathode and electrical conductivity while overcoming manufacturing challenges associated with the production of one single large air cathode. Another approach used to overcome manufacturing challenges of large air cathodes was to stack the electrodes in a horizontal arrangement to minimize water pressure and avoid cathode failure [23]. Overall, the limitations associated with engineering large air cathodes resulted in only a few studies utilizing this configuration in pilot MFCs for wastewater treatment.

From the data reported in Table 7.1, air cathode MFCs produced an area-normalized power density (47 ± 30 mW m^{-2}) 224% larger than that of liquid catholyte MFCs (21 ± 14 mW m^{-2}) and a volumetric power density 138% larger (458 ± 462 mW m^{-3} for air cathode compared to 332 ± 361 mW m^{-3} for liquid catholyte), indicating that the air-cathode configuration allows to maximize power production. Despite enabling larger electrochemical performance, treatment efficiency of air-cathode MFCs is sub-par, resulting in an average COD removal rate of $51 \pm 18\%$ compared to $73 \pm 25\%$ of liquid catholyte MFCs at a larger HRT (32 ± 44 hours for air cathode compared to 15 ± 9 hours for liquid catholyte). The larger organic matter removal rate in liquid catholyte reactors was likely due to aeration of the catholyte but energy consumption for this process has not always been reported, making it difficult to identify an optimal approach. From the comparison of COD removal and power production in pilot MFCs, it is apparent that there is a trade-off between treatment capacity and electrochemical performance, and it is not yet clear whether the additional power produced in the air cathode configuration could offset the energy needed for additional effluent treatment downstream of the MFC. Technoeconomic analysis could help identify whether one architecture is to be preferred in specific environmental conditions and identify specific metrics for pilot MFC performance evaluation to be reported in future studies.

7.5 Internal Resistance in Large-Scale MFCs

A critical limitation in the design and development of pilot-scale MFCs is that the area-specific electrode resistance (mΩ m^2 – independent of the electrode size) increases as the electrode area is scaled up, resulting in the same electrode material and architecture having much lower performance at scale. The factors controlling this increase in resistance remain to be elucidated. A larger electrode resistance at scale will reduce the maximum power density and the energy recovered during wastewater treatment. A modelling study indicated that a portion of the resistance increase during scale-up is due to the inherently poor conductivity of the carbonaceous materials used as electrodes (copper: $0.1\,\Omega\,cm^{-1}$; carbon paper: $0.8\,\Omega\,cm^{-1}$; graphite fiber: $1.6\,\Omega\,cm^{-1}$; carbon cloth: $2.2\,\Omega\,cm^{-1}$) [51]. In that study, it was reported that increasing the electrode area of a carbon felt anode in an MFC from $10\,cm^2$ to $1\,m^2$ increased the electrode resistance from 0.16 to $0.46\,\Omega$, resulting in power losses as high as 4.19 W at scale. The adverse impact of electrode dimensions on performance has also been shown experimentally. For example, it was reported that increasing electrode size from 7 to $6200\,cm^2$ ($0.62\,m^2$) while maintaining the same cathode material and architecture increased the electrode resistance from $66 \pm 17\,m\Omega\,m^2$ to $555 \pm 24\,m\Omega\,m^2$, indicating that at the same voltage output, the large-scale cathode will generate 8.4× less current density. The same study reported that the area-normalized resistance of the anode was less affected by the larger size of the electrodes. Graphite fiber brush anodes were scaled up from $7\,cm^2$ to an array of 22 brushes, each 5.1 cm in diameter and 61 cm long, resulting in a total area of $0.68\,m^2$. This larger area resulted in a 3.3× increase in the electrode resistance from $71 \pm 3\,m\Omega\,m^2$ for the $7\,cm^2$ brush to $238 \pm 18\,m\Omega\,m^2$ for the larger anode [63].

Several factors can contribute to the increase in the area-specific resistance during scale-up and low electrical conductivity is only one of them. As the electrode size is increased, the water pressure on the electrodes becomes larger (Figure 7.2). In air cathodes, this translates in larger flooding of the activated carbon catalyst layer where water pressure is larger, which in turn reduces the oxygen availability. A potential solution to this problem that has been previously proposed, even though it has never been tested, is to modulate the hydrophobicity of the polymer layer in the air cathode as a function of the water pressure, producing catalyst layers more hydrophobic where the water pressure is larger in order to maintain a constant water/air distribution across the whole height of the cathode. In the case that liquid catholytes are used, a large water pressure will increase the energy needed to sparge air in the catholyte.

The impact of water pressure on the anodic biofilm performance has not been extensively investigated. One study reported high variability in the power production (from $697 \pm 65\,mW\,m^{-2}$ to

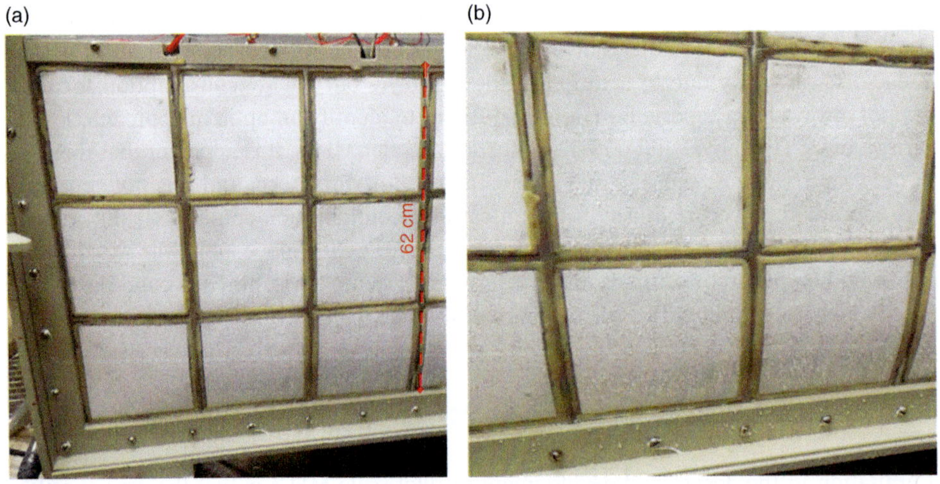

Figure 7.2 (a) Impact of water height on air cathode water leakage. The air cathode showed excellent resistance to water heights up to 20 cm. At the bottom of the cathode, where the water pressure was the largest, water from the liquid chamber leaked through the hydrophobic polymer layer, forming droplets on the air side. (b) Zoom in on the bottom of the cathode.

$1260 \pm 24\,mW\,m^{-2}$) of MFCs operating at different water pressure (from 100 to 2000 mmH_2O) [64]. The reactors operating at large water pressure (2000 mmH_2O, $697 \pm 65\,mW\,m^{-2}$) produced almost half the power density of the reactors at small water pressure (100 mmH_2O, $1260 \pm 24\,mW\,m^{-2}$). The microbial community was similar across the different water pressures, and the large difference in performance was ascribed to a metabolic inhibition of the microorganisms. To date, despite several studies reporting an increase in electrode resistance and lower performance during scale-up, the underlying phenomena triggering this change have yet to be identified. More work will be needed in the future to unveil the impact of scale on the electrochemical, microbiological, and biological characteristics of MFCs in order to rationally identify the optimal routes for scaling up.

Together with anode and cathode resistance, solution resistance due to the transport of ions can play a role in dictating the performance of pilot MFCs. The solution resistance is due to the transport of ions in the media between the electrodes to balance the electrons exchanged in the external circuit. Two approaches are typically used to decrease the solution resistance in microbial electrochemical technologies: (i) increasing the solution conductivity and ion concentration or (ii) minimizing the electrode spacing to diminish the distance that the ions must travel. Amending wastewater with salts to reduce the solution resistance does not appear to be a practical approach. However, wastewaters with high solution conductivity such as oilfield wastewater, seafood processing wastewater, and seawater-based domestic wastewater sludge are available and can be investigated in large-scale reactors [65, 66]. Reducing electrode spacing is the most effective approach to diminish solution resistance in MFCs. However, caution must be used to not short-circuit the cell or allow for an excessive intrusion of oxygen from the cathode near the anode, which adversely impacts the microbial community and the anodic performance. Short circuiting can be prevented by using membranes, which will then increase the cost of the reactor. The impact of the solution resistance on pilot-scale MFCs is minor compared to anode and cathode resistance. While the electrode resistance (in $m\Omega\,m^2$, i.e. normalized by the reactor cross-sectional area) increases during scale-up, the solution resistance remains constant, and its contribution to the whole internal resistance diminishes compared to anode and cathode resistances. For example, it was reported that the solution resistance increased only from 87 to 93 $m\Omega\,m^2$ during scale-up of an MFC from a projected area of 7–6200 cm^2 as the electrode spacing was maintained similar. In the same reactor, the anode resistance increased by 3.3 times, while cathode resistance increased by 8.4 times [63]. Nevertheless, minimizing electrode spacing should be prioritized in the design of microbial electrochemical systems in order to diminish energy losses, reactor footprint, and optimize ion transport in the cell.

7.6 Large-Scale MFCs for Wastewater Treatment – Substrate

Substrate is a critical factor in MFCs, primarily controlling electricity generation and microbial ecology. Pilot MFCs have been operated with domestic and industrial wastewaters, swine wastewaters, and anaerobic digestion filtrate (Table 7.1). The water chemistry and substrate content and composition vary as a function of the wastewater type and are also subject to spatiotemporal variation within the same wastewater type. Therefore, general conclusions on the optimal wastewater type for MFCs cannot be immediately drawn. Nevertheless, some general characteristics of the wastewater composition can be obtained by the typical requirements for MFCs. High strength wastewater with a large concentration of organic matter and high conductivity should generally be preferred in order to provide plenty of substrate for the exoelectrogenic bacteria and simple substrates such as acetate, lactate, or formate are to be preferred over more complex organics such as sugars or oligomers.

7.7 Treatment Efficiency

From the data reported in Table 7.1 and from each individual publication, it is possible to analyze the efficiency of pilot-scale MFCs for wastewater treatment in terms of treatment efficiency and energy recovery. As shown in Table 7.1, the average COD removal was 60 ± 23%, indicating that 60% of the influent COD was consumed in the MFC. This COD removal was significantly lower than that typically obtained from conventional activated sludge (~90%) and MBRs (~95%) [67]. The average effluent COD was 322 ± 513 mg L^{-1} (191 ± 205 mg L^{-1} excluding Walter et al. [35]due to the large influent COD of 6000 ± 1000 mg L^{-1}) which is again larger than that typically obtained in conventional activated sludge processes (~100 mg L^{-1}) and in MBRs (~50 mg L^{-1}). Such effluent COD is exceeding the maximum threshold for direct discharge in the United States and Europe, indicating that additional treatment units will likely be needed downstream of the MFC to further remove organic matter from the effluent and meet stringent requirements for direct discharge (~30 mg L^{-1} BOD, corresponding to ~60 mg L^{-1} COD in the United States).

To reduce the effluent organic content to a level suitable for discharge, MFCs were coupled with membrane filtration, resulting in final effluent COD as low as 28 ± 11 mg L^{-1} [68] or with an anaerobic fluidized bed MBR to produce a final effluent COD of 16 ± 3 mg L^{-1} [69] and 36 ± 6 mg L^{-1} [70] (Figure 7.3). Other approaches included bioelectrochemical MBRs (BEMR) comprising a hybrid MFC/MBR, producing an effluent with only 21.5 ± 8.4 mg L^{-1} of COD [71]. These processes, while being effective in reducing some pollutant concentration to levels suitable for discharge or even water reuse, require additional energy that should be computed in the overall energy demand of the MFC. The COD in the MFC effluent cannot be typically reduced below a certain threshold (~150 mg L^{-1}) without a large reduction in electrochemical performance due to the development of mass-transfer limitations [72]. As the COD in solution is consumed by the exoelectrogenic bacteria, a lower amount of organic matter is transported to the anode, leading to a decrease in current production, which further slows down COD removal and compromises the integrity of the microbial biofilm.

Despite the lower average COD removal and effluent quality in MFCs compared to other well-established wastewater treatment processes, some MFC configurations were able to achieve excellent COD removal (~90%) with minimal effluent COD (70 ± 17 mg L^{-1}, Feng et al. [23]; 50 mg L^{-1}, Blatter et al. [29]; 25 mg L^{-1}, He et al. [30]) with effluents suitable for direct discharge, indicating that engineering solutions are available to obtain high-quality effluents directly in the MFC alone. In one study, a low COD in the effluent was achieved by using a long retention time (144 hours, Feng et al.[23]), which will likely not be practical in most conditions. In two studies, a low COD effluent was obtained by employing the effluent from the anode chamber as an influent in the cathode chamber following aeration (Blatter et al. [29]; He et al. [30]). Aerating the catholyte was necessary to increase the oxygen concentration in the catholyte to facilitate the reduction of oxygen at the cathode, and some oxygen likely contributed to accelerating the consumption of the residual COD from the anode. While enabling low COD effluents, energy will be consumed to aerate the anolyte effluent, which could offset the energy advantages of MFCs compared to conventional aerobic wastewater treatment processes.

The average HRT (21 ± 28 hours) for the pilot MFCs was approximately three times larger than that of conventional activated sludge (~6 hours), indicating that the reactor footprint of MFCs could be substantially larger than conventional treatment technologies for the same amount of wastewater processed. HRTs in pilot MFCs were not always optimized to maximize power production or organic removal and were in a few studies limited by the low availability of wastewater [26, 35]. HRT optimization in pilot MFCs will require the use of technoeconomic analysis tools

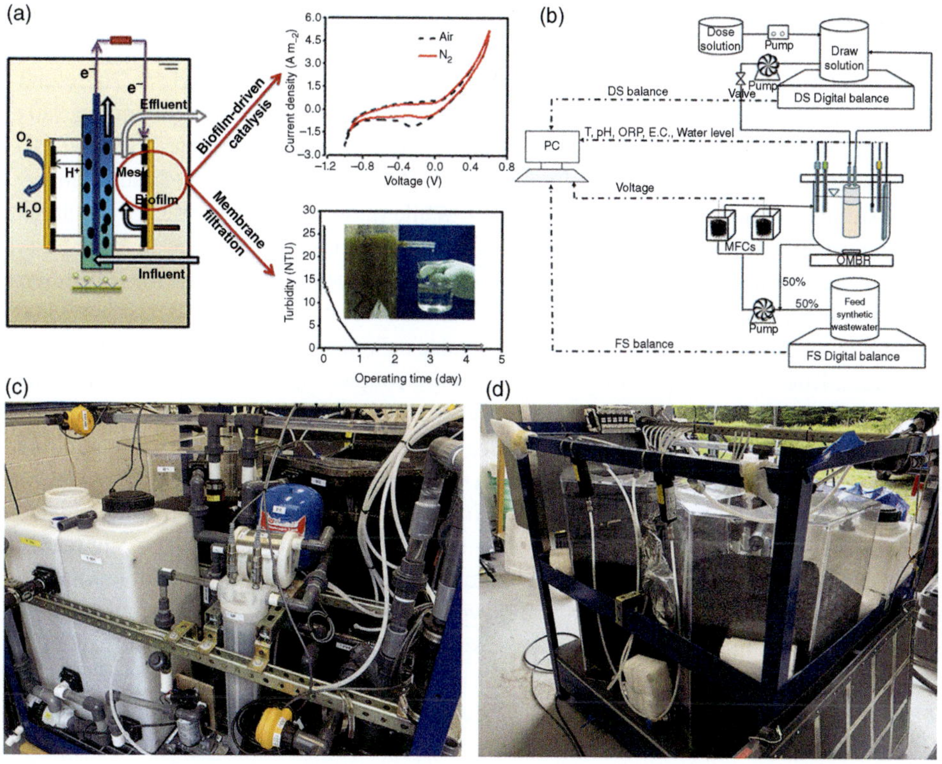

Figure 7.3 Water filtration units for polishing MFC effluents. (a) Schematic of a BEMR. *Source:* Reprinted from Wang et al. [71] with permission from the American Chemical Society. Copyright 2011 American Chemical Society. (b) Schematic of an osmotic membrane bioreactor operating in conjunction with an MFC to diminish the COD content in the treated effluent. *Source:* Reprinted from Hou et al. [68] with permission from Elsevier. Copyright 2016 Elsevier. (c) Biofilter unit used to treat the effluent of the pilot MFC described by [26]. (d) Biofilter tanks containing granular activated carbon. *Source:* Reproduced from Rossi et al. [26]. Copyright 2022 Elsevier.

in order to balance the reactor efficiency with a suitable COD effluent to diminish the burden on downstream treatment units. Despite the lower wastewater treatment capacity and larger HRTs in pilot MFCs, both activated sludge and MBRs require large energy input (and for MBRs, chemicals are consumed) that can be avoided in MFCs. Most of the studies reported in Table 7.1 used a simple pump for wastewater delivery to the reactor, resulting in minimal energy consumption. For example, activated sludge typically requires approximately 0.6 kWh m^{-3} and MBR 2.3 kWh m^{-3} compared to only 0.3 kWh m^{-3} reported by Rossi et al. [26] in an 850-L reactor or 0.02 kWh m^{-3} reported by Dong et al. [31] in a 1500-L MFC.

Sludge production has not been extensively investigated in pilot MFC studies, likely due to the limited operation time of less than one year for most of the published works. In one study, it was reported a sludge accumulation rate of 0.16 kg per kg of COD removed, half of that typically reported for conventional aerobic processes (0.32 kg per kg of COD removed) [17, 26]. Assuming an average sludge disposal cost of \$200 t^{-1} [73], reducing sludge by half can dramatically decrease expenditures during wastewater treatment.

7.8 Energy Recovery

A critical feature of MFCs is to recover energy from wastewater and produce electricity. Therefore, since the inception of this technology, a large focus has been on maximizing power generation. This has been accomplished by optimizing the reactor architecture in zero-gap configurations [74, 75], by selecting efficient materials and catalysts [76], substrate [77, 78], and through genetic engineering of specific strains and microbial communities [79]. While these approaches resulted in a large increase in performance in laboratory and bench-scale reactors, their application in large-scale MFCs for wastewater treatment is limited. Large-scale MFCs have typically been manufactured using inexpensive configurations with carbon-based electrodes at both the anode and the cathode, limiting the use of membranes and separators to minimize capital costs and using wastewaters as inoculum sources. Due to the simple design and operation, coupled with the inherent limitations (low substrate concentration, poor solution conductivity, and buffer capacity) of wastewater, area or volume normalized power production of large-scale systems is typically orders of magnitude lower than that of optimized lab-scale reactors with a volume of a few milliliters.

The average power density produced during continuous operation by the large-scale reactors (both air cathode and submerged cathode) reported in Table 7.1 was $40 \pm 32\,\mathrm{mW\,m^{-2}}$ based on the electrode area and $429 \pm 439\,\mathrm{mW\,m^{-3}}$ based on liquid volume. These power densities were 20 (based on the electrode area) and 50 (based on liquid volume) times smaller than maximum power densities produced in optimized lab scale reactors using wastewater as a feed ($800 \pm 30\,\mathrm{mW\,m^{-2}}$, $21{,}600 \pm 810\,\mathrm{mW\,m^{-3}}$) [80], indicating that (i) there is a lack of understanding in how to efficiently scale up MFCs (i.e. the power cannot be maintained during scale-up) and (ii) some of the strategies (media, substrate, architectures) used to maximize performance in small-scale systems cannot be immediately translated in pilot systems. The CE of the pilot MFCs has not always been reported, likely due to the challenges associated with its analysis in continuous feeding mode with wastewater. The COD content of wastewater varies with time and while the current density allows for a continuous measurement of coulombs generated in an MFC, integrating a continuous analysis of the COD is not immediately feasible. Therefore, CE measurements for pilot-scale reactors are generally based on average COD removal efficiency and average current production over variable timeframes. The CE reported in large-scale systems was $10 \pm 13\%$, indicating that up to 10% of the electrons contained in the reduced organic matter were recovered and utilized for current generation. This could seem a small number but it is in line with lab-scale studies [72, 81, 82], and if extended to the calculated amount of energy available in the wastewater worldwide, it would result in 1.7 GW of potential power, corresponding to approximately 4200 million photovoltaic panels.

7.9 Comparison of MFCs with Conventional Wastewater Treatment Technologies

Only one study used the results from a pilot-scale MFC as a basis for comparing performance and cost with a current state-of-the-art MBR, which is an established technology for wastewater treatment. Results from that work were not published in a peer-reviewed journal but were included in a final report, which is available online [83]. The MFC was an 850-L active volume (1400 L total volume) reactor with brush anodes (11 $\mathrm{m^2}$ of total anode projected area) and air cathodes with activated carbon catalysts (20 $\mathrm{m^2}$ of total cathode projected area, two cathodes per module, each anode was connected with two cathodes) fed ~1500 L $\mathrm{day^{-1}}$ of domestic wastewater from a primary clarifier. The reactor was coupled with a biofilter using granular activated carbon for intermittently operated filtration to polish the MFC effluent and achieve a final $91 \pm 6\%$ COD removal ($49 \pm 15\%$

removed in the MFC) and 91% BOD removal while producing 43 mW m^{-2} (0.46 ± 0.35 W at 1.54 ± 0.90 A) and 18 Wh m^{-3} on average during nine months of operation [26]. The MFC size and operation resembled a small, decentralized reactor for wastewater treatment and was, therefore, compared to an MBR with similar wastewater treatment capacity rather than to a conventional centralized large-scale activated sludge plant. Based on the results obtained during the demonstration, the MFC treatment capacity was projected to 18,500 L day^{-1} and compared to a commercial MBR.

The capital cost analysis revealed large expenditure for the MFC system ($322,000), contributing to 23% of the overall cost ($1,310,000 – of which $730,000 direct costs). These estimated costs were likely inflated by the early stage of research on pilot-scale MFCs that did not allow to leverage the economy of scale for minimizing production costs. For example, in the early stage of that research, only one air-cathode vendor among the three tested was able to provide an electrode capable of withstanding the large water pressure from the anolyte and the electrode was in its largest past assembled by hand. Cathodes were the most expensive part of the reactor and mainly contributed to the large initial cost of the reactor. It can be expected that, as the MFC technology matures and more interest toward commercialization is attracted, the spectrum of potential vendors will increase while mass manufacturing will diminish costs for MFC components.

Operational costs ($133,500 per year) in the pilot MFC were dominated by chemical use ($26,200 per year), in particular, low-concentration acids for restoring cathode performance. The MFC power production rapidly deteriorated due to the formation of carbonate salt precipitates and a thick biofilm on the cathode. Early work demonstrated that the performance can be restored following cathode scrubbing and an acid wash to remove carbonate precipitates [39, 84]. This limitation was common to other MFC reactors with similar architecture and performance. Other direct costs included replacement for anodes ($217,600 every 10 years) and cathodes ($33,900 every 3 years) and for the granular activated carbon for the biofilter ($13,600 every 5 years). Labor and equipment represented a large portion of operational costs ($17,700 per year) and were primarily due to the time-consuming cathode cleanup. Future studies will need to investigate the stability of anodes and cathodes in pilot MFCs and clarify whether electrode replacement is needed and how often. Typically, carbon-based anodes are stable in MFCs and are not degraded over time, while cathode acid wash has been proved to be an effective approach for cathode regeneration. Results from the direct cost analysis clearly indicate that the most pressing challenge in MFC application is the stability of operation and prevention of cathode performance degradation. Some approaches have been explored at the lab scale to prevent cathode fouling but their application in pilot-scale reactors has yet to be tested [57, 85].

Cost comparison between the pilot MFC and the MBR returned similar direct costs ($730,300 for the MFC and $668,300 for the MBR) with a much larger estimated annual energy cost for the MBR ($5500) compared to the MFC ($100). These additional expenditures are primarily driven by the energy consumption associated with wastewater aeration in the MBR. This analysis clearly indicates that MFCs are already cost-competitive with commercially available wastewater treatment technologies. More work will be needed primarily in the development of effective strategies toward the scale-up in order to increase predictability at scale from lab and bench scale studies and in the minimization of capital and operational expenditures by enhancing performance stability to avoid electrode replacement costs and chemical consumption for cathode cleaning.

7.10 Treatment Technologies Other than MFCs: Microbial Electrolysis Cells

MECs couple the oxidation of organic matter in used water at the anode with hydrogen gas production at the cathode. In MECs, the ORR at the cathode used in MFCs is replaced by the hydrogen evolution reaction (HER), shifting the whole cell voltage from positive with the ORR in MFCs (electricity is produced) to negative with the HER in MECs (electricity is consumed), and an additional voltage input is required to drive hydrogen production at the cathode. Thermodynamically, the minimum voltage requirement is <0.2 V, but kinetic and mass-transfer limitations typically increase the applied voltage to 0.8–1.0 V, which is still lower than the thermodynamic voltage required for operating conventional water electrolyzers fed ultrapure water (1.2 V). Hydrogen gas is an essential building block of the chemical industry and MECs are of particular interest for the treatment of fermentation effluent wastewaters due to the large concentration of substrate and the possibility of maximizing the hydrogen yield of the overall process [86]. Current densities in lab- and bench-scale MECs are now approaching 100 A m^{-2} at applied voltages lower than 1.0 V [87] with hydrogen production rates exceeding 50 L L^{-1} day^{-1}. Despite these advancements, and the technology being first demonstrated at bench-scales almost 20 years ago [88, 89], only a few MECs have been successfully operated at pilot-scale (>100 L), treating real waste streams (Table 7.2).

The configuration of pilot MECs resembles the MFC architecture previously described with a few differences. Two chambers are typically used in MECs to separate anode from cathode and limit the growth of microorganisms in the cathode chamber capable of consuming the hydrogen gas produced. In only one of the pilot MEC studies reported in Table 7.2, a separator was not included, resulting in large methane production rates and no appreciable hydrogen gas production after less than two months of operation. To avoid methane production in MEC with no separators, inhibitors of methanogenesis such as chloroform can be added to the media [97, 98]. While effective at bench scale, this approach is not practical at pilot scale, as the inhibitor needs to be removed from the effluent prior to discharge. AEMs and polyethylene separators were preferred in the pilot MEC studies reported in Table 7.1, likely due to their lower cost compared to cation exchange membranes such as Nafion. AEMs have been previously shown to boost MEC performance by optimizing ion transport efficiency and pH management across the electrodes [87, 99, 100].

Carbon-based materials were the preferred choice for the anode, which is in line with pilot MFC studies. In fact, the anodic community of MFCs and MECs tends to be similar, as the same reaction is occurring on the anode [101]. Compared to MFCs, where carbon brushes were preferentially used as anodes, carbon felt was used in five out of seven pilot MEC studies, likely due to the fact that carbon felt allows for the development of compact modules, which can be filled with a catholyte and immersed in wastewater. The preferred cathode material was stainless steel, which has shown good efficiency for hydrogen generation [102] for a limited material cost. The current being delivered from the anode protects the stainless-steel cathode and avoids corrosion. Catholytes were used in most of the studies containing a buffer or sodium chloride to minimize ohmic losses. Hydrogen gas can be captured from the headspace of the reactors, but caution must be used as hydrogen is a small molecule and can easily diffuse through tubing and connectors, resulting in experimental production rates much lower than theoretical [91, 92].

The average COD removal of pilot MECs was similar to that obtained in pilot MFC studies with air-cathodes, averaging 51 ± 24% indicating that also in the case of MECs, additional treatment technologies will be needed downstream to reduce the organic matter concentrations to a level suitable for discharge. The CE in pilot MECs (18 ± 12%) was slightly larger than that of pilot MFCs (10 ± 13%), which is in line with results from lab-scale reactors. The oxygen diffusing through the cathode in MFCs can diminish CE by providing an alternative electron acceptor to the anode for the microorganisms. In MECs, no oxygen is provided to the cathode and hydrogen is evolved from water. The hydrogen production rates in the pilot MEC studies reported in Table 7.2 are low (0.020 ± 0.013 $L_{hydrogen}$ $L_{reactor}{}^{-1}$ day^{-1}), approximately one-to-two orders of magnitude smaller than that obtained in lab-scale reactors. This could be due to (i) the low packing density

Table 7.2 Summary of the performance and configuration of pilot-scale MECs with volume larger than 100 L.

Wastewater type	Volume	Flow rate ($L\ h^{-1}$)	Configuration	Anode	Cathode	Separator[a]	Catholyte	COD removal (%)	CE (%)[b]	Hydrogen production rate ($L\ L^{-1}\ day^{-1}$)	Current density ($A\ m^{-2}$)	Applied voltage (V)	Packing density ($m^2\ m^{-3}$)	Study
Domestic	1000	42	Modular	Carbon brushes	SS 304	None	–	62 ± 20	n.r.	0.027	0.41	0.9	18.1	[90]
Domestic	120	5	Cassette	Carbon felt	SS wool	Polyethylene separator	50 mM phosphate buffer pH 7	34	n.r.	0.015	0.0076	1.1	2.8	[91]
Domestic	120	5	Cassette	Carbon felt	SS wool	Polyethylene separator	50 mM phosphate buffer pH 7	44	2	0.007	0.01	1.1	2.8	[92]
Domestic	181	36	Cassette	Carbon felt	SS 316 mesh	Polyethylene separator	NaCl 0.1 M	64	10	0.003	0.23	0.9	13	[93]
Domestic	130	2.7	Cassette	Carbon fibers	SS wool	AEM	NaCl 0.07 M	25	28	0.032	0.31	1.5	13	[94]
Domestic	150	6.3	Modular	Carbon felt	Carbon felt	AEM	Same as anolyte	100	–	–	0.22	1.0	16	[95]
Domestic	135	2	Cassette	Carbon felt	SS wool	AEM	NaCl 0.07 M	30	30	0.038	0.42	0.9	11	[96]

a) AEM, anion exchange membrane; CEM, cation exchange membrane.
b) CE, Coulombic efficiency.

($10 \pm 6\,m^2\,m^{-3}$) of the pilot MECs tested to date compared to lab-scale reactors (>$100\,m^2\,m^{-3}$) [103, 104], which indicates that larger electrodes can be fit in the same MEC volume, boosting the volume-normalized hydrogen production rates; or (ii) a lower productivity at scale due to an increase in the electrode resistance during scale-up, as previously described for pilot-scale MFC reactors.

7.11 Outlook and Future Directions

Microbial electrochemical technologies have been demonstrated at large scale (>100 L), treating domestic and industrial wastewaters while producing electricity in MFCs or hydrogen and methane gas in MECs. Approximately half of the influent organic matter fed to the pilot bioelectrochemical reactor was removed from the wastewater, and 10–30% of the chemical energy in the influent was recovered either as electricity or valuable chemicals. Increasing organic matter removal will be at the expense of the electrochemical performance of the reactor. Technologies capable of removing the remaining contaminants in the MFC effluent downstream of the electrochemical reactors need to be demonstrated at the relevant scale to meet treated effluent quality for discharge. In most of the studies reported in the literature, only the COD influent, effluent, and removal rate were investigated and reported. Future work should include the analysis of the efficiency of pilot-scale bioelectrochemical reactors in the removal of contaminants other than COD and BOD, including pathogens, metals, nitrogen, and phosphorous. While data are available at the lab and bench scale for some of these contaminants, these results cannot be immediately translated into pilot-scale systems.

Energy recovery in MFCs and MECs is limited by the intrinsic characteristics of wastewater, including a large substrate variability, low solution conductivity, and poor buffer capacity. The substrate range that can be consumed by common exoelectrogens is limited, and the degradation of complex and recalcitrant organics will require synergy with other microbial strains, which could in turn compete for a limited space on the electrode while not directly contributing to electricity generation. A low conductivity of the solution increases the internal resistance of the cell, while poor buffer capacity facilitates the development of large pH gradients across the electrodes, limiting the microbial activity at low pH while facilitating solid precipitation on the cathode at high pH. A poor wastewater solution chemistry is likely the main driver in the large and rapid decrease in electrochemical performance common in all the previous studies on pilot MFCs and MECs published so far, primarily due to cathode and membrane fouling. Biofouling and solid precipitation cannot be avoided, and the only effective approach used so far to overcome this challenge was removing the fouled parts and soaking them in diluted acid washes. However, this procedure is time-consuming, can pose a hazard for wastewater treatment operators, and can dramatically increase the operational costs. Despite the use of inexpensive electrode materials in most of the pilot reactors tested, frequent replacement of the materials due to fouling is not practical.

There are avenues for improving electrochemical performance in future studies. Using high strength wastewater can minimize internal resistance while providing plenty of substrate for the microbial anode to generate electricity. The modularity of bioelectrochemical systems allows to scale up or down the reactor dimension and operate with low and variable flow rates, which is not always feasible in anaerobic digesters. Diminishing anode and cathode resistance during scale-up is a research priority, as well as identifying the drivers for the larger resistance at scale. Currently, despite several studies pointing out that the area-normalized resistance of the electrodes increases as the electrode is scaled up, the underlying processes driving this phenomenon remain unclear [63, 105, 106]. Minimizing the electrode spacing, which is critical at lab and bench scale, is relatively ineffective at pilot scale due to the limited contribution of the solution resistance on the overall internal resistance compared to anode and cathode resistances. Lab-scale studies have demonstrated that minimizing spacing to zero-gap configurations can diminish anode and cathode resistances, as well as the solution resistance, by optimizing the transport of ions in the cell [107, 108]. The feasibility of such an approach is questionable at large scale due to the potential problems associated with clogging due to the large solid content of typical wastewaters.

Advancements in the engineering of pilot-scale systems should be accompanied by the identification of key performance indicators (KPIs) through life cycle and technoeconomic analysis, to guide future research in the field and aid in selecting the most promising pilot configurations and optimal avenues for scale-up. As of today, despite a broad range of reactor configurations have been tested, there is not a preferred choice, due to the lack of KPIs and the large variability in the wastewater composition, operational procedures, and architectures tested [109]. Nevertheless, preliminary analysis indicated that pilot-scale bioelectrochemical reactors are already competitive with commercial wastewater treatment technologies such as MBR and several for-profit companies currently commercialize MFCs and related technologies (Sentry, Aquacycl, Cambrian Innovation, MicroOrganic, among others). Larger competition in this space can be expected following government subsidies for green hydrogen and electricity that can further enhance the profitability of MFCs and MECs. As more interest is drawn to the development of pilot-scale studies and demonstrations, a better understanding of the technology and its application in wastewater treatment is expected.

References

1 Viessman, W. (2013). *Water Supply and Pollution Control*. Upper Saddle River, NJ: Pearson Education Limited.

2 Heidrich, E.S., Curtis, T.P., and Dolfing, J. (2011). Determination of the internal chemical energy of wastewater. *Environ. Sci. Technol.* 45: 827–832. https://doi.org/10.1021/es103058w.

3 McCarty, P.L., Bae, J., and Kim, J. (2011). Domestic wastewater treatment as a net energy producer – can this be achieved? *Environ. Sci. Technol.* 45: 7100–7106. https://doi.org/10.1021/es2014264.

4 Logan, B.E. and Rabaey, K. (1979). Conversion of wastes into bioelectricity and chemicals by using microbial electrochemical technologies. *Science* 337 (2012): 686–690.

5 Shizas, I., Bagley, D.M., and Asce, M. (2004). Experimental determination of energy content of unknown organics in municipal wastewater streams. *J. Energy Eng.* 130: 45–53. https://doi.org/10.1061/ASCE0733-94532004130:245.

6 Liu, H., Liu, H., Ramnarayanan, R. et al. (2004). Production of electricity during wastewater treatment using a single chamber microbial fuel cell. *Environ. Sci. Technol.* 38: 2281–2285. https://doi.org/10.1021/es034923g.

7 Potter, M.C. (1911). Electrical effects accompanying the decomposition of organic compounds. *Proc. R. Soc. B* 84: 260–276. https://doi.org/10.1098/rspb.1911.0073.

8 Logan, B.E., Hamelers, B., Rozendal, R. et al. (2006). Microbial fuel cells: methodology and technology. *Environ. Sci. Technol.* 40: 5181–5192. https://doi.org/10.1021/es0605016.

9 Ucar, D., Zhang, Y., and Angelidaki, I. (2017). An overview of electron acceptors in microbial fuel cells. *Front. Microbiol.* 8: https://doi.org/10.3389/fmicb.2017.00643.

10 Lawson, K., Rossi, R., Regan, J.M., and Logan, B.E. (2020). Impact of cathodic electron acceptor on microbial fuel cell internal resistance. *Bioresour. Technol.* 316: https://doi.org/10.1016/j.biortech.2020.123919.

11 Caccavo, F., Lonergan, D.J., Lovley, D.R. et al. (1994). Geobacter sulfurreducens sp. nov., a hydrogen-and acetate-oxidizing dissimilatory metal-reducing microorganism. *Appl. Environ. Microbiol.* 60: 3752–3759.

12 Speers, A.M. and Reguera, G. (2012). Electron donors supporting growth and electroactivity of geobacter sulfurreducens anode biofilms. *Appl. Environ. Microbiol.* 78: 437–444. https://doi.org/10.1128/AEM.06782-11.

13 Christgen, B., Spurr, M., Milner, E.M. et al. (2023). Does pre-enrichment of anodes with acetate to select for Geobacter spp. enhance performance of microbial fuel cells when switched to more complex substrates? *Front. Microbiol.* 14: https://doi.org/10.3389/fmicb.2023.1199286.

14 Logan, B.E., Rossi, R., Ragab, A., and Saikaly, P.E. (2019). Electroactive microorganisms in bioelectrochemical systems. *Nat. Rev. Microbiol.* 17: https://doi.org/10.1038/s41579-019-0173-x.

15 Yates, M.D., Kiely, P.D., Call, D.F. et al. (2012). Convergent development of anodic bacterial communities in microbial fuel cells. *ISME J.* 6: 2002–2013. https://doi.org/10.1038/ismej.2012.42.

16 Chen, H., Simoska, O., Lim, K. et al. (2020). Fundamentals, applications, and future directions of bioelectrocatalysis. *Chem. Rev.* 120: 12903–12993. https://doi.org/10.1021/acs.chemrev.0c00472.

17 Ginestet, P. and Camacho, P. (2006). *Technical Evaluation of Sludge Production and Reduction. Comparative Evaluation of Sludge Reduction Routes - EU project number: EVK1-CT2000-00050*, vol. 5. London: IWA Publishing. https://doi.org/10.2166/9781780402352.

18 Peccia, J. and Westerhoff, P. (2015). We should expect more out of our sewage sludge. *Environ. Sci. Technol.* 49: 8271–8276. https://doi.org/10.1021/acs.est.5b01931.

19 Westerhoff, P., Lee, S., Yang, Y. et al. (2015). Characterization, recovery opportunities, and valuation of metals in municipal sludges from U.S. wastewater treatment plants nationwide. *Environ. Sci. Technol.* 49: 9479–9488. https://doi.org/10.1021/es505329q.

20 Xie, L., Tanaka, F., Yagi, T. et al. (2024). Multi-anode enhanced the bioelectricity generation in air-cathode microbial fuel cells towards energy self-sustaining wastewater treatment. *Environ. Res.* 243: https://doi.org/10.1016/j.envres.2023.117744.

21 Heinrichmeier, J., Littfinski, T., Vasyukova, E. et al. (2023). On-site performance evaluation of a 1,000-litre microbial fuel cell system using submergible multi-electrode modules with air-cathodes for sustainable municipal wastewater treatment and electricity generation. *Water Sci. Technol.* 87: 1969–1981. https://doi.org/10.2166/wst.2023.106.

22 Shimidzu, A., Tanaka, F., Matsumura, T. et al. (2023). Optimizing low-voltage boosting for an air-cathode microbial fuel cell with an anion exchange membrane in a 246 L wastewater treatment reactor. *Environ. Sci. (Camb.)* 10: 296–303. https://doi.org/10.1039/d3ew00448a.

23 Feng, Y., He, W., Liu, J. et al. (2014). A horizontal plug flow and stackable pilot microbial fuel cell for municipal wastewater treatment. *Bioresour. Technol.* 156: 132–138. https://doi.org/10.1016/j.biortech.2013.12.104.

24 Hiegemann, H., Littfinski, T., Krimmler, S. et al. (2019). Performance and inorganic fouling of a submergible 255 L prototype microbial fuel cell module during continuous long-term operation with real municipal wastewater under practical conditions. *Bioresour. Technol.* 294: https://doi.org/10.1016/j.biortech.2019.122227.

25 Babanova, S., Jones, J., Phadke, S. et al. (2020). Continuous flow, large-scale, microbial fuel cell system for the sustained treatment of swine waste. *Water Environ. Res.* 92: 60–72. https://doi.org/10.1002/wer.1183.

26 Rossi, R., Hur, A., Page, M.A. et al. (2022). Pilot scale microbial fuel cells using air cathodes for producing electricity while treating wastewater. *Water Res.* 118208: https://doi.org/10.1016/j.watres.2022.118208.

27 Das, I., Ghangrekar, M.M., Satyakam, R. et al. (2020). On-site sanitary wastewater treatment system using 720-L stacked microbial fuel cell: case study. *J. Hazard. Toxic Radioact. Waste* 24: 04020025. https://doi.org/10.1061/(asce)hz.2153-5515.0000518.

28 Sugioka, M., Yoshida, N., Yamane, T. et al. (2022). Long-term evaluation of an air-cathode microbial fuel cell with an anion exchange membrane in a 226L wastewater treatment reactor. *Environ. Res.* 205: https://doi.org/10.1016/j.envres.2021.112416.

29 Blatter, M., Delabays, L., Furrer, C. et al. (2021). Stretched 1000-L microbial fuel cell. *J. Power Sources* 483: https://doi.org/10.1016/j.jpowsour.2020.229130.

30 He, W., Dong, Y., Li, C. et al. (2019). Field tests of cubic-meter scale microbial electrochemical system in a municipal wastewater treatment plant. *Water Res.* 155: 372–380. https://doi.org/10.1016/j.watres.2019.01.062.

31 Dong, Y., He, W., Liang, D. et al. (2019). Operation strategy of cubic-meter scale microbial electrochemistry system in a municipal wastewater treatment plant. *J. Power Sources* 441: 227124. https://doi.org/10.1016/j.jpowsour.2019.227124.

32 Leininger, A., Yates, M.D., Ramirez, M., and Kjellerup, B. (2021). Biofilm structure, dynamics, and ecology of an upscaled biocathode wastewater microbial fuel cell. *Biotechnol. Bioeng.* 118: 1305–1316. https://doi.org/10.1002/bit.27653.

33 Liu, R., Tursun, H., Hou, X. et al. (2017). Microbial community dynamics in a pilot-scale MFC-AA/O system treating domestic sewage. *Bioresour. Technol.* 241: 439–447. https://doi.org/10.1016/j.biortech.2017.05.122.

34 Mohamed, A., Zmuda, H.M., Ha, P.T. et al. (2021). Large-scale switchable potentiostatically controlled/microbial fuel cell bioelectrochemical wastewater treatment system. *Bioelectrochemistry* 138: 107724. https://doi.org/10.1016/j.bioelechem.2020.107724.

35 Walter, X.A., You, J., Winfield, J. et al. (2020). From the lab to the field: self-stratifying microbial fuel cells stacks directly powering lights. *Appl. Energy* 277: https://doi.org/10.1016/j.apenergy.2020.115514.

36 Liang, P., Duan, R., Jiang, Y. et al. (2018). One-year operation of 1000-L modularized microbial fuel cell for municipal wastewater treatment. *Water Res.* 141: 1–8. https://doi.org/10.1016/j.watres.2018.04.066.

37 Liang, P., Duan, R., Jiang, Y. et al. (2019). Corrigendum to "One-year operation of 1000-L modularized microbial fuel cell for municipal wastewater treatment". *Water Res.* 166: 114878. https://doi.org/10.1016/j.watres.2019.114878.

38 Ge, Z. and He, Z. (2016). Long-term performance of a 200 liter modularized microbial fuel cell system treating municipal wastewater: Treatment, energy, and cost. *Environ. Sci. (Camb.)* 2: 274–281. https://doi.org/10.1039/c6ew00020g.

39 Hiegemann, H., Herzer, D., Nettmann, E. et al. (2016). An integrated 45 L pilot microbial fuel cell system at a full-scale wastewater treatment plant. *Bioresour. Technol.* 218: 115–122. https://doi.org/10.1016/j.biortech.2016.06.052.

40 He, Z. (2017). Development of microbial fuel cells needs to go beyond "power density". *ACS Energy Lett.* 2: 700–702. https://doi.org/10.1021/acsenergylett.7b00041.

41 Bird, H., Heidrich, E.S., Leicester, D.D., and Theodosiou, P. (2022). Pilot-scale microbial fuel cells (MFCs): A meta-analysis study to inform full-scale design principles for optimum wastewater treatment. *J. Cleaner Prod.* 346: https://doi.org/10.1016/j.jclepro.2022.131227.

42 Santoro, C., Serov, A., Stariha, L. et al. (2016). Iron based catalysts from novel low-cost organic precursors for enhanced oxygen reduction reaction in neutral media microbial fuel cells. *Energy Environ. Sci.* 9: 2346–2353. https://doi.org/10.1039/c6ee01145d.

43 Peera, S.G., Maiyalagan, T., Liu, C. et al. (2021). A review on carbon and non-precious metal based cathode catalysts in microbial fuel cells. *Int. J. Hydrogen Energy* 46: 3056–3089. https://doi.org/10.1016/j.ijhydene.2020.07.252.

44 Zhang, F., Pant, D., and Logan, B.E. (2011). Long-term performance of activated carbon air cathodes with different diffusion layer porosities in microbial fuel cells. *Biosens. Bioelectron.* 30: 49–55. https://doi.org/10.1016/j.bios.2011.08.025.

45 Zhang, F., Cheng, S., Pant, D. et al. (2009). Power generation using an activated carbon and metal mesh cathode in a microbial fuel cell. *Electrochem. Commun.* 11: 2177–2179. https://doi.org/10.1016/j.elecom.2009.09.024.

46 Rossi, R., Jones, D., Myung, J. et al. (2019). Evaluating a multi-panel air cathode through electrochemical and biotic tests. *Water Res.* 148: 51–59. https://doi.org/10.1016/j.watres.2018.10.022.

47 Pant, D., Van Bogaert, G., De Smet, M. et al. (2010). Use of novel permeable membrane and air cathodes in acetate microbial fuel cells. *Electrochim. Acta* 55: 7710–7716. https://doi.org/10.1016/j.electacta.2009.11.086.

48 Bowman, L., Van Houtven, D., Pant, D. et al. 2019). Carbon based electrode with large geometric dimensions WO2019068488A1.

49 Rossi, R. and Logan, B.E. (2022). Impact of reactor configuration on pilot-scale microbial fuel cell performance. *Water Res.* 225: https://doi.org/10.1016/j.watres.2022.119179.

50 Santoro, C., Arbizzani, C., Erable, B., and Ieropoulos, I. (2017). Microbial fuel cells: From fundamentals to applications. A review. *J. Power Sources* 356: 225–244. https://doi.org/10.1016/j.jpowsour.2017.03.109.

51 Cheng, S., Ye, Y., Ding, W., and Pan, B. (2014). Enhancing power generation of scale-up microbial fuel cells by optimizing the leading-out terminal of anode. *J. Power Sources* 248: 631–638. https://doi.org/10.1016/j.jpowsour.2013.10.014.

52 Gorby, Y.A., Yanina, S., McLean, J.S. et al. (2006). Electrically conductive bacterial nanowires produced by Shewanella oneidensis strain MR-1 and other microorganisms. *Proc. Natl. Acad. Sci.* 103: 11358–11363. https://doi.org/10.1073/pnas.0604517103.

53 Sun, D., Chen, J., Huang, H. et al. (2016). The effect of biofilm thickness on electrochemical activity of Geobacter sulfurreducens. *Int. J. Hydrogen Energy* 41: 16523–16528. https://doi.org/10.1016/j.ijhydene.2016.04.163.

54 Guo, K., Freguia, S., Dennis, P.G. et al. (2013). Effects of surface charge and hydrophobicity on anodic biofilm formation, community composition, and current generation in bioelectrochemical systems. *Environ. Sci. Technol.* 47: 7563–7570. https://doi.org/10.1021/es400901u.

55 Li, B., Zhou, J., Zhou, X. et al. (2014). Surface modification of microbial fuel cells anodes: approaches to practical design. *Electrochim. Acta* 134: 116–126. https://doi.org/10.1016/j.electacta.2014.04.136.

56 Li, C., Liang, D., Tian, Y. et al. (2024). Sorting out the latest advances in separators and pilot-scale microbial electrochemical systems for wastewater treatment: concomitant development, practical application, and future perspective. *Environ. Sci. Technol.* https://doi.org/10.1021/acs.est.4c03169.

57 Li, C., Yi, K., Hu, S., and Yang, W. (2023). Cathodic biofouling control by microbial separators in air-breathing microbial fuel cells. *Environ. Sci. Ecotechnol.* 15: https://doi.org/10.1016/j.ese.2023.100251.

58 Yuan, Y., Zhou, S., and Tang, J. (2013). In situ investigation of cathode and local biofilm microenvironments reveals important roles of OH- and oxygen transport in microbial fuel cells. *Environ. Sci. Technol.* 47: 4911–4917. https://doi.org/10.1021/es400045s.

59 An, J., Li, N., Wan, L. et al. (2017). Electric field induced salt precipitation into activated carbon air-cathode causes power decay in microbial fuel cells. *Water Res.* 123: 369–377. https://doi.org/10.1016/j.watres.2017.06.087.

60 Popat, S.C., Ki, D., Rittmann, B.E., and Torres, C.I. (2012). Importance of OH− transport from cathodes in microbial fuel cells. *ChemSusChem* 5: 1071–1079. https://doi.org/10.1002/cssc.201100777.

61 Rossi, R., Wang, X., Yang, W., and Logan, B.E. (2019). Impact of cleaning procedures on restoring cathode performance for microbial fuel cells treating domestic wastewater. *Bioresour. Technol.* 290: 121759. https://doi.org/10.1016/j.biortech.2019.121759.

62 Noori, M.T., Rossi, R., Logan, B.E., and Min, B. (2024). Hydrogen production in microbial electrolysis cells with biocathodes. *Trends Biotechnol.* https://doi.org/10.1016/j.tibtech.2023.12.010.

63 Rossi, R., Cario, B.P., Santoro, C. et al. (2019). Evaluation of electrode and solution area-based resistances enables quantitative comparisons of factors impacting microbial fuel cell performance. *Environ. Sci. Technol.* 53: 3977–3986. https://doi.org/10.1021/acs.est.8b06004.

64 Cheng, S., Liu, W., Guo, J. et al. (2014). Effects of hydraulic pressure on the performance of single chamber air-cathode microbial fuel cells. *Biosens. Bioelectron.* 56: 264–270. https://doi.org/10.1016/j.bios.2014.01.036.

65 Grattieri, M., Suvira, M., Hasan, K., and Minteer, S.D. (2017). Halotolerant extremophile bacteria from the Great Salt Lake for recycling pollutants in microbial fuel cells. *J. Power Sources* 356: 310–318. https://doi.org/10.1016/j.jpowsour.2016.11.090.

66 Grattieri, M. and Minteer, S.D. (2018). Microbial fuel cells in saline and hypersaline environments: Advancements, challenges and future perspectives. *Bioelectrochemistry* 120: 127–137. https://doi.org/10.1016/j.bioelechem.2017.12.004.

67 Pauwels, B., Fru Ngwa, F., Deconinck, S., and Verstraete, W. (2006). Effluent quality of a conventional activated sludge and a membrane bioreactor system treating hospital wastewater. *Environ. Technol.* 27: 395–402. https://doi.org/10.1080/09593332708618651.

68 Hou, D., Lu, L., and Ren, Z.J. (2016). Microbial fuel cells and osmotic membrane bioreactors have mutual benefits for wastewater treatment and energy production. *Water Res.* 98: 183–189. https://doi.org/10.1016/j.watres.2016.04.017.

69 Ren, L., Ahn, Y., and Logan, B.E. (2014). A two-stage microbial fuel cell and anaerobic fluidized bed membrane bioreactor (MFC-AFMBR) system for effective domestic wastewater treatment. *Environ. Sci. Technol.* 48: 4199–4206. https://doi.org/10.1021/es500737m.

70 Kim, K.Y., Yang, W., Ye, Y. et al. (2016). Performance of anaerobic fluidized membrane bioreactors using effluents of microbial fuel cells treating domestic wastewater. *Bioresour. Technol.* 208: 58–63. https://doi.org/10.1016/j.biortech.2016.02.067.

71 Wang, Y.-K., Sheng, G.-P., Li, W.-W. et al. (2011). Development of a novel anaerobic membrane bioelectrochemical reactor for wastewater treatment. *Environ. Sci. Technol.* 45: 9256–9261.

72 Zhang, X., He, W., Ren, L. et al. (2015). COD removal characteristics in air-cathode microbial fuel cells. *Bioresour. Technol.* 176: 23–31. https://doi.org/10.1016/j.biortech.2014.11.001.

73 Domini, M., Abbà, A., and Bertanza, G. (2022). Analysis of the variation of costs for sewage sludge transport, recovery and disposal in Northern Italy: a recent survey (2015-2021). *Water Sci. Technol.* 85: 1167–1175. https://doi.org/10.2166/wst.2022.040.

74 Fan, Y., Han, S.K., and Liu, H. (2012). Improved performance of CEA microbial fuel cells with increased reactor size, Energy. *Environ. Sci.* 5: 8273–8280. https://doi.org/10.1039/c2ee21964f.

75 Rossi, R. and Logan, B.E. (2021). Using an anion exchange membrane for effective hydroxide ion transport enables high power densities in microbial fuel cells. *Chem. Eng. J.* 422: https://doi.org/10.1016/j.cej.2021.130150.

76 Cao, B., Zhao, Z., Peng, L. et al. (1979). Silver nanoparticles boost charge-extraction efficiency in Shewanella microbial fuel cells. *Science* 373 (2021): 1336–1340. https://www.science.org.

77 Pant, D., Van Bogaert, G., Diels, L., and Vanbroekhoven, K. (2010). A review of the substrates used in microbial fuel cells (MFCs) for sustainable energy production. *Bioresour. Technol.* 101: 1533–1543. https://doi.org/10.1016/j.biortech.2009.10.017.

78 Pandey, P., Shinde, V.N., Deopurkar, R.L. et al. (2016). Recent advances in the use of different substrates in microbial fuel cells toward wastewater treatment and simultaneous energy recovery. *Appl. Energy* 168: 706–723. https://doi.org/10.1016/j.apenergy.2016.01.056.

79 Simoska, O., Cummings, D.A., Gaffney, E.M. et al. (2023). Enhancing the performance of microbial fuel cells via metabolic engineering of Escherichia coli for phenazine production. *ACS Sustainable Chem. Eng.* 11: 11855–11866. https://doi.org/10.1021/acssuschemeng.3c01593.

80 Yang, W. and Logan, B.E. (2016). Immobilization of a metal-nitrogen-carbon catalyst on activated carbon with enhanced cathode performance in microbial fuel cells. *ChemSusChem* 9: 2226–2232. https://doi.org/10.1002/cssc.201600573.

81 He, Z., Minteer, S.D., and Angenent, L.T. (2005). Electricity generation from artificial wastewater using an upflow microbial fuel cell. *Environ. Sci. Technol.* 39: 5262–5267. https://doi.org/10.1021/es0502876.

82 Lobo, F.L., Wang, X., and Ren, Z.J. (2017). Energy harvesting influences electrochemical performance of microbial fuel cells. *J. Power Sources* 356: 356–364. https://doi.org/10.1016/j.jpowsour.2017.03.067.

83 Cropek, D. (2021). Hybrid microbial fuel cell-biofiltration system for energy-neutral wastewater treatment ESTCP project #ER-201635 environmental restoration projects. https://www.serdp-estcp.org/Program-Areas/Environmental-Restoration/Wastewater-and-Drinking-Water/ER-201635/ER-201635#factsheet-25820-publications (accessed 29 July 2022).

84 Zhang, E., Wang, F., Yu, Q. et al. (2017). Durability and regeneration of activated carbon air-cathodes in long-term operated microbial fuel cells. *J. Power Sources* 360: 21–27. https://doi.org/10.1016/j.jpowsour.2017.05.119.

85 Yang, W., Rossi, R., Tian, Y. et al. (2018). Mitigating external and internal cathode fouling using a polymer bonded separator in microbial fuel cells. *Bioresour. Technol.* 249: 1080–1084. https://doi.org/10.1016/j.biortech.2017.10.109.

86 Srivastava, P., García-Quismondo, E., Palma, J., and González-Fernández, C. (2024). Coupling dark fermentation and microbial electrolysis cells for higher hydrogen yield: technological competitiveness and challenges. *Int. J. Hydrogen Energy* 52: 223–239. https://doi.org/10.1016/j.ijhydene.2023.04.293.

87 Wang, L., Du, H., Elsyed, A.F.N. et al. (2024). Impact of reactor architecture and design parameters on the performance of microbial electrolysis cells revealed by the electrode potential slope analysis. *Electrochim. Acta* 144072. https://doi.org/10.1016/j.electacta.2024.144072.

88 Liu, H., Grot, S., and Logan, B.E. (2005). Electrochemically assisted microbial production of hydrogen from acetate. *Environ. Sci. Technol.* 39: 4317–4320. https://doi.org/10.1021/es050244p.

89 Rozendal, R.A., Hamelers, H.V.M., Euverink, G.J.W. et al. (2006). Principle and perspectives of hydrogen production through biocatalyzed electrolysis. *Int. J. Hydrogen Energy* 31: 1632–1640. https://doi.org/10.1016/j.ijhydene.2005.12.006.

90 Cusick, R.D., Bryan, B., Parker, D.S. et al. (2011). Performance of a pilot-scale continuous flow microbial electrolysis cell fed winery wastewater. *Appl. Microbiol. Biotechnol.* 89: 2053–2063. https://doi.org/10.1007/s00253-011-3130-9.

91 Heidrich, E.S., Dolfing, J., Scott, K. et al. (2013). Production of hydrogen from domestic wastewater in a pilot-scale microbial electrolysis cell. *Appl. Microbiol. Biotechnol.* 97: 6979–6989. https://doi.org/10.1007/s00253-012-4456-7.

92 Heidrich, E.S., Edwards, S.R., Dolfing, J. et al. (2014). Performance of a pilot scale microbial electrolysis cell fed on domestic wastewater at ambient temperatures for a 12 month period. *Bioresour. Technol.* 173: 87–95. https://doi.org/10.1016/j.biortech.2014.09.083.

93 Cotterill, S.E., Dolfing, J., Jones, C. et al. (2017). Low temperature domestic wastewater treatment in a microbial electrolysis cell with 1 m^2 anodes: Towards system scale-up. *Fuel Cells* 17: 584–592. https://doi.org/10.1002/fuce.201700034.

94 Baeza, J.A., Martínez-Miró, À., Guerrero, J. et al. (2017). Bioelectrochemical hydrogen production from urban wastewater on a pilot scale. *J. Power Sources* 356: 500–509. https://doi.org/10.1016/j.jpowsour.2017.02.087.

95 Isabel San-Martín, M., Mateos, R., Carracedo, B. et al. (2018). Pilot-scale bioelectrochemical system for simultaneous nitrogen and carbon removal in urban wastewater treatment plants. *J. Biosci. Bioeng.* 126: 758–763. https://doi.org/10.1016/j.jbiosc.2018.06.008.

96 Guerrero-Sodric, O., Baeza, J.A., and Guisasola, A. (2023). Exploring key operational factors for improving hydrogen production in a pilot-scale microbial electrolysis cell treating urban wastewater. *Chem. Eng. J.* 469: https://doi.org/10.1016/j.cej.2023.144001.

97 Singh, L., Miller, A.G., Wang, L., and Liu, H. (2021). Scaling-up up-flow microbial electrolysis cells with a compact electrode configuration for continuous hydrogen production. *Bioresour. Technol.* 331: 125030. https://doi.org/10.1016/j.biortech.2021.125030.

98 He, K., Li, W., Tang, L. et al. (2022). Suppressing methane production to boost high-purity hydrogen production in microbial electrolysis cells. *Environ. Sci. Technol.* 56: 11931–11951. https://doi.org/10.1021/acs.est.2c02371.

99 Rozendal, R.A., Hamelers, H.V.M., and Buisman, C.J.N. (2006). Effects of membrane cation transport on pH and microbial fuel cell performance. *Environ. Sci. Technol.* 40: 5206–5211. https://doi.org/10.1021/es060387r.

100 Rozendal, R.A., Sleutels, T.H.J.A., Hamelers, H.V.M., and Buisman, C.J.N. (2008). Effect of the type of ion exchange membrane on performance, ion transport, and pH in biocatalyzed electrolysis of wastewater. *Water Sci. Technol.* 57: 1757–1762. https://doi.org/10.2166/wst.2008.043.

101 Kiely, P.D., Cusick, R., Call, D.F. et al. (2011). Anode microbial communities produced by changing from microbial fuel cell to microbial electrolysis cell operation using two different wastewaters. *Bioresour. Technol.* 102: 388–394. https://doi.org/10.1016/j.biortech.2010.05.019.

102 Tang, J., Bian, Y., Jin, S. et al. (2022). Cathode material development in the past decade for H2 production from microbial electrolysis cells. *ACS Environ. Au* 2: 20–29. https://doi.org/10.1021/acsenvironau.1c00021.

103 Wang, L., Linowski, K., and Liu, H. (2023). Scalable membrane-less microbial electrolysis cell with multiple compact electrode assemblies for high performance hydrogen production. *Chem. Eng. J.* 454: 140257. https://doi.org/10.1016/j.cej.2022.140257.

104 Logan, B.E., Wallack, M.J., Kim, K.Y. et al. (2015). Assessment of microbial fuel cell configurations and power densities. *Environ. Sci. Technol. Lett.* 2: 206–214. https://doi.org/10.1021/acs.estlett.5b00180.

105 Fan, Y., Sharbrough, E., and Liu, H. (2008). Quantification of the internal resistance distribution of microbial fuel cells. *Environ. Sci. Technol.* 42: 8101–8107. https://doi.org/10.1021/es801229j.

106 Miller, A., Singh, L., Wang, L., and Liu, H. (2019). Linking internal resistance with design and operation decisions in microbial electrolysis cells. *Environ. Int.* 126: 611–618. https://doi.org/10.1016/j.envint.2019.02.056.

107 Torres, C.I., Marcus, A.K., and Rittmann, B.E. (2008). Proton transport inside the biofilm limits electrical current generation by anode-respiring bacteria. *Biotechnol. Bioeng.* 100: 872–881. https://doi.org/10.1002/bit.21821.

108 Rossi, R., Wang, X., and Logan, B.E. (2020). High performance flow through microbial fuel cells with anion exchange membrane. *J. Power Sources* 475: https://doi.org/10.1016/j.jpowsour.2020.228633.

109 Jadhav, D.A., Chendake, A.D., Vinayak, V. et al. (2022). Scale-up of the bioelectrochemical system: Strategic perspectives and normalization of performance indices. *Bioresour. Technol.* 363: https://doi.org/10.1016/j.biortech.2022.127935.

8

Electrosynthesis

Marcos Pita, Gabriel García-Molina, Kavita Jayakumar, Jose María Abad, and Antonio L. De Lacey

Instituto de Catalisis y Petroleoquimica, CSIC, Department of Biocatalysis, Madrid, Spain

8.1 Introduction

A major category of chemical reactions is the reduction–oxidation (redox) processes. They imply the participation of at least two species, in which one of them is oxidized transferring electrons to the other one, which is reduced. The expression of a redox reaction can be comprehensively split into two half-reactions, the reduction reaction and the oxidation reaction, which should balance the positive and negative charges transferred:

$$O^{n} \rightarrow O^{n+m} + me^{-} \tag{8.1}$$

$$R^{k} + me^{-} \rightarrow R^{k-m} \tag{8.2}$$

The development of knowledge related to electricity during the nineteenth century blossomed into many applications of general use for society and was used in combination with redox processes such as batteries, but it was not until the 1920s that the concept of electrosynthesis was introduced. It is focused on splitting the two half-reactions of a redox process and performing each of them in combination with an anode and a cathode while controlling the electron transfer, reaction rate, and yield. An early example of electrosynthesis was the analysis of Kolbe's reaction, which is the oxidation of acetate to CO_2 [1, 2]. A polarized anode managed to gather the electrons from acetate, yielding its oxidation to CO_2. In the early decades after this work, plenty of oxidative reactions assisted with electrochemistry were studied, which are considered electrosynthetic reactions. However, in this chapter, we will regard the electrosynthetic approach used for reactions to generate new chemicals instead of degrading them to the final CO_2. An early example showing a high degree of selectivity is the electro-assisted reduction of trichloroacetic acid to dichloroacetic acid [3], where a negative potential yielded the single hydrogenation of the substrate. Although the monosubstitution is favored at the selected potential, it is recommended to check in the last transformation stages to stop the reaction before the over-reduction to chloroacetic acid takes place. This comment in the educational publication highlights one of the challenges to overcome when using electrochemistry for reduction reactions: its lack of selectivity. Other transformations that imply oxidation can also produce more complex products, as it might be exemplified by the electrodimerization of aromatic ring derivatives. Several products can be achieved by electro-oxidation of aniline into dimers or asymmetric combinations of several anilinium radicals [4] in what

Bioelectrocatalysis: From Electron Transfer Processes to Emerging Technological Applications, First Edition.
Edited by Shelley D. Minteer and Matteo Grattieri.

might be considered the earliest research toward electropolymerization. Another early example of this synthetic route is the formation of combined aromatic rings in trifluoroacetic acid from dialkylarene molecules [5]. Soon after the establishment of the electrochemical reactions for synthesis purposes, the research field took off, and many other examples became available.

An interesting feature has been the use of catalysts for improving the electrosynthetic reactions. The addition of catalysts to the electrolyte solution has been the subject of research, and thus also the modification of electrodes with such catalysts. Many reviews can be found in this field [6], covering organic electrosynthesis [7, 8], polymer electrosynthesis [9], and microbial electrosynthesis (MES) [10]. In this chapter, we will focus mainly on the knowledge involving the use of biological catalysts together with electrochemistry to enhance electrosynthetic reactions, in what is known as bioelectrosynthesis. In particular, the catalysts to accomplish bioelectrosynthetic reactions are mainly enzymes and microorganisms.

Enzymes play a major role in physiological redox reactions. When combined with electrochemical systems, they can interact in different ways: (i) a substrate needed for the enzymatic reaction can be produced by electrochemical means and later used by the enzyme allocated in the electrolyte solution and (ii) redox enzymes can either give or take electrons to or from the electrode by means of a mediator (or directly) and use it to complete its redox reaction. This possibility was already described in the late 1970s [11], where the field was theorized and kinetically analyzed, suggesting many possible applications of bioelectrochemical systems, such as sensors, or electrosynthesis. Since then, there has been a significant growth and broadening of the research field, covering many different applications, which are too numerous to review in this work. Therefore, in this chapter, we will present the applications of bioelectrosynthesis for some paradigmatic energy-related examples: ammonia and hydrogen synthesis, CO_2 abatement, and organic synthesis.

8.2 Bioelectrosynthesis for CO_2 Reduction

Addressing the challenge of worldwide net-zero carbon emissions has become imperative due to the escalating levels of greenhouse gases (GHGs) in the atmosphere. C1 molecules, particularly carbon dioxide (CO_2), are a focal point, constituting nearly 70% of the GHGs responsible for intensifying global warming and climate change [12]. The current atmospheric CO_2 concentration, a consequence of the nineteenth-century industrial revolution, has surged to approximately 50% higher than preindustrial levels, averaging 412 parts per million (ppm) [13]. This environmental challenge, however, presents an opportunity: CO_2 serves as a substrate for constructing the carbon framework and producing valuable materials like fuels and chemicals. Consequently, attention has shifted toward CO_2 sequestration, encompassing carbon capture, utilization, and storage (CCUS). Various techniques, including metal- and nanomaterials-fused electrochemical catalysis, photocatalysis, and biological catalysis, contribute to studies on CO_2 mitigation and utilization [14–16].

Biological CO_2 conversion – a sustainable, substrate-specific, and reusable method that transforms CO_2 into value-added products – is of particular interest, with seven carbon fixation pathways identified in nature (Figure 8.1). Ecosystems efficiently reduce CO_2 emissions through biological CO_2 assimilation by plants and microorganisms, instigating research into biomimetic strategies, such as introducing partial heterologous carbon fixation pathways into *in vivo* and *in vitro* models. However, the limitations of biological CO_2 reduction by enzymes and microbes become evident in the face of the vast emissions from human activities, surpassing the natural ecosystem's assimilation capacity and resulting in excessive global warming [18]. Additionally, enzymes and microbes involved in CO_2 reduction may not operate optimally under industrial conditions, leading

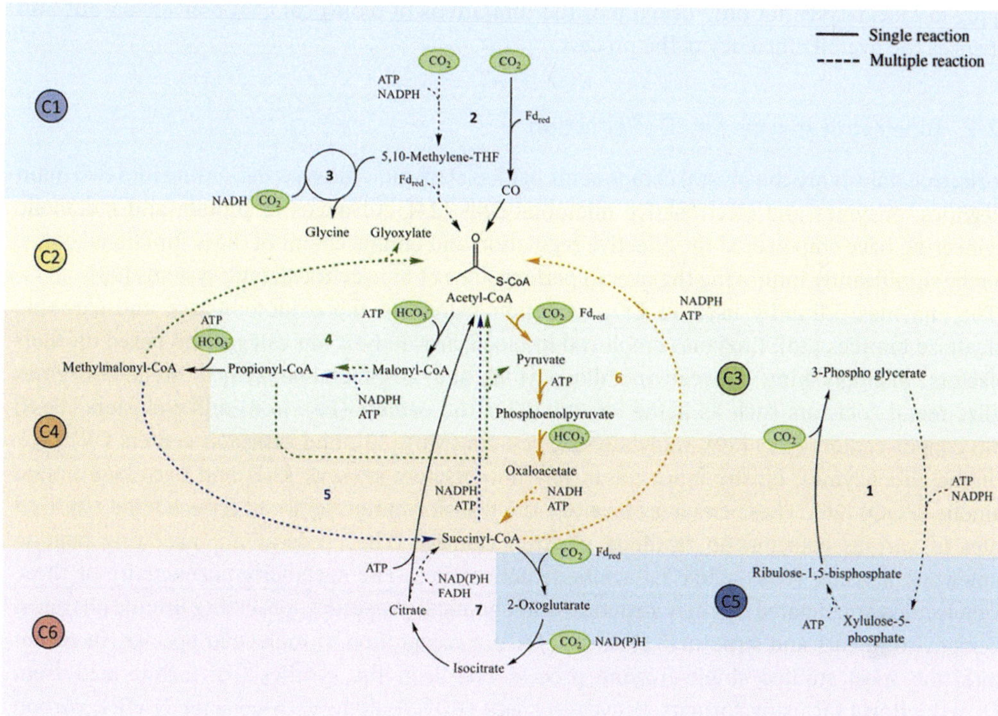

Figure 8.1 Major CO_2 assimilation pathways in nature. (1) Calvin-Benson-Basham cycle; (2) Wood-Ljungdahl pathway; (3) reductive glycine pathway; (4) 3-hydroxypropionate bicycle; (5) 3-hydroxypropionate/4-hydroxybutyrate cycle; (6) dicarboxylate/4-hydroxybutyrate cycle; and (7) reductive citric acid cycle. The 3-HP bicycle, 3-HP/4-HB cycle, and DC/4-HB cycle are represented by green, blue, and orange lines, respectively. *Source:* Adapted from Kang et al. [17], Copyright 2023 by the Authors, distributed under the terms and conditions of the Creative Commons Attribution (CC BY) license https://creativecommons.org/licenses/by/4.0/.

to suboptimal performance in scale-up. Moreover, the rates of biological processes are inherently constrained, and they may struggle to keep pace with the large-scale and rapid reduction of CO_2 emissions required for meaningful environmental impact.

In traditional biocatalysis, redox reactions require two substrates (electron donor and electron acceptor), and the transfer of electrons between them is facilitated by various electron mediators or enzyme-bound cofactors, such as nicotinamide adenine dinucleotide [NAD(H)] and its phosphorylated form [NADP(H)], flavin mononucleotide (FMN), flavin adenine dinucleotide (FAD), heme, and other metal cofactors [17]. Biological processes are often limited due to their dependence on cofactors for redox equivalents. To overcome these limitations, the integration of electrochemistry proves instrumental, with the electrode substituting the second substrate and acting as a source of electrons for the CO_2 reduction process. This approach combines the merits of both biocatalysis and electrocatalysis. Electrochemical reactions can safely provide the necessary redox equivalents for biocatalysis with the consumption of electricity, which can be produced from renewable resources (e.g. solar and wind). For instance, bioelectrochemical regeneration of redox cofactors has been widely employed in CO_2 fixation [19–21]. This breakthrough in cofactor regeneration involves a hybrid system integrating biological or chemical materials with electrochemically regenerated cofactor [22, 23]. Utilizing electric power to replenish redox cofactors or provide a direct electron supply to biocatalysts not only overcomes the limitations of biological CO_2 conversion but also enhances the overall efficiency of the process.

8.2.1 Bioelectrocatalysts for CO_2 Reduction

Bioelectrocatalysts are the pivotal components of bioelectrocatalytic systems, falling into two main categories: enzymes and electroactive microbial cells [24]. Advances in protein and metabolic engineering have empowered the effective regulation and enhancement of these functional cores, thereby significantly improving the overall performance of bioelectrocatalytic systems [25].

The enzymes primarily used in CO_2 reduction belong to the oxidoreductase and carbonic anhydrase families [26]. Enzymes employed in bioelectrocatalysis are categorized based on their cofactors, distinguishing between metalloenzymes and non-metalloenzymes. Metalloenzymes utilize metal cofactors such as heme centers [27], iron centers (Fe), iron–sulfur clusters (Fe-S) [28], copper centers (Cu) [29], molybdenum centers (Mo) [30], and tungsten centers (W) [31]. Non-metalloenzymes, on the other hand, rely on cofactors such as FAD and pyrroloquinoline quinone (PQQ) [26]. These enzymes leverage the transformation between reduced and oxidized states of various cofactors to facilitate electron transfer. They possess an insulating peptide framework housing redox-active cofactors or active sites. The specificity and activity of these biomolecules are dictated by their structure and chemical properties, providing unique chemical microenvironments and structural platforms for the recognition of molecular species. In recent times, the most studied single-enzyme process has been the synthesis of formic acid from CO_2 reduction [32] using formate dehydrogenases (FDHs) along with cofactor NADH, carbon monoxide dehydrogenase (CODH) for CO_2 conversion to CO [33], and a multienzyme approach to transfer electrons in a cascade of reactions to generate biofuel [34].

In contrast to enzymes, electroactive microbial cells offer a broader catalytic spectrum due to their ability to function as miniature, self-reproducing bioreactors [24]. The complex metabolic network within microbial cells results in less substrate specificity compared to isolated enzymes. The metabolic activity of the microbial cell enables the regeneration of reduced equivalents, while unique electron transport mechanisms evolved by electroactive microbial cells facilitate electronic communication with electrodes. The intricate and flexible metabolic pathways in microbial systems allow for less specific but more complex multistep transformations, making them capable of catalyzing processes such as the conversion of CO_2 to alcohols or fatty acids [35, 36]. Although the use of whole-cell microbial catalysts offers some important advantages, it also brings along several metabolism and process-related challenges. For instance, it can result in less productivity due to the mass transport losses during the translocation of redox mediators across the cellular membrane and activation losses during the long electron transfer pathway.

8.2.2 CO_2 Bioelectrocatalytic Pathways

The mechanisms governing the reception of redox equivalents by bioelectrocatalysts are pivotal in the design of effective bioelectrocatalytic systems. The mode of charge transfer dictates how the biocatalyst communicates with the electrode, influencing sensitivity, efficiency, and overall system viability.

As previously discussed in Chapters 1 and 2, direct electron transfer (DET) involves the direct transfer of electrons from the electrode to a redox cofactor through electron tunneling [37]. While DET is extremely fast, it is limited to short-distance electron transfer, typically within 14 Å [38]. Active sites buried within the protein structure are often not capable of directly receiving redox equivalents from the electrode because the distance through the insulating peptide material is too far for electron tunneling to occur. Metal-centered enzymes (W-FDH, molybdenum-dependent FDH (Mo-FDH), etc.) are capable of DET. In microbial DET, microbes engage in extracellular electron transfer (EET) mechanisms for electrical communication with electrodes. Microbes facilitate extracellular DET through direct contact or by forming electron-conductive chains extending from the cell surface, utilizing redox-active species such as c-type cytochromes [39].

Mediated electron transfer (MET) employs a redox molecule (mediator) to transport charge to and from active sites that are inaccessible to DET [40]. Mediators can be freely diffusing or part of a redox polymer, and their redox potentials must closely match those of the active site. Common redox mediators include metal complexes, polymers (e.g. osmium) [40], metallocene molecules (e.g. ferrocene or cobaltocene) [19], quinones [41], phenazines [42], and viologens [43].

Naturally occurring mediators, including hydrogen (H_2), hydrogen peroxide (H_2O_2), and cofactors like NAD(P)H [44–46], often act as co-substrates of enzymes. In the context of CO_2 reduction, NADH and NADPH are relevant examples but present challenges in mediated systems due to their expense and difficult *in situ* regeneration. Efficient regeneration or replacement of these cofactors is crucial for advancing bioelectrosynthesis.

8.2.3 CO_2 Reduction Products and Recent Advances

The choice between enzymatic and microbial bioelectrosynthesis plays a pivotal role in dictating the spectrum of products obtained from CO_2 reduction. Enzymatic systems, often featuring immobilized enzymes on electrodes, offer a controlled and selective approach. Specific enzymatic cascades, such as those involving formate dehydrogenase or nitrogenase, can lead to the formation of distinct carbon products. The localized environment created by immobilized enzymes allows for enhanced selectivity, combating mass transport issues encountered in homogeneous systems.

On the other hand, microbial bioelectrosynthesis, driven by various bacterial species, introduces a broader range of pathways and metabolic capabilities. Microbes can utilize complex pathways like the Calvin-Benson-Bassham cycle, Wood-Ljungdahl pathway, and others, enabling the reduction of CO_2 to multicarbon organic molecules (Figure 8.1). The EET mechanisms employed by microbes, including direct contact with redox proteins or MET, contribute to the versatility of the products formed.

8.2.4 Formic Acid Production: Enzymatic Approach

Formate dehydrogenase (FDH) is an enzyme capable of catalyzing the reversible reduction and oxidation between CO_2 and formic acid. However, FDHs often require an expensive coenzyme (NADH) for the CO_2 fixation reaction, necessitating reduced equivalent supplementation for effective reaction progress. Innovative enzymatic CO_2 reduction to formic acid involved optimizing electrode potential for FDH by Srikanth et al., achieving 12.74% Faradaic efficiency [32]. Reda et al. employed a tungsten-containing FDH adsorbed on glassy carbon, achieving exceptional Faradaic efficiencies of 97% [47]. These advances demonstrate the continual exploration of electrode materials and enzyme immobilization techniques for enhanced performance.

Focusing on the DET pathway and NADH-free enzymatic CO_2 conversion, Kuk et al. used a conductive PANi hydrogel to immobilize FDH (W) from *Clostridium ljungdahlii* (ClFDH), achieving a faradaic efficiency of 92.7% and a rate of $1.42\,\mu mol\,h^{-1}\,cm^{-2}$ at a low potential of −0.6 V (vs. Ag|AgCl) [22]. The Zhang group achieved a faradaic efficiency of 99.3% after the application of −0.6 V (vs. SHE) for two hours, utilizing the same ClFDH enzyme via DET, creating a new benchmark for electroenzymatic CO_2 reduction to formate [31]. Reisner et al. investigated the effects of the local environment on the enzyme using meso-indium tin oxide (ITO) electrodes and W-FDH from *Desulfovibrio vulgaris* Hildenborough (*Dv*H-FDH). They discovered that proper buffer selection was critical, as local pH concentrations may increase by 2 pH units, and determined that using a mixed bicarbonate-MOPS buffer resulted in a twofold increase in current density and produced $21 \pm 0.1\ \mu mol\ cm^{-2}\ h^{-1}$. Further improvements were achieved by a macroporous electrode architecture that improved both catalyst loading and mass transfer, yielding $55 \pm 4.3\ \mu mol\ cm^{-2}\ h^{-1}$. These studies highlight the balance between electrode architecture and the local environment of the enzymatic catalyst and the importance of optimizing these parameters to give improved devices [48].

Investigating a MET pathway, Yuan et al. developed a redox polymer-based system utilizing Mo-FDH embedded in a redox polymer with a poly(allylamine) backbone with cobaltocene pendants. This system, with low redox potential centered at −0.66 V (vs. SHE), efficiently reduced CO_2 to formate with high faradaic efficiency (99%), while simultaneously indicating the potential of the cobaltocene redox polymer for application in H^+ reduction to molecular hydrogen (H_2) and dinitrogen (N_2) reduction to ammonia (NH_3) [19]. Another interesting approach was the use of gas diffusion electrodes with W-dependent formate dehydrogenase by Schuhmann et al. Using a methyl viologen-modified poly(methacrylate)-based redox polymer, the system is capable of direct delivery of the gaseous substrate to the enzyme, avoiding mass transport issues. The electroenzymatic transformation, carried out at a mild potential of −0.59 V (vs. SHE) under ambient temperature, yielded between 1.9 and 2.6 mg L^{-1} after 45 hours of run time in a single-compartment gas diffusion cell, while system activity held at 80% of its initial value. This innovative approach addressed mass transport issues in enzymatic transformations, showcasing the feasibility of using gas diffusion electrodes for enhanced activity [49].

The Reisner group demonstrated a paired photoelectrochemical system that converted polyethylene terephthalate (PET) to glycolic acid while simultaneously reducing CO_2 to formate. Lead halide perovskites served as a photocathode, achieving simultaneous conversion of waste PET to glycolic acid (a feedstock within the pharmaceutical industry) and CO_2 to formate with a yield of $121 \pm 87\ \mu mol\ cm^{-2}$ after 10 hours with a 95% faradaic efficiency [50].

8.2.5 Acetate and Butyrate Production: MES

MES has been instrumental in producing acetate, with certain bacteria such as *Clostridium ljungdahlii* and *Sporomusa* spp., utilizing electrons from graphite electrodes to efficiently reduce CO_2 to acetate and other by-products. Strategies involving selective enrichment and acclimation to CO_2 electroreduction have achieved significant acetate production rates [51–53]. Chong Liu's group developed a novel biological–inorganic hybrid system utilizing a cobalt phosphate anode to split O_2, producing H_2 at a cobalt–phosphorous alloy cathode, which is selectively consumed by the hydrogenase of *Sporomusa ovata* and powering the reduction of CO_2 to acetate. They addressed limitations in acetate production by introducing a biocompatible perfluorocarbon nanoemulsion as an H_2 carrier, resulting in a 190% increase in acetate production. The average acetate yield of 6.4 g L^{-1} was achieved in four days with close to 100% Faradaic efficiency [54]. Ganigué and coworkers demonstrated butyrate production from CO_2, using a two-chambered H-type bioelectrochemical system with a cathode made of commercial carbon cloth and an enriched carboxydotrophic mixed culture from a syngas fermenting reactor, producing high concentrations of butyrate (20.2 mM), offering insights into the potential of bioelectrochemical systems for producing longer-chain organic compounds [55].

8.2.6 Alcohol Production

Bajracharya's group utilized a mixed microorganism culture to produce ethanol through CO_2 reduction. The cathode potential was adjusted to shift the reaction from acetate to ethanol production. This hybrid inorganic–biological system can potentially produce various multi-carbon alcohols through the metabolism of different microbial cells [53]. Torella and coworkers achieved isopropanol production using a triple-junction amorphous silicon solar cell, showcasing the synergy of bioelectrocatalysis and metabolic engineering. CO_2 was continuously sparged into the engineered *Ralstonia eutropha* cell in which the polyhydroxybutyrate synthesis pathway is disrupted, and four genes were introduced into the cell to construct a pathway to redirect acetyl-CoA toward the synthesis of isopropanol. In a subsequent study, they replaced the cathode material with an ROS-resistant cobalt–phosphorus alloy, enhancing isopropanol production. These advancements represent novel strategies for efficiently coupling electrochemical processes with metabolic engineering for alcohol production [56].

In 2017, the Lee and Park groups developed a photoelectrochemical cell to reduce CO_2 to methanol [57]. Using an enzyme cascade of NADH-dependent dehydrogenase enzymes, FDH-FaldDH-ADH, the sequential reduction of CO_2 to formate, to formaldehyde, and finally, methanol was produced at a rate of $220\,\mu M\,h^{-1}$. Reduction of formate to formaldehyde is the rate-limiting step due to the low activity of FaldDH. NADH mediation in this dehydrogenase cascade was achieved via NADH regeneration at a bismuth ferrite photocathode with a rhodium catalyst. This integrated system highlights the potential of coupling enzymatic cascades with photoelectrochemical processes (Figure 8.2).

8.2.7 Olefin and Methane Production: MES and Nitrogenases

Some studies show that nitrogenases, including MoFe, FeFe, and VFe nitrogenases, can also perform CO_2 fixation. Previous studies had demonstrated the potential of both the MoFe- and FeFe-nitrogenases to drive CO_2 reduction to formic acid when paired with cobaltocene redox polymer [59]. The Minteer group investigated the ability of nitrogenases to produce olefins using vanadium nitrogenase (VFe) originating from *Azotobacter vinelandii* [20]. Employing two cobaltocene electron mediators (1,1′-dicarboxy-cobaltocenium and 1-carboxy-cobaltocenium), this reaction yielded 25 ± 5 nmol ethylene (C_2H_4) and 42 ± 7 nmol propene (C_3H_6) per µmol VFe after application of −0.86 V (vs. SHE). The electroenzymatic C—C bond formation from CO_2 catalyzed by VFe nitrogenase bypasses the requirement of CO as the substrate for the production of olefin.

Nichols and coworkers developed a nanoparticle-based nickel sulfide hydrogen-evolving cathode for methane production from CO_2. The system, using methanogenic bacterium *Methanosarcina barkeri* as the biocatalyst for CO_2 fixation, achieved significant CH_4 production (4.3 mmol) with high Faradaic efficiency (74%), establishing a synergistic coupling of inorganic and biological components for solar-driven CH_4 generation [60].

8.2.8 Bioelectrochemical CO_2 Reduction to Complex Carbon-Based Products

Plumeré et al. reported a stereoselective CO_2 fixation to crotonyl-CoA using a viologen-based redox-active hydrogel, co-immobilizing ferredoxin $NADP^+$ reductase and crotonyl-CoA carboxylase/reductase [58]. This system demonstrated significant activity and Faradaic efficiency (91%) with a yield of 37% on application of −0.575 V (vs. SHE) for 25 hours, showcasing the potential of enzymatic cascades for producing complex carbon-based products.

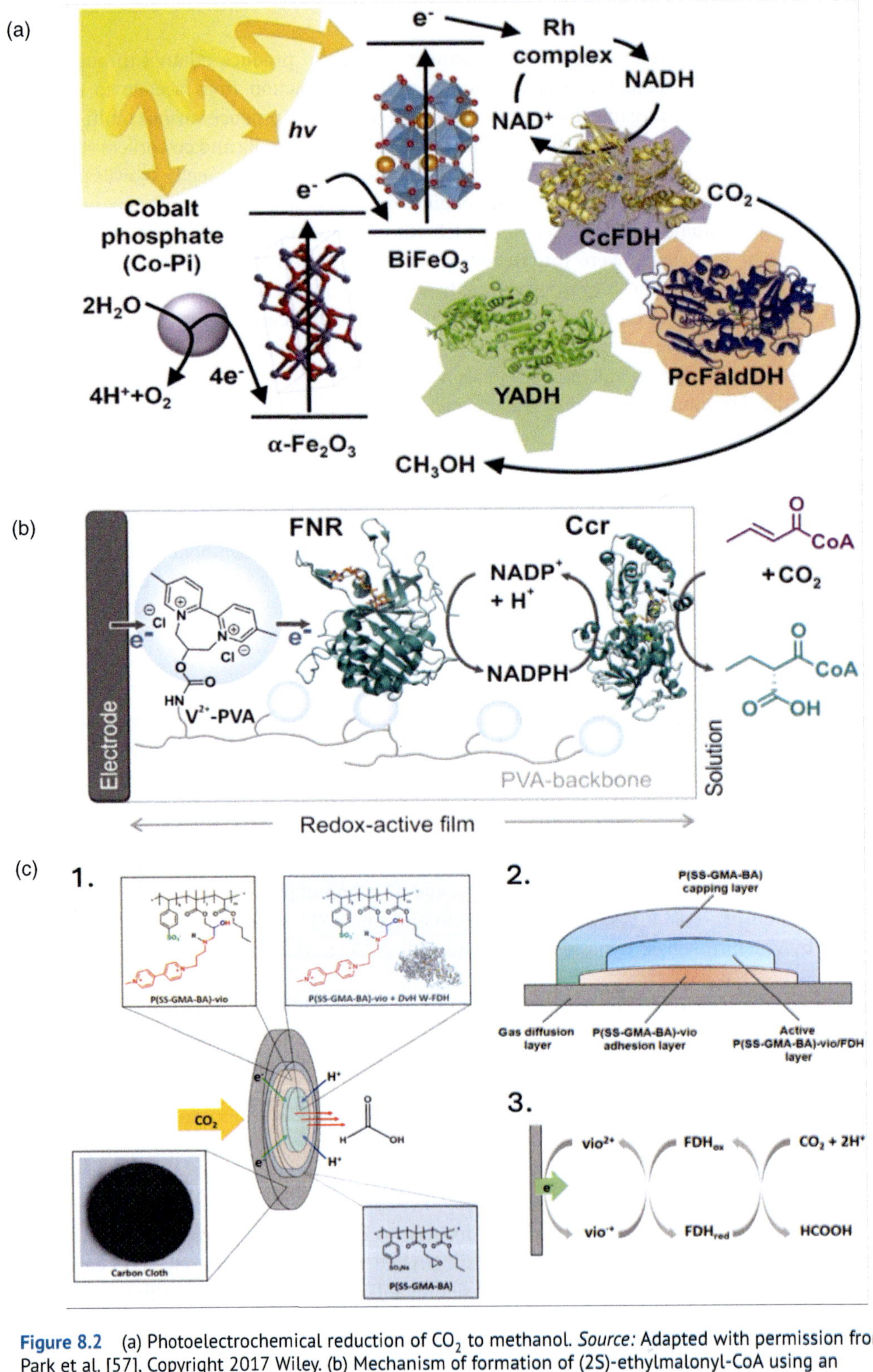

Figure 8.2 (a) Photoelectrochemical reduction of CO_2 to methanol. *Source:* Adapted with permission from Park et al. [57], Copyright 2017 Wiley. (b) Mechanism of formation of (2S)-ethylmalonyl-CoA using an electroenzymatic cascade. *Source:* Adapted with permission from Castañeda-Losada et al. [58], Copyright Wiley 2021. (c) polymer/enzyme-based gas diffusion layer for electrochemical CO_2 fixation. *Source:* Adapted with permission from Szczesny et al. [49], Copyright 2019 American Chemical Society.

Armstrong et al. demonstrated an ITO-Ti electrode containing nanoconfined enzyme cascades for CO_2 fixation to pyruvate. The system, integrated into a solar water splitting/CO_2 reduction device, achieved high Faradaic efficiency (70%) and low potential (−0.5 V vs. SHE) [61]. This innovative approach offers a solution to mass transport issues, enhancing enzymatic cascade efficiency.

8.3 Bioelectrocatalytic Production of H_2

8.3.1 Introduction

Molecular hydrogen is considered the ideal fuel from the environmental point of view because its oxidation only produces water. However, H_2 is a very scarce gas on our planet and thus has to be synthesized for its industrial applications. Currently, most of the H_2 used is still obtained from fossil fuels [62]. Therefore, the goal for sustainable use of H_2 as an energy vector is to obtain it from renewable sources, such as water or biomass. In particular, the electrocatalytic production of H_2 is a very attractive strategy because it can use electricity obtained from clean energy devices, such as solar panels or wind turbines, for reducing protons in an aqueous medium. In this way, the energy obtained from natural and abundant sources, such as the sun, wind, and water, is stored in the form of chemical energy that can be converted again to electrical power when required by electrochemical oxidation of H_2 in fuel cells [63]. Thus, it is in theory a perfectly sustainable process. However, the main drawback of electrochemical reduction of protons to H_2 is that the more efficient electrocatalysts for this reaction are based on Pt, a noble metal that is scarce and expensive [63]. Biocatalysts are then an alternative to noble metals for obtaining H_2 from water or biomass, either by their direct use [64, 65] or as inspiration for synthesizing biomimetic electrocatalysts based on abundant metals [66, 67]. In bioelectrocatalytic systems, we can differentiate between those using whole cells and those based on purified enzymes, each of them having advantages and disadvantages. We review both types of bioelectrocatalytic systems applied to H_2 synthesis.

8.3.2 Electroenzymatic Production of H_2

Hydrogenases are the enzymes that specifically catalyze the reversible reaction of H_2 production/uptake with almost no overpotential [68]. Hydrogenases are classified according to the metal content of their active site; thus, they either belong to the FeFe-hydrogenases, NiFe-hydrogenases, or Fe-hydrogenases groups [69]. These metals are abundant in nature; thus, much research has been devoted to synthesizing biomimetic electrocatalysts based on Fe or Ni complexes for both oxidation and production of H_2. Hydrogenases have the advantages of their specificity, high turnover, and low overpotential under mild conditions. Their main disadvantage is their O_2 sensitivity, which differs between the different types of hydrogenases. The more O_2-tolerant membrane-bound NiFe-hydrogenases are catalytically biased toward H_2-oxidation and are only capable of H_2 production at quite acidic pH values [70]. A completely aerobic hydrogenase has been reported recently, but its catalytic activity is very low. On the other hand, FeFe-hydrogenases have very high activity for H_2 production at neutral pH, and contrary to NiFe-hydrogenases, they are hardly inhibited by product accumulation [71, 72]. However, their main weakness is the sensitivity to light and O_2, both of which cause their irreversible inhibition [73, 74].

Since the 1980s of the last century, many reports of DET of hydrogenases on electrodes have been published [64, 75]. DET is possible because all hydrogenases have a redox relay formed by a sequence of iron–sulfur clusters that connects the protein surface with the active site, allowing for fast intramolecular electron transfer [69]. If the most exposed iron–sulfur cluster faces the electrode surface, then it can sequentially receive the two electrons required at the active site for forming an H_2 molecule, which is then released to the medium by the protein's hydrophobic channel [76]. There are two strategies to enhance DET of hydrogenases on an electrode: (i) chemical modification of the electrode surface to favor oriented immobilization of the hydrogenase with its distal redox site facing the surface by electrostatic interactions [77]; (ii) immobilization on carbon materials with high roughness or modified with conductive nanoentities, to obtain a high number of enzyme molecules with their distal redox near the electrode due to the roughness of the surface at the nanometric scale [78]. The immobilization of hydrogenase on nanostructured electrodes has the advantage of increasing the number of enzyme molecules per geometric area able to establish efficient DET [79]. Kihara et al. reported the immobilization of *T. roseopersicina* Ni-Fe hydrogenase on an electrode modified with a single-walled carbon nanotube (SWCNT) forest. The vertically aligned SWCNTs allowed stable confinement of the hydrogenase, but the current measured at −0.5 V vs. *S*HE was limited to $0.13\,mA \times cm^{-2}$ with the faradaic yield of H_2 production at 32% [80]. Higher DET-based currents and faradaic yields near 100% were obtained when immobilizing Fe-Fe hydrogenases directly adsorbed on nanostructured TiO_2 [81] or ITO electrodes [82]. Milton and coworkers have reported electrocatalytic current densities of H_2 production around $-8\,mA \times cm^{-2}$ at −0.8 V vs. *S*HE with 100% faradaic yield and great stability. Ninety-two percent of the initial electrocatalytic current was obtained after 120 hours of continuous operation [82].

An alternative to DET-based electroenzymatic H_2 production is the MET-based approach, in which the hydrogenase is entrapped within redox polymers of low redox potential. The main advantages are that a higher number of electroactive enzyme molecules can be immobilized on the electrode, as multilayers of hydrogenase are wired by the redox moieties of the polymers, and their higher stability compared to direct adsorption on the electrode surface due to the entrapment within the polymer structure. However, the potential of the redox polymer has to be low enough to drive H_2 production, as is the case with cobaltocene [83] and some viologen-based polymers [84]. Very high catalytic currents of H_2 production ($6.2\,mA \times cm^{-2}$ at −0.62 V vs. SHE) with almost 100% faradaic yield have been reported by Plumeré and coworkers by immobilizing the Fe-Fe hydrogenase from *Desulfovibrio desulfuricans* integrated into a 2,2′-viologen-based redox hydrogel [84]. An additional advantage of viologen-based redox polymers is that their redox groups can scavenge O_2 in the medium by its reduction, thus protecting the hydrogen from inhibition, although at the expense of decreasing the faradaic yield [85].

Recent strategies studied for increasing the stability of hydrogenases toward the presence of O_2, which is one of the main drawbacks for applications of this type of enzyme, are based on genetic engineering. Rutz et al. have replaced a glutamate residue with a methionine to enhance the structural flexibility of the Fe-Fe hydrogenase from *Clostridium beijerinckii*, causing an increased resistance of the enzyme to irreversible inhibition by oxygen, as it favors fast formation of the inactive oxygen-protected state, which can be reactivated at low redox potentials [86]. Fasano et al. have studied the oxygen tolerance of NiFe hydrogenase chimeras formed by combining subunits from the more O_2-tolerant Hyd 1 with the O_2-sensitive Hyd 2, both from *Escherichia coli* [87]. A completely aerobic hydrogenase has been discovered recently from *Mycobacterium smegmatis*. However, it has the disadvantage of very low catalytic activity [88].

Photoelectrocatalytic systems take advantage of light energy to decrease the driving force required to reduce protons to hydrogen. A photosensitive compound is required to excite low-energy electrons to high-energy ones, which are then transferred to the catalyst for producing hydrogen. This concept has been applied by combining photosystem I (PSI) as a biological photochemical complex with hydrogenase as a biocatalyst. Tapia et al. were the first to implement this strategy for photoelectrocatalytic hydrogen production by immobilizing both biological elements on a glassy carbon electrode and wiring them with two redox polymers: an Os complex-based polymer with positive redox potential for transferring electrons from the electrode to PSI and a cobaltocene-based polymer with negative redox potential to transfer the excited electrons from PSI to *D. gigas* NiFe-hydrogenase [89]. Conzuelo, Schuhmann, and coworkers optimized this strategy by oriented immobilization of a PSI monolayer on the electrode surface [90] and later by layer-by-layer deposition of PSI and Ni-Fe-Se hydrogenase from *D. vulgaris* [91].

Semi-artificial photoelectrocatalytic hydrogen production systems directly connect the hydrogenase to inorganic p-type semiconductors. Several materials capable of absorbing visible light have been tested as support for hydrogenase immobilization, and their application as photocathodes of hydrogen production has been studied. A layer-by-layer combination of two p-type semiconductors, ZnO and Cu_2O, has been reported as a successful configuration for photobioelectrochemical H_2 production. Cu_2O is capable of harvesting light energy within the visible range but is prone to oxidation, whereas ZnO has a wide bandgap that allows only UV light absorption. The layer-by-layer deposition of both oxides allows both visible light absorption and protection against corrosion, as well as hosting hydrogenases to catalyze the hydrogen production. The photobiocathode produced $-0.8\,mA \times cm^{-2}$ upon illumination with $100\,mW\,cm^{-2}$ visible light, although the faradaic efficiency was only 1% [92].

Sulfides have also been studied as semiconductors for photobioelectrochemical H_2 production. In_2S_3 has been shown to be a suitable material for Ni-Fe-Se hydrogenase immobilization and their electronic coupling [93], but it is an n-type semiconductor and thus only adequate for developing photoanodes. Luna-López et al. reported a photobiocathode based on a p-n junction using In_2S_3 and NiO; the former allowed the coupling with the hydrogenase activity of H_2 production, while the latter allowed fast hole transport to the electrode to outcompete with hole–electron recombination after photochemical excitation [94]. Nevertheless, higher photoelectrocatalytic currents were measured with the same Ni-Fe-Se hydrogenase when adsorbed on an electrode modified with electrodeposited $CuGaS_2$, which is a p-type semiconductor [95].

Nanoporous p-type silicon has allowed a high loading of adsorbed Fe-Fe hydrogenase and afforded an onset potential of 0 V *vs.* SHE at pH 6.8 and a turnover frequency of H_2 production of 1300 $second^{-1}$ for a photocurrent density of $3\,mA \times cm^{-2}$ at −0.3 V vs. SHE under $50\,mW \times cm^{-2}$ irradiation of visible light [96]. Silicon can also be covered with a protective layer of nanostructured TiO_2, which can act as support for hydrogenase immobilization for H_2 evolution. The p-Si-TiO_2-hydrogenase photocathode produced H_2 with an onset potential of +0.35 V vs. RHE and a faradaic yield of 95% [97]. Reisner's group has more recently established the benchmark for photoelectroenzymatic H_2 production of $-5\,mA \times cm^{-2}$ at 0 V vs. RHE by integrating the TiO_2-hydrogenase construction in a lead halide perovskite photocathode [98].

Further improvements required to implement hydrogenase-based cathodes as a viable strategy of electrocatalytic or photoelectrocatalytic H_2 production are to increase their O_2 tolerance and to develop methods for their mass production. An interesting approach is the use of genetic engineering for rapid and robust production of recombinant of O_2-tolerant *D. baculatum* Ni-Fe-Se hydrogenase [99]. Other recent strategies include the immobilization of the O_2-tolerant Ni-Fe hydrogenase from *Cupriavidus necator* on tin-rich indium oxide, which alters the catalytic bias of the enzyme from H_2 oxidation to proton reduction [100], and the construction of chimera Ni-Fe hydrogenases from *E. coli* to tune both the catalytic bias and oxygen tolerance [87].

8.3.3 Microbial Production of H_2

Microbial cells can also be used as electrocatalysts for H_2 production. Since the beginning of this century, it was discovered that some bacteria were able to exchange electrons directly with electrodes; thus, the number of articles on microbial electrochemical cells has experienced a significant increase [101–103]. The main advantages of the use of enzymes, which normally require lengthy and expensive processes to be isolated and are less stable *in vitro* than *in vivo*, are that microbial cells are much more robust, are capable of growth and self-repair, and the corresponding bioprocesses can be easily scaled up. On the other hand, the main disadvantages of microbial electrochemical systems are a lower coverage of catalytic sites per electrode area, high kinetic barriers for electron exchange between the cells and the electrode due to the presence of the cellular membrane, and diffusional constraints of substrates and products through the biofilms formed on the electrode surface. These disadvantages lead to lower current densities measured with microbial cathodes than with enzymatic biocathodes [104].

In the literature, microbial electrolysis cells (MECs) are defined as those in which at least one of the electrodes is biotic. Typically, all MECs have in the anode exoelectrogenic bacteria that oxidize organic substrates, such as acetate, glucose, lactate, etc. The anodic bacteria transfer the generated electrons to the electrode surface, either via outer-membrane cytochromes, conductive pili, or self-produced redox mediators [105]. In this chapter, we will only consider those MECs that have as a cathodic electrocatalyst a pure or mixed bacterial culture.

Some MECs use pure bacterial cultures as biocathodes for H_2 production, predominating species that contain highly active hydrogenases, such as *Desulfovibrio* ones [65]. Perona-Vico et al. studied the electrocatalytic H_2 production by biofilms grown on cathodes from pure cultures of 10 different strains from *Rhodobacter*, *Rhodopseudomonas*, *Rhodocyclus*, *Desulfovibrio*, and *Sporomusa*. The highest H_2 production rate corresponded to *D. paquesii*, reaching $46 \pm 20\,\mu mol\ H_2\ min^{-1}$ with a current density of $0.68\,mA\,cm^{-2}$ at 0 V vs. SHE [106]. Other exoelectrogenic bacteria, such as *Geobacter sulfurreducens* [107] and *Shewanella oneidensis* [108], which are widely used in bioanodes of microbial electrochemical systems, give much lower H_2 production rates when used as pure cultures on biocathodes.

The use of mixed cultures has some advantages over the use of pure cultures in MECs. The preparation and operation of the MEC is simplified as it does not require a specific medium or growth conditions. Furthermore, syntrophic relationships between the bacterial consortium may enhance the H_2 production rate, although other products can be obtained [65]. The microbial consortia can be established starting from different sources, such as sediments [109], wastewater [110], or from the inoculum of microbial fuel cell anodes. In the latter case, Dai et al. reported an inoculum of sulfate-reducing bacteria that was first used in a microbial fuel cell configuration, and then, by reversing the polarity, the bioanode was changed into a biocathode for H_2 production [111]. Common species found in microbial consortium used on biocathodes are *Desulfovibrio* sp., *Clostridium* sp., *E. coli*, *S. oneidensis*, *G. sulfurreducens*, and *Pseudomonas aeruginosa* [112, 113].

8.4 Bioelectrosynthesis of Ammonia

Ammonia is a product of high added value for the food, chemical, and pharmaceutical industries [114, 115]. Nowadays, most of the ammonia production (~90%) is carried out by the Haber-Bosch process that was devised at the beginning of the twentieth century [116]. Despite being quite an efficient process—mainly due to the enthalpy of dissociation of the N_2 triple bond $+945\,kJ\,mol^{-1}$, it requires high temperature (~700 K) and high pressure (above 100 atm) [114, 117]. As a result, this process accounts for approximately 1% of the annual energy consumed in the world. Moreover, this process uses metal catalysts and requires N_2 and H_2 as substrates, the latter usually coming from fossil fuels [114, 116]. These demands in energy and raw materials make this industrial process responsible for approximately 1–3% of the total CO_2 emitted to the atmosphere in a year [114–119].

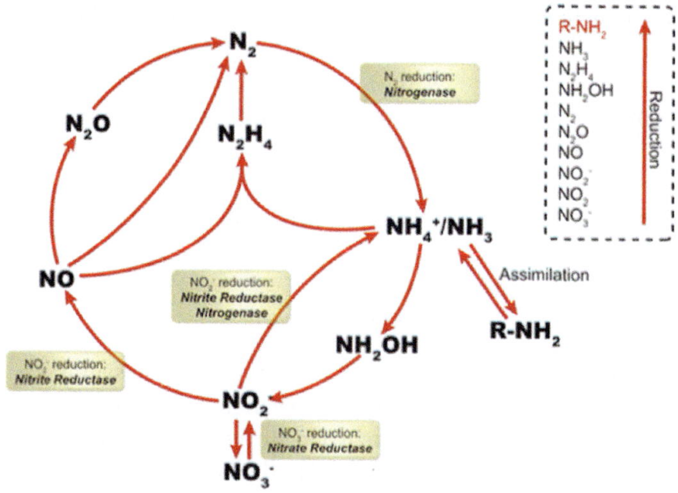

Figure 8.3 Scheme of the nitrogen cycle. *Source:* Adapted with permission from Stein and Klotz [120], Copyright 2016 Elsevier.

It is of utmost importance to obtain NH_3 *via* more environmentally friendly synthetic routes that, although a full replacement of the Haber-Bosch process is unbearable in the short and mid-term, can offer an alternative or a complement.

Ammonia is an essential molecule for life and a precursor of key biomolecules (proteins, nucleic acids, etc.) [118]. Only one enzyme is known to carry out the reduction reaction of atmospheric N_2 to NH_3 (NH_4^+ at physiological pH); this enzyme is nitrogenase, present in certain archaea, bacteria, and some plants in symbiosis with these prokaryotic microorganisms, also known as diazotrophic microorganisms [118]. This N_2 reduction is carried out under mild conditions of near-ambient temperature and pressure; in nature, this fixation of atmospheric nitrogen is part of the nitrogen cycle (Figure 8.3, right). Both sources of ammonia synthesis, industrial and by microorganisms, account for 50% of total atmospheric nitrogen fixation each [121].

As mentioned above, the only known biological fixation of N_2 in nature goes through the enzyme nitrogenase, so insight into its structure and mechanism is of vital importance. Nitrogenase is an oxidoreductase enzyme that catalyzes the reduction of N_2 to NH_3 using adenosine triphosphate (ATP) as an energy source. This enzyme works at near atmospheric temperature and pressure; on the other hand, its kinetics is slow, and it is sensitive to oxygen, becoming permanently inactivated in the presence of oxygen within a few minutes [30, 122].

Nitrogenase is a complex system formed by two proteins: (i) a reductase protein (Fe protein), which has the function of transferring electrons from a donor to the second protein (in nature those electron donors are flavin and ferredoxins that supply low-potential electrons to the reductase, but the reductase protein can also take electrons from another protein, a molecule, or an electrode, and (ii) a catalytic protein containing the active metal center, where the reduction of N_2 to NH_3 takes place. In nature, nitrogenase appears as three isozymes whose main difference lies in these active centers, where in each of them, three different metals can be found. Nitrogenases are classified as follows:

- Nitrogenases of molybdenum iron (MoFe), whose active center has Mo as the metal.
- Nitrogenases of vanadium iron (VFe), whose active center has V as the metal.
- Nitrogenases of iron (FeFe), in which only Fe is found in its active center.

The stoichiometry of the N_2 to NH_3 reduction reaction is different in each nitrogenase [123, 124].

FeFe nitrogenase:

$$N_2 + 40\,MgATP + 20e^- + 20H^+ \rightarrow 2\,NH_3 + 7\,H_2 + v40\,MgADP + 40\,Pi \quad (8.3)$$

VFe nitrogenase:

$$N_2 + 24\,MgATP + 12e^- + 12\,H^+ \rightarrow 2\,NH_3 + 3\,H_2 + 24\,MgADP + 24\,Pi \quad (8.4)$$

MoFe nitrogenase:

$$N_2 + 16\,MgATP + 8\,e^- + 8\,H^+ \rightarrow 2\,NH_3 + H_2 + 16\,MgADP + 16\,Pi \quad (8.5)$$

All diazotrophs can synthesize Mo nitrogenase, and only some of them are able to synthesize the other two. VFe and FeFe nitrogenases are less efficient for this reaction, and thus they are considered alternative nitrogenases that contribute to N_2 fixation when Mo (essential for the active center of the other "main" nitrogenase) is scarce in the environment [125]. The stoichiometry also shows that hydrogen production, which is mandatory to balance the reaction, is higher for the alternative enzymes. In addition, nitrogenase is also capable of reducing CO_2, C_2H_2, N_3, HCN, NO_2, N_2H_4, CO, R-CN, and R-NC, as well as producing CH_4 in the case of FeFe nitrogenase [126].

The reductase is a protein consisting of two identical subunits forming a homodimer of about ~66 kDa, with a binding site for MgATP and MgADP in each of the subunits and between the two subunits. This protein has a sulfoferric cluster 4Fe-4S (Figure 8.4) [127] and transfers electrons one at a time to the MoFe protein in dissociation-association events, consuming two molecules of ATP for each event [128]. The structure of the second protein (MoFe, VFe, or FeFe) is different among the nitrogenases, with VFe and FeFe nitrogenases having two more subunits. MoFe nitrogenase forms a heterotetramer composed of two α-subunits and two β-subunits structured in a 2 αβ shape; it has a molecular weight of about ~240 kDa, and it is where the reduction of N_2 to NH_3 takes place. This protein has four metal centers, two of which are called P clusters [8Fe-7S]. These P clusters are the sites where the electrons coming from the reductase are received, and the electron transfer takes place with a rate constant of about 13 second^{-1} [122]. The P clusters are embedded between

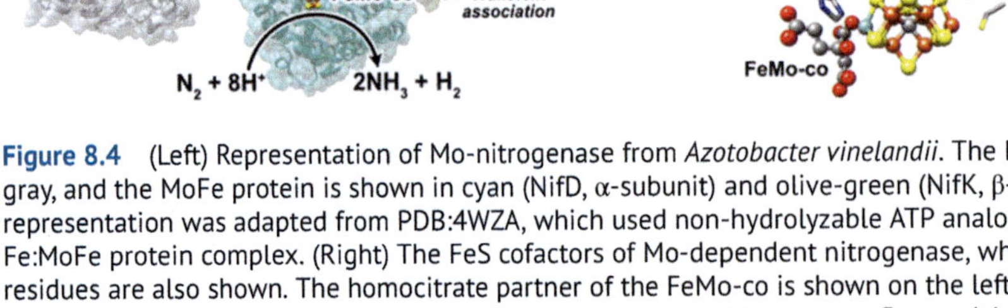

Figure 8.4 (Left) Representation of Mo-nitrogenase from *Azotobacter vinelandii*. The Fe protein is shown in gray, and the MoFe protein is shown in cyan (NifD, α-subunit) and olive-green (NifK, β-subunit). This representation was adapted from PDB:4WZA, which used non-hydrolyzable ATP analogues to form a tight Fe:MoFe protein complex. (Right) The FeS cofactors of Mo-dependent nitrogenase, where the coordinating residues are also shown. The homocitrate partner of the FeMo-co is shown on the left of the FeMo-co. Fe = rust, S = yellow, C = gray, N = blue, O = red, Mo = cyan, Mg = green. *Source:* Adapted from Gu and Milton [121], Copyright 2020 The Authors Creative Commons Attribution (CC-BY) license (http://creativecommons.org/licenses/by/4.0/).

the α and β units (Figure 8.4). The other two metal centers are responsible for the reduction to ammonia. [7Fe-Mo-9S-C-homocitrate] FeMo-cofactor (FeMo-co) receives the electrons from the P clusters and is located in the α-subunits (Figure 8.4) [119].

The mechanism behind the reduction of nitrogen to ammonia by nitrogenase has been a matter of research for decades, with the most widely studied nitrogenase being the MoFe from *Azotobacter vinelandii*. The sequence of major events in the catalytic cycle of this nitrogenase comprises (i) the binding event, (ii) the transfer of electrons from the reductase to the FeMo protein, (iii) hydrolysis of MgATP to MgADP, (iv) the release of Pi and the dissociation of the reductase protein from the FeMo protein, with the release of the Pi being the limiting step in the cycle [122]. As for the electronic transfer events, first, the transfer from the P cluster to the FeMo-co takes place, then the electron transfer between the 4Fe-4S reductase and the P cluster occurs, which is independent of the hydrolysis of MgATP, and this process is called "deficit spending" [129, 130].

The reduction mechanism of nitrogenase comprises eight steps. Six electrons and six protons are needed to generate two molecules of NH_3 and a H_2 molecule. H_2 production takes place during step four, which is a crossroads because the reaction can either continue toward the binding and reduction of N_2 or, if the necessary conditions are not met, return to the initial step, producing more H_2 (Figure 8.5) [131–133].

In vivo, the nitrogenase enzyme relies on the molecular machinery of the cell to overcome the obstacles of stability. The cell is able to replenish and generate alternative nitrogenases and provide the ATP needed for consumption. Inside the cell, the O_2 sensitivity is palliated by the oxidative phosphorylation; besides, when the O_2 concentration starts to be elevated, nitrogenase forms a reversible complex with a protein called FeSII, or "Shethna." This protein is a ~13 kDa ferredoxin [2Fe-2S] containing a [2Fe-2S] cluster that *in vivo* is in a homodimer form of ~26 kDa. When O_2 concentrations reach approximately 2%, FeSII undergoes a conformational change to its "open" state, forming a catalytically inactive triprotein complex with nitrogenase, which protects the two proteins (reductase and MoFe) from oxygen. The interactions between the FeSII protein and nitrogenase are electrostatic; when O_2 levels drop, the triprotein complex dissociates, FeSII returns to its "closed" state, and nitrogenase regains its activity [134]. Furthermore, this increase can stimulate the expression of nitrogenase and N_2 fixation [135]. When the enzyme is purified without the cellular machinery at work, it is necessary to devise new strategies to counteract these drawbacks.

Having briefly described this complex enzyme, we will focus on the bioelectrosynthesis of NH_3, one of the possible alternatives to current industrial synthesis. For this alternative production route, nature is used as a source of inspiration and raw material, which, as mentioned at the beginning, is responsible for approximately 50% of the amount of N_2 fixed in the world. Bioelectrocatalysis for

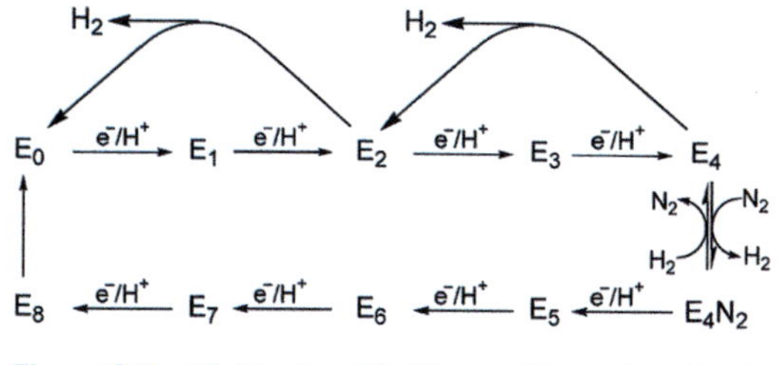

Figure 8.5 Highly simplified Lowe–Thorneley kinetic scheme, highlighting (i) the correlated electron/proton delivery in eight steps, (ii) some of the possible pathways for decay by H_2 release, and (iii) N_2 binding and H_2 release at the E4 level. LT also denotes the protons added to FeMo-co (e.g. E1H1); for clarity, we have omitted this. *Source:* Reprinted with permission from Hoffman et al. [131], Copyright 2013 American Chemical Society.

the synthesis of products of interest is relatively new because many of the applications for which bioelectrosynthesis can be used are reductive processes, which are a challenge due to the mediating systems for extreme reducing potentials (beyond the reduction of oxygen) being unstable [130].

One approach for the bioelectrosynthesis of ammonia is to make direct use of diazotrophic microorganisms for NH_3 production. This approach uses all the available metabolic machinery of the bacteria, allowing greater stability and protection of the enzyme from O_2. However, on the other hand, it is more expensive to extract pure ammonia from the bacterial population. This requires ammonia assimilation inhibitors, such as methionine sulfoximine and phosphinothricin, which slow down the incorporation of ammonia into amino acids in the glutamine synthetase and glutamate synthase pathway and thus help to produce extracellular ammonia. Another complementary way is to eliminate the glutamine synthetase and glutamate synthase genes or limit their expression [136].

Studies using this approach for NH_3 bioelectrosynthesis focus on microbial fuel cells, taking advantage of other microorganisms to synthesize or reduce other products of interest, such as H_2 production or CO_2 reduction. In a work by Chong Liu et al., a bioelectrochemical system was designed for the formation of NH_3 from N_2 and H_2O at ambient conditions in a single reactor by coupling hydrogen generation from catalytic water splitting to an H_2-oxidizing bacterium, *Xanthobacter autotrophicus*, which performs the reduction of N_2 and CO_2 to solid biomass, obtaining free NH_3 after adding a glutamate synthetase inhibitor [137]. In subsequent studies, the system was improved by the group by constructing silicon-based microwire array electrodes to mimic a symbiotic root nodule associative diazotroph and achieved 6.6 mg N_2 per gram dry biomass per hour of free ammonia. This is two orders of magnitude more than their natural counterparts [138].

Another option to achieve extracellular ammonium in the absence of assimilatory inhibitors or genetic manipulation was demonstrated [139]. Using an electrode as the sole electron donor and *Pseudomonas stutzeri* A1501 (as a cathodic catalyst). *P. stutzeri* produced $2.32 \pm 0.25\ mg \cdot L^{-1}$ of extracellular NH_4^+ at a potential of −300 mV vs. SHE without the addition of inhibitors or expensive catalysts. Endogenous excretion of phenazine-1-carboxylic acid by *P. stutzeri* played a key role in EET, mediating electron transfer from the electrode to the inner membrane. Next, the proton motive force moves electrons into the respiratory chain with complex III. These are used to synthesize reduced cofactors (NADH and $FADH_2$) to transfer energy. Finally, they are used by nitrogenase to reduce N_2 to NH_3.

A slightly different approach using bacteria for ammonia production was performed by Bo Wang et al. In this study, biocompatible cadmium sulfide nanoparticles were used to coat the membrane of a photoheterotrophic bacterium, *Rhodopseudomonas palustris*; the photoinduced electrons from the CdS nanoparticles significantly enhanced biological nitrogen fixation in pure cultures of *R. palustris* (Figure 8.6a) [140]. Using the biohybrid cells, an increase in nitrogenase activity and an additional 153% of solid biomass, with a photosynthetic efficiency of 6.73% was observed. Following the same line as this work, but this time starting from the non-photosynthetic bacterium *Azotobacter chroococcum*, a photosynthetic biohybrid based on a conductive polymer was constructed to enhance biological nitrogen fixation by increasing nitrogenase activity [141]. The light-scavenging cationic poly(fluorene-alt-phenylene) was electrostatically bound to the bacterial surface, which facilitated electron transfer to the bacterium (Figure 8.6b), promoting the nitrogen fixation pathway through redox proteins on the bacterial surface when under illumination. With this system, an increase in nitrogenase activity by 260% and an increase in NH_4^+ production by 44% were recorded.

Another approach to bioelectrosynthesizing NH_3 is the direct use of purified nitrogenase enzyme for this purpose. Practically, all the studies carried out on ammonia bioelectrosynthesis have used

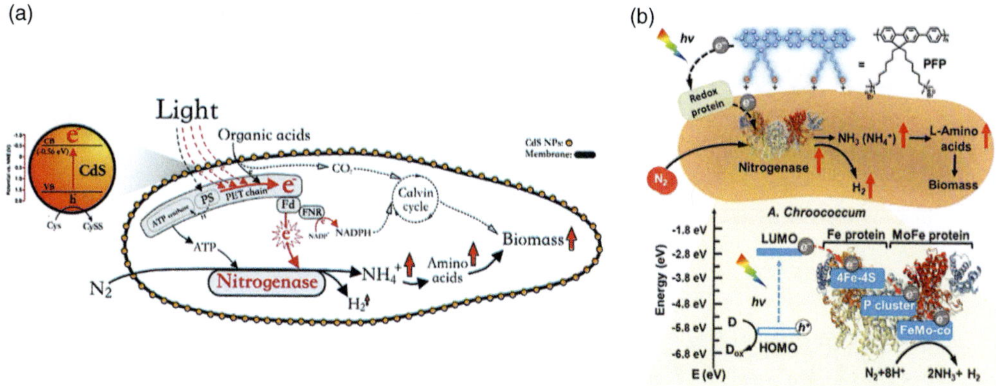

Figure 8.6 (a) Schematic illustration of enhanced N_2 fixation and S2C conversion in the CdS-coated *R. palustris* cell (PS: photosystem; PET chain: photosynthetic electron transfer chain; Fd: ferredoxin; FNR: ferredoxin–NADP$^+$ oxidoreductase; Cys: cysteine; CySS: cystine). *Source:* Adapted from Wang et al. [140]; with permission from Energy & Environmental Science, Copyright 2019. (b) Schematic diagram of PFP/*A. Chroococcum* photosynthetic hybrid biosystems and how the photoexcited electrons are generated from PFP under irradiation. *Source:* Adapted from Zeng et al. [141], Copyright 2023 John Wiley and Sons.

the MoFe nitrogenase enzyme, since it is the most efficient enzyme for this purpose. This type of approach, like the previous one, has its advantages, such as the ease of extraction of the pure product and greater simplicity than a system with interactions among living organisms, and as disadvantages, the purification of the enzyme, its lower stability, and its greater sensitivity to oxygen.

In this type of system, nitrogenase interconnects with electrodes to exchange the electrons needed for the bioelectrosynthesis of ammonia. The interconnection between the electrode and nitrogenase can occur in either DET [142] or MET [143], analogously to other enzymes already mentioned previously in this chapter, in a direct way, as DET. However, the high consumption of ATP (16 MgATP per 2 NH_3) prompts the inhibition of nitrogenase caused by an increasing presence of the inhibitor ADP. This limitation has been addressed in different ways, either by adding an enzymatic ATP regeneration system to keep its concentration constant or by devising new strategies to replace the dependence on MgATP. Some of the studies that have used these methods for NH_3 bioelectrosynthesis are described below, showing a wide range of different variations that exist within the studies performed with DET and MET.

In a very interesting study by Shelley Minteer et al. [43], methyl viologen (N,N′-dimethyl-4,4′-bipyridinium) was used to interconnect the surface of an electrode with the nitrogenase enzyme. Bulk electrolysis at a mild applied potential of −610 mV vs. SHE resulted in the production of NH_3 with a faradaic efficiency of 59% at room temperature and under 1 atm N_2 at pH 7. This faradaic efficiency was quite remarkable, since the maximum efficiency achievable is 75%. Furthermore, they coupled this nitrogenase cathode with an anode that used the enzyme hydrogenase and MV to oxidize H_2, developing an enzymatic fuel cell (EFC) capable of producing NH_3 from H_2 and N_2 while producing an electric current. To demonstrate this, a 60 mC charge was passed through the H_2/N_2 EFCs, resulting in the formation of 286 nmol NH_3 mg^{-1} of MoFe protein, corresponding to a faradaic efficiency of 26.4%. Using the same MET principle, the coupling of ammonia bioelectrosynthesis to an enzymatic cascade was achieved, using nitrogenase, diaphorase, and alanine dehydrogenase to electrochemically drive the transaminase away from its reactant-favored equilibrium to produce chiral amines. Methyl viologen mediates both nitrogenase and diaphorase regeneration of the cofactor [144]. To maintain a constant ATP concentration in both experiments,

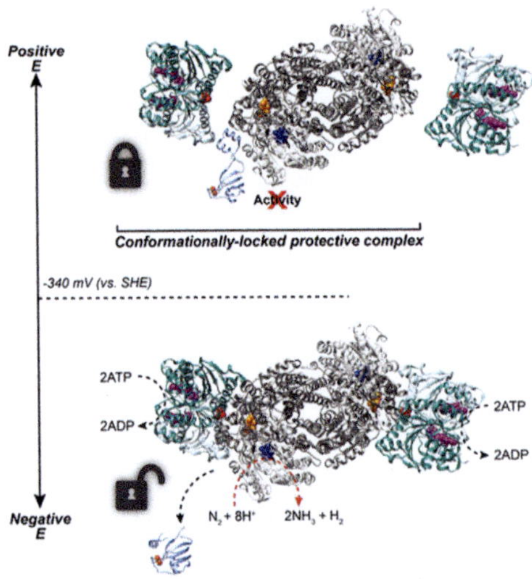

Figure 8.7 Conformational locked MoFe/Fe/FeSII protein complex vs. the reduction potential of the FeSII protein for illustrational purposes; the FeSII protein is shown as a monomer. *Source:* Adapted with permission from Milton et al. [145], Copyright 2017 American Chemical Society.

another enzyme system was added to the nitrogenase solution to regenerate ATP, consisting of the enzyme creatine phosphokinase and its corresponding substrate creatine phosphate.

In another study, one of the crucial problems for the purification of nitrogenases, its high sensitivity to O_2, was addressed. In this work, the enzyme nitrogenase and FeSII were used together to protect nitrogenase from inactivation by the presence of O_2 (Figure 8.7). For this purpose, methyl viologen was used, which in previous work was shown to be able to generate some protection against O_2 [146, 147]. A reduction potential of −760 mV vs. SHE was applied, and air was injected to test nitrogenase activity in the presence of 21% O_2. 586 ± 197 nmol NH_3 mg^{-1} were obtained after 4 hours of experiment [145].

In addition, spectroelectrochemistry of the FeSII protein under electrochemical control was used to determine the reduction potential of the [2Fe-2S] cluster, which was determined to be −340 ± 6 mV vs. SHE. This determination is an important insight for further learning about the control of the "open and closed" states of this protein.

The reduction potential of the $[4Fe\text{-}4S]^{1+/2+}$ reductase protein cluster is approximately −430 mV vs. SHE after association with MgATP, indicating that the reduction potential of the redox mediators of the MoFe protein should be more negative [148]. In addition, the one-electron reduction of Mo^N to the Mo^R state is believed to have an approximate reduction potential of −465 mV vs. SHE. One of the first studies using this approach was performed by Lance C. Seefeldt's group [149]. In this work, it was reported that a point modification of an amino acid of the MoFe protein, $\beta\text{-}98^{Tyr \rightarrow His}$, allows the resulting protein to reduce hydrazine (N_2H_4) to two ammonia molecules when provided with a low-potential reductant, polyaminocarboxylate-ligated EuII (Em −1.1 V vs. NHE). A rate of up to 170 nmol NH_3 min^{-1} mg^{-1} MoFe protein was obtained.

In an approach similar to the studies performed in intact bacteria, cadmium sulfide nanocrystals were used to supply electrons to the MoFe protein nitrogenase, bypassing ATP and protein reductase utilization [150]. The CdS:MoFe protein biohybrids provide a photochemical model for achieving light-driven reduction of N_2 to NH_3. Turnover rates of 75 $minute^{-1}$ were observed, with 63% of the nitrogenase working with ATP under optimal conditions.

The strategy of dispensing with ATP was used in the work of David P. Hickey et al. [151]; in this study, a material was developed to facilitate DET with the enzymes and was tested for the reduction of N_2 to NH_3 (Figure 8.8). A method based on linear poly(ethylenimine) (LPEI) covalently modified with pyrene particles was developed. MoFe protein was immobilized using this method

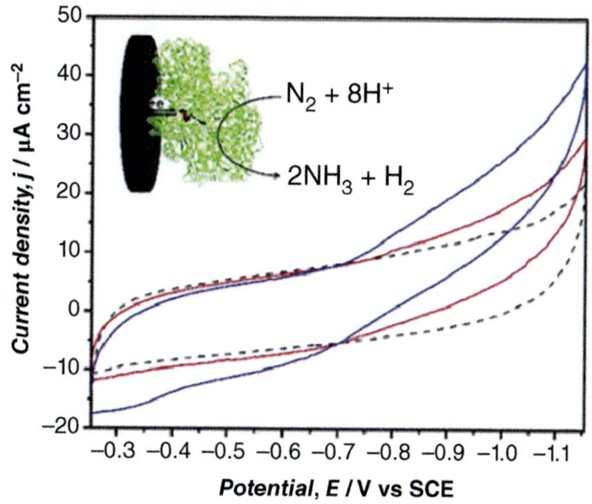

Figure 8.8 Cyclic voltammograms of pyrene-LPEI films containing MoFe protein and 5 mg mL^{-1} MWCNT–COOH under Ar (dashed line) and after 5 minutes (red line) or 10 minutes (blue line) of bubbling ultrahigh purity N_2. *Source:* Adapted with permission from Hickey et al. [151], Copyright 2018 Royal Society of Chemistry.

by crosslinking on a carbon electrode. The redox peak of the P cluster of the MoFe protein was identified at a potential of −510 mV vs. SHE using a combination of cyclic voltammetry and square wave voltammetry, and 180 ± 30 nmol of NH_3 were produced by the reduction of N_2.

In another study carried out by Ross D. Milton et al. [116], two types of MoFe proteins were immobilized: a wild-type one and the other with a β-98Tyr His amino acid point substitution. Previous studies had shown that the modified MoFe protein was more capable of accepting electrons from non-natural reducing agents [118]. MoFe proteins were immobilized within a polymer on the surface of a glassy carbon electrode, and cobaltocene (Cc+, bis(cyclopentadienyl)cobalt(III)) was used as a mediator (E^0 = −963 mV vs. SHE). In the experiments, chronoamperometries (CAs) were performed by applying a potential of −1 V vs. SHE. Current densities of 370 ± 14 µA cm^2 for the wild-type MoFe protein and 725 ± 22 µA cm^2 for the modified MoFe protein were obtained. The modified MoFe protein produced 70 ± 9 and 234 ± 62 nmol of NH_3 by reducing N_3 and NO_2^-, respectively. The faradaic yield obtained was approximately 100%.

Another electron transfer system based on redox polymers was developed using cobaltocene-functionalized poly(allylamine) (Cc-PAA), which is able to directly reduce MoFe nitrogenase with a low redox potential of −580 mV vs. SHE [152]. These types of redox polymers form a water-soluble skeleton with the bound redox mediators, which act both to exchange electrons and form a matrix for the immobilization of the enzyme, thus preventing its leaching [40]. The MoFe protein of nitrogenase was immobilized within this polymer, and the synthesis of ammonia from N_3, NO_2^- and even the reduction of N_2 to NH_3 could be observed. Although methyl viologen has given very good results in solution, no attempt was made to perform these experiments with a polymer with MV as a mediator because they are highly unstable [19].

Quantification of the ammonia produced is also a challenge, particularly when dealing with low-volume systems with slow turnover frequency. A typical option to measure ammonia is a fluorescence assay based on *o*-phthalaldehyde sulfite, as shown in Figure 8.9:

In conclusion, it can be stated that, although ammonia bioelectrosynthesis technology is still far from being applicable at the industrial level, it appears as a feasible alternative toward zero-emission GHGs due to ammonia synthesis. Great steps have been taken to achieve this, including the bioelectrosynthesis of ammonia under aerobic conditions and the production of ammonia without the high energy consumption of ATP. However, accomplishing this goal will undoubtedly require a multidisciplinary approach that encompasses not only the fields of biology

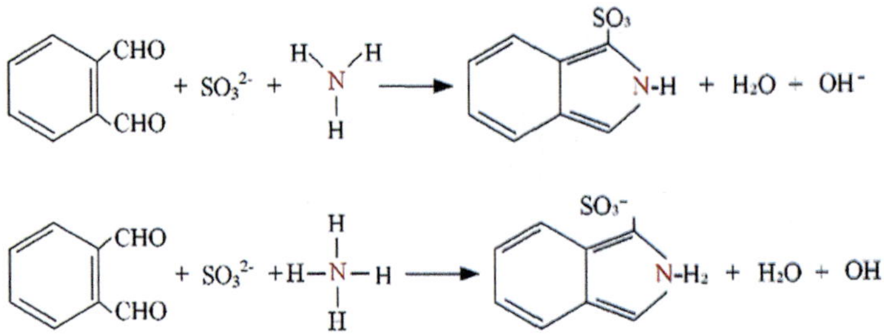

Figure 8.9 Mechanism of ammonia detection based on OPA-sulfite. *Source:* Adapted from Li et al. [153], Copyright 2020 Elsevier.

and chemistry but also materials science, synthetic biology [154], and the new technologies that are emerging today, such as the use of machine learning applied to bioelectrocatalysis [155].

8.5 Bioelectrocatalysis for Organic Synthesis

Enzymatic (EESs) or MESs systems are bioelectrochemical approaches that can be used for transforming electricity into valuable organic compounds [156, 157].

In EES, isolated enzymes or multienzymatic cascades can be immobilized on the electrode surfaces as bioelectrocatalysts for the synthesis of organic compounds. EES bears some excellent properties of enzymes, such as high activity, selectivity, and specificity to synthesize a variety of useful chemicals, especially fine chemicals with high added value, avoiding side reactions. They are carried out mainly by oxidoreductases applied to various oxidation and reduction systems in combination with electrochemical cofactor regeneration and in situ substrate generation. One of them has been the highly challenging hydroxylation of inert C-H bonds, oxyfunctionalizing them with a high catalytic activity and chemo-, site/regio-, and stereoselectivity [158]. This reaction is a broadly useful transformation for the efficient synthesis of complex organic molecules and pharmaceuticals. It constitutes a challenge for chemists since it is frequently accompanied by poor selectivity, unlike for oxidative enzymes existing in nature, such as monooxygenases and peroxygenases, which can accomplish this reaction with suitable activity and selectivity [159–161].

In this way, Minteer and coworkers have reported a bioelectrosynthetic system employing alkane monooxygenase from *Pseudomonas putida* (alkB) in a biocathode. With the proposed approach, it was possible to catalyze the terminal oxyfunctionalization of alkanes, producing alcohols, epoxides, and sulfoxides through bioelectrochemical hydroxylation, epoxidation, sulfoxidation, and demethylation in an H_2-fueled biofuel cell with a coupled hydrogenase bioanode [162].

Peroxygenases are considered ideal catalysts for CH-activation without the need for flavoproteins and cofactors to be regenerated. They can also be employed for efficient selective biocatalytic C–H bond hydroxylation. Holtmann and coworkers obtained the electro-enzymatic hydroxylation of ethylbenzene by the evolved unspecific peroxygenase of *Agrocybe aegerita*, promoting catalyst efficiency *via* electrochemical *in situ* supply of H_2O_2 [163].

Similarly, a recombinant peroxygenase of *Agrocybe aegerita* has also been employed to catalyze the hydroxylation of 4-ethylbenzoic acid in an all-in-one electrode system [164].

EES also offers a novel approach to the production of organic compounds through effective cathodic CO_2 reduction, such as formic acid, methane, or methanol. Enzymatic electrosynthesis (EES) of formic acid is based on the use of FDH as an enzymatic electrocatalyst at the cathode, which shows a high selectivity and catalytic efficiency under mild conditions [32, 165].

However, most of the approaches employ nicotinamide adenine dinucleotide-dependent (NAD^+) FDHs, which constitutes a limitation due to the low cofactor stability and high cost [166, 167].

To overcome this limitation, recently reported NAD^+-independent Mo- and W-FDHs have higher reaction rates in the reduction of CO_2 to formate ($HCOO^-$) [22, 168].

In the case of methanol, its conversion from CO_2 is more complicated, requiring the use of a multienzyme cascade incorporating, besides FDH, formaldehyde dehydrogenase (FLDH), and alcohol dehydrogenase (ADH) [169].

In another study, electrochemical conversion of carbon dioxide to methanol was achieved with the assistance of formate dehydrogenase and methanol dehydrogenase as biocatalysts in the presence of PQQ as the electron mediator [170].

Another novel EES approach has been reported for the reductive conversion of CO_2 to hydrocarbons, such as ethylene (C_2H_4) and propene (C_3H_6), by the electroenzymatic formation of C-C bonds using a vanadium iron nitrogenase from *Azotobacter vinelandii* mediated by derivatives of cobaltocene/cobaltocenium [20].

Also, Minteer's group has described the EES of alkanes by bioelectrocatalytic decarbonylation of fatty aldehydes. The EES is formed by combining an aldehyde deformylating oxygenase (ADO) from cyanobacteria that catalyzes the decarbonylation of fatty aldehydes to alkanes and formic acid. This system is able to produce a range of alkanes (octane to propane) from aldehydes and alcohols [171].

EES can also be employed for the enzymatic epoxidation of olefins using flavin-dependent monooxygenases, which constitute a useful synthetic intermediate. Schmid and coworkers reported the conversion of styrene to S-styrene oxide employing $FADH_2$-dependent styrene monooxygenase from *Pseudomonas* sp. VLB120, which catalyzes the enantioselective epoxidation of C=C double bonds [172]. Other EESs have also been reported for the reduction of carbonyl to produce alcohol, especially chiral alcohols employing carbonyl reductases. They can catalyze the asymmetric reduction of carbonyl, consuming NAD(P)H, which is regenerated in the cathode by an electrochemical mediator [173]. Finally, enantioselective reductive amination of carbonyls can also be accomplished by EES. In this approach, chiral amines are obtained via the asymmetric amination of α-keto acid by amino acid dehydrogenases that catalyze the reductive amination to L-amino acids in the presence of the cofactor NADH acting as an electron donor. Minteer et al. designed an EES to transform N_2 into nitrogenous organic compounds based on the coupling of a bioelectrochemical Haber-Bosch mimic with a H_2/N_2 biofuel cell. The reduction of N_2 to NH_3 and oxidation of H_2 were catalyzed by MoFe nitrogenase and hydrogenase, respectively, mediated by methyl viologen. The ammonia generated was combined with prochiral alpha-keto acids and NADH to synthesize amino acids, such as L-norleucine, catalyzed by leucine dehydrogenase with a high conversion rate [174].

The case of MES relies on the use of whole-cell microbes as biocatalysts to realize electricity-driven synthesis based on microbial EET with cathodes, allowing lower overpotential, longer operational stability, and milder operational conditions [175].

The MES can be used to reduce carbon dioxide to multicarbon, extracellular long-chain organic compounds. Until now, methane, formate, acetate, butyrate, 2-oxobutyrate, propionate, ethanol, 1-butanol, isobutanol, 3-methyl-1-butanol, and wax esters have all been produced by MES with CO_2 as the sole carbon source, providing an alternative green approach for the reduction of global warming gases. Typically, in MES approaches, electrons are derived at the anode either from water, reduced sulfur, or organic wastes and delivered at the biocathode, where microorganisms can directly accept electrons or indirectly through H_2 evolved at the electrode, for carrying out the reduction of carbon dioxide to organic compounds that are excreted from the cell [176].

Acetate is the most common bioproduct generated by MES, and although low in value, this compound can serve as a precursor for the synthesis of many other organic chemicals like ethanol, n-butanol, butyrate, or longer-chain fatty acids of higher economic value [177, 178]. Acetogens, especially *Sporumusa ovata*, *Clostridium ljungdahlii*, *Acetobacterium woodii*, and *Moorella thermoacetica*. *Sporomusa* strains are the microorganisms commonly employed due to their efficiency in producing acetate via the central intermediate acetyl-CoA from CO_2 through reactions of the Wood-Ljungdahl pathway. They have been used either as pure or mixed cultures, although the productivity of mixed cultures is relatively higher than that of pure cultures. It is associated with the formation of a biofilm on the cathode in a mixed microbiological culture that favors direct contact with the electrode surface and thus an increased DET [179].

In a first study, Nevin et al. reported the possibility of providing the acetogenic microorganism *Sporomusa ovata* with electrons delivered directly to the cells with a graphite electrode for the reduction of carbon dioxide to organic compounds. Biofilms of *S. ovata* growing on graphite cathode surfaces consumed electrons with the reduction of carbon dioxide to acetate and small amounts of 2-oxobutyrate, exhibiting coulombic efficiencies up to $86 \pm 21\%$ [10].

Recently, Krige et al. have reported great improvements in the acetate production by the development of a synthetic biofilm containing *Sporomusa ovata* using 3D bio-printing. This denser biofilm printed on a carbon cloth promotes an efficient electron transfer from the cathode to microbes. Also, a wider range of acetogenic microorganisms have been employed to catalyze the reduction of CO_2 for the production of organic acids. *Sporomusa silvacetic* and *Sporomusa sphaeroides* were both reported as biocatalysts in MES, although their electrical conversion efficiency was much lower than that of *Sporomusa ovata* [180].

On the other side, Xie et al. obtained enhanced acetate and butyrate production from CO_2 reduction employing engineered *Clostridium ljungdahlii* by electrochemically depositing a nickel phosphide (Ni – P) catalyst on the cathode [181]. Also, Zhang et al. studied the performance of different *Sporomusa* species for the MES of acetate from carbon dioxide, obtaining different efficiencies in function of the strain [182]. MES of organic chemicals such as acetic acid, butyric acid, and ethanol was also obtained from CO_2 reduction by *Clostridium scatologenes* ATCC 25775 [183].

Another of the main organic products produced by MES is methane, which can also be obtained by conversion of CO_2 using electrochemically active methanogenic microorganisms in the cathode (electromethanogenesis) [184–189]. These microbes use CO_2 and H_2 and/or small organic molecules, such as acetate, formate, and methylamine, and convert them into methane. Different MES have been developed to produce CH_4. MES are able to catalyze methane production either by direct EET, taking the electrons from the electrode to directly reduce CO_2 (reaction), or by an indirect EET initially producing hydrogen that is further used together with CO_2 to produce methane. Logan et al. described for the first time the conversion of CO_2 into methane using *Methanobacterium palustre* by DET without an organic fuel such as acetate [190].

Majone et al. reported the performance of a microbial biocathode based on a hydrogenophilic methanogenic culture, capable of reducing carbon dioxide to methane via direct and indirect EET at high rates and electron capture efficiencies [187].

Although EES and MES are considered promising technologies for the production of value-added chemicals, current levels of cathodic productivity in both cases are too low for any application at the industrial level. Still, there is a large potential to increase the productivity, efficiency, and robustness of EES and MES systems. Future developments and perspectives on bioelectrocatalysis for electrosynthesis in both EES and MES are focused on some key issues to be improved. These are the constructions of highly electroactive biocathodes using biocompatible support materials with high conductivity, chemical stability, and porous three-dimensional structure carrying high enzyme or bacteria coverages and showing high rates and efficiencies of electron transfer. Also, the use of inexpensive enzymes and microorganisms with new-to-nature reactivities and better stability obtained by protein engineering, besides new artificial electron mediators with broad redox potentials for MET. Designs and strategies to enhance the reactor configuration beyond traditional "H-cell" reactors and their operational conditions are required to allow their scale-up and long-term performance. A better fundamental understanding of mechanisms of charge transfer occurring across the biological/material interface is fundamental [191].

At the current rate of progress, all these issues should turn EES and MES into widely used technologies for industrial applications in the near future.

8.6 Conclusions

Using electrochemical assistance for either facilitating or causing redox processes has been a growing matter of research for over a century, but it was not until its combination with enzymatic or bacterial catalysts that the field could open the ground of selectivity and specificity in the production yield. The advantages of biological catalysts are enhanced when they can be linked to an artificial electron donor, opening a research field with applications toward biotechnological, energy, and remediation. It is expected that new applications for commodities and complex biosynthetic approaches will appear in the near future, overcoming some specific classic synthetic and biosynthetic routes limited by a redox transfer process.

References

1 Kolbe, A.W.H. (1848). Zersetzung der Valeriansäure durch den elektrischen Strom. *Justus Liebigs Ann. Chem.* 64 (3): 339–341.

2 Fleischmann, M., Mansfield, J.R., Thirsk, H.R. et al. (1967). The investigation of the kinetics of electrode reactions by the application of repetitive square pulses of potential. *Electrochim. Acta* 12 (8): 967–982.

3 Iversen, P.E. (1971). Reduction of trichloroacetic acid to dichloroacetic acid. An organic electrosynthesis experiment. *J. Chem. Educ.* 48 (2): 136.

4 Hand, R.L. and Nelson, R.F. (1974). Anodic oxidation pathways of N-alkylanilines. *JACS* 96 (3): 850–860.

5 Ronlan, A. and Parker, V.D. (1974). Electrosynthesis of medium and large sized rings by oxidative cyclization of bis(3,4-dimethoxyphenyl)alkanes. *J. Org. Chem.* 39 (7): 1014–1016.

6 Leech, M.C. and Lam, K. (2022). A practical guide to electrosynthesis. *Nat. Rev. Chem.* 6 (4): 275–286.

7 Frontana-Uribe, B.A., Little, R.D., Ibanez, J.G. et al. (2010). Organic electrosynthesis: a promising green methodology in organic chemistry. *Green Chem.* 12 (12): 2099–2119.

8 Francke, R. and Little, R.D. (2014). Redox catalysis in organic electrosynthesis: basic principles and recent developments. *Chem. Soc. Rev.* 43 (8): 2492–2521.

9 Li, C., Bai, H., and Shi, G. (2009). Conducting polymer nanomaterials: electrosynthesis and applications. *Chem. Soc. Rev.* 38 (8): 2397–2409.

10 Nevin, K.P., Woodard, T.L., Franks, A.E. et al. (2010). Microbial electrosynthesis: feeding microbes electricity to convert carbon dioxide and water to multicarbon extracellular organic compounds. *mBio* 1 (2): https://doi.org/10.1128/mbio.00103-10.

11 Varfolomeev, S.D. and Berezin, L.V. (1978). Enzymes as catalysts of electrochemical reactions. *J. Mol. Catal.* 4 (6): 387–399.

12 Intergovernmental Panel on Climate Change (IPCC) (ed.) (2023). Summary for policymakers. In: *Climate Change 2021 – The Physical Science Basis: Working Group I Contribution to the Sixth Assessment Report of the Intergovernmental Panel on Climate Change [Internet]*, 3–32. Cambridge: Cambridge University Press [cited 2024 Apr 8] Available from: https://www.cambridge.org/core/books/climate-change-2021-the-physical-science-basis/summary-for-policymakers/8E7A4E3AE6C364220F3B76A189CC4D4C.

13 Vats, G. and Mathur, R. (2022). A net-zero emissions energy system in India by 2050: an exploration. *J. Cleaner Prod.* 352: 131417.

14 Derrick, J.S., Loipersberger, M., Chatterjee, R. et al. (2020). Metal–ligand cooperativity via exchange coupling promotes iron- catalyzed electrochemical CO_2 reduction at low overpotentials. *JACS* 142 (48): 20489–20501.

15 Zhang, X., Guo, S.X., Gandionco, K.A. et al. (2020). Electrocatalytic carbon dioxide reduction: from fundamental principles to catalyst design. *Mater. Today Adv.* 7: 100074.

16 Sahoo, A., Chowdhury, A.H., Islam, S.M., and Bala, T. (2022). Successful CO_2 reduction under visible light photocatalysis using porous NiO nanoparticles, an atypical metal oxide. *New J. Chem.* 46 (22): 10806–10813.

17 Kang, D.K., Kim, S.H., Sohn, J.H., and Sung, B.H. (2023). Insights into enzyme reactions with redox cofactors in biological conversion of CO_2. *J. Microbiol. Biotechnol.* 33 (11): 1403–1411.

18 D'Amario, B., Pérez, C., Grelaud, M. et al. (2020). Coccolithophore community response to ocean acidification and warming in the Eastern Mediterranean Sea: results from a mesocosm experiment. *Sci. Rep.* 10 (1): 12637.

19 Yuan, M., Sahin, S., Cai, R. et al. (2018). Creating a low-potential redox polymer for efficient electroenzymatic CO_2 reduction. *Angew. Chem. Int. Ed.* 57 (22): 6582–6586.

20 Cai, R., Milton, R.D., Abdellaoui, S. et al. (2018). Electroenzymatic C–C bond formation from CO_2. *JACS* 140 (15): 5041–5044.

21 Mateos, R., Escapa, A., Vanbroekhoven, K. et al. (2019). Microbial electrochemical technologies for CO_2 and its derived products valorization. In: *Microbial Electrochemical Technology [Internet]*, Biomass, Biofuels and Biochemicals (ed. S.V. Mohan, S. Varjani, and A. Pandey), 777–796. Elsevier [cited 2024 Apr 8], Available from: https://www.sciencedirect.com/science/article/pii/B9780444640529000327.

22 Kuk, S.K., Gopinath, K., Singh, R.K. et al. (2019). NADH-free electroenzymatic reduction of CO_2 by conductive hydrogel-conjugated formate dehydrogenase. *ACS Catal.* 9 (6): 5584–5589.

23 Song, H., Ma, C., Liu, P. et al. (2019). A hybrid CO_2 electroreduction system mediated by enzyme-cofactor conjugates coupled with cu nanoparticle-catalyzed cofactor regeneration. *J. CO_2 Util.* 34: 568–575.

24 Freguia, S., Virdis, B., Harnisch, F., and Keller, J. (2012). Bioelectrochemical systems: microbial versus enzymatic catalysis. *Electrochim. Acta* 82: 165–174.

25 Milton, R.D. and Minteer, S.D. (2017). Direct enzymatic bioelectrocatalysis: differentiating between myth and reality. *J. R. Soc. Interface* 14 (131): 20170253.

26 Flexer, V. and Brun, N. (2017). Fundamentals of enzymatic electrochemical systems. In: *Functional Electrodes for Enzymatic and Microbial Electrochemical Systems*, 3–50. World Scientific (Europe) [Internet, cited 2024 Apr 8], Available from: https://www.worldscientific.com/doi/abs/10.1142/9781786343543_0001.

27 Poulos, T.L. (2014). Heme enzyme structure and function. *Chem. Rev.* 114 (7): 3919–3962.

28 Schilter, D., Camara, J.M., Huynh, M.T. et al. (2016). Hydrogenase enzymes and their synthetic models: the role of metal hydrides. *Chem. Rev.* 116 (15): 8693–8749.

29 Liu, J., Chakraborty, S., Hosseinzadeh, P. et al. (2014). Metalloproteins containing cytochrome, iron–sulfur, or copper redox centers. *Chem. Rev.* 114 (8): 4366–4469.

30 Burgess, B.K. and Lowe, D.J. (1996). Mechanism of molybdenum nitrogenase. *Chem. Rev.* 96 (7): 2983–3012.

31 Li, W., Gao, Y., Sun, X. et al. (2023). Direct detection of a single [4Fe-4S] cluster in a tungsten-containing enzyme: electrochemical conversion of CO_2 into formate by formate dehydrogenase. *Carbon Energy* 5 (5): e304.

32 Srikanth, S., Alvarez-Gallego, Y., Vanbroekhoven, K., and Pant, D. (2017). Enzymatic electrosynthesis of formic acid through carbon dioxide reduction in a bioelectrochemical system: effect of immobilization and carbonic anhydrase addition. *ChemPhysChem* 18 (22): 3174–3181.

33 Hansen, H.A., Varley, J.B., Peterson, A.A., and Nørskov, J.K. (2013). Understanding trends in the electrocatalytic activity of metals and enzymes for CO_2 reduction to CO. *J. Phys. Chem. Lett.* 4 (3): 388–392.

34 Addo, P.K., Arechederra, R.L., Waheed, A. et al. (2011). Methanol production via bioelectrocatalytic reduction of carbon dioxide: role of carbonic anhydrase in improving electrode performance. *Electrochem. Solid-State Lett.* 14 (4): E9.

35 Chu, N., Hao, W., Wu, Q. et al. (2022). Microbial electrosynthesis for producing medium chain fatty acids. *Engineering* 16: 141–153.

36 Srikanth, S., Singh, D., Vanbroekhoven, K. et al. (2018). Electro-biocatalytic conversion of carbon dioxide to alcohols using gas diffusion electrode. *Bioresour. Technol.* 265: 45–51.

37 Ghindilis, A.L., Atanasov, P., and Wilkins, E. (1997). Enzyme-catalyzed direct electron transfer: fundamentals and analytical applications. *Electroanalysis* 9 (9): 661–674.

38 Marcus, R.A. and Sutin, N. (1985). Electron transfers in chemistry and biology. *Biochim. Biophys. Acta Rev. Biomembr.* 811 (3): 265–322.

39 Choi, O. and Sang, B.I. (2016). Extracellular electron transfer from cathode to microbes: application for biofuel production. *Biotechnol. Biofuels* 9 (1): 11.

40 Ruff, A. (2017). Redox polymers in bioelectrochemistry: common playgrounds and novel concepts. *Curr. Opin. Electrochem.* 5 (1): 66–73.

41 Milton, R.D., Hickey, D.P., Abdellaoui, S. et al. (2015). Rational design of quinones for high power density biofuel cells. *Chem. Sci.* 6 (8): 4867–4875.

42 Rhodes, Z., Simoska, O., Dantanarayana, A. et al. (2021). Using structure-function relationships to understand the mechanism of phenazine-mediated extracellular electron transfer in *Escherichia coli*. *iScience* 24 (9): 103033.

43 Milton, R.D., Cai, R., Abdellaoui, S. et al. (2017). Bioelectrochemical Haber–Bosch process: an ammonia-producing H_2/N_2 fuel cell. *Angew. Chem. Int. Ed.* 56 (10): 2680–2683.

44 Blanchet, E., Duquenne, F., Rafrafi, Y. et al. (2015). Importance of the hydrogen route in up-scaling electrosynthesis for microbial CO_2 reduction. *Energy Environ. Sci.* 8 (12): 3731–3744.

45 Kim, J., Jang, J., Hilberath, T. et al. (2022). Photoelectrocatalytic biosynthesis fuelled by microplastics. *Nat. Synth.* 1 (10): 776–786.

46 Elving, P.J., Bresnahan, W.T., Moiroux, J., and Samec, Z. (1982). 524—NAD/NADH as a model redox system: mechanism, mediation, modification by the environment. *J. Electroanal. Chem. Interfacial Electrochem.* 141 (3): 365–378.

47 Reda, T., Plugge, C.M., Abram, N.J., and Hirst, J. (2008). Reversible interconversion of carbon dioxide and formate by an electroactive enzyme. *Proc. Natl. Acad. Sci.* 105 (31): 10654–10658.

48 Edwardes Moore, E., Cobb, S.J., Coito, A.M. et al. (2022). Understanding the local chemical environment of bioelectrocatalysis. *Proc. Natl. Acad. Sci.* 119 (4): e2114097119.

49 Szczesny, J., Ruff, A., Oliveira, A.R. et al. (2020). Electroenzymatic CO_2 fixation using redox polymer/enzyme-modified gas diffusion electrodes. *ACS Energy Lett.* 5 (1): 321–327.

50 Bhattacharjee, S., Rahaman, M., Andrei, V. et al. (2023). Photoelectrochemical CO_2-to-fuel conversion with simultaneous plastic reforming. *Nat. Synth.* 2 (2): 182–192.

51 Batlle-Vilanova, P., Ganigué, R., Ramió-Pujol, S. et al. (2017). Microbial electrosynthesis of butyrate from carbon dioxide: production and extraction. *Bioelectrochemistry.* 117: 57–64.

52 Nevin, K.P., Hensley, S.A., Franks, A.E. et al. (2011). Electrosynthesis of organic compounds from carbon dioxide is catalyzed by a diversity of acetogenic microorganisms. *Appl. Environ. Microbiol.* 77 (9): 2882–2886.

53 Bajracharya, S., Yuliasni, R., Vanbroekhoven, K. et al. (2017). Long-term operation of microbial electrosynthesis cell reducing CO_2 to multi-carbon chemicals with a mixed culture avoiding methanogenesis. *Bioelectrochemistry* 113: 26–34.

54 Rodrigues, R.M., Guan, X., Iñiguez, J.A. et al. (2019). Perfluorocarbon nanoemulsion promotes the delivery of reducing equivalents for electricity-driven microbial CO_2 reduction. *Nat. Catal.* 2 (5): 407–414.

55 Ganigué, R., Puig, S., Batlle-Vilanova, P. et al. (2015). Microbial electrosynthesis of butyrate from carbon dioxide. *Chem. Commun.* 51 (15): 3235–3238.

56 Torella, J.P., Gagliardi, C.J., Chen, J.S. et al. (2015). Efficient solar-to-fuels production from a hybrid microbial–water-splitting catalyst system. *Proc. Natl. Acad. Sci.* 112 (8): 2337–2342.

57 Kuk, S.K., Singh, R.K., Nam, D.H. et al. (2017). Photoelectrochemical reduction of carbon dioxide to methanol through a highly efficient enzyme cascade. *Angew. Chem. Int. Ed.* 56 (14): 3827–3832.

58 Castañeda-Losada, L., Adam, D., Paczia, N. et al. (2021). Bioelectrocatalytic cofactor regeneration coupled to CO_2 fixation in a redox-active hydrogel for stereoselective C–C bond formation. *Angew. Chem. Int. Ed.* 60 (38): 21056–21061.

59 Hu, B., Harris, D.F., Dean, D.R. et al. (2018). Electrocatalytic CO_2 reduction catalyzed by nitrogenase MoFe and FeFe proteins. *Bioelectrochemistry* 120: 104–109.

60 Nichols, E.M., Gallagher, J.J., Liu, C. et al. (2015). Hybrid bioinorganic approach to solar-to-chemical conversion. *Proc. Natl. Acad. Sci.* 112 (37): 11461–11466.

61 Morello, G., Siritanaratkul, B., Megarity, C.F., and Armstrong, F.A. (2019). Efficient electrocatalytic CO_2 fixation by nanoconfined enzymes via a C3-to-C4 reaction that is favored over H_2 production. *ACS Catal.* 9 (12): 11255–11262.

62 Navarro, R.M., Peña, M.A., and Fierro, J.L.G. (2007). Hydrogen production reactions from carbon feedstocks: fossil fuels and biomass. *Chem. Rev.* 107 (10): 3952–3991.

63 You, B. and Sun, Y. (2018). Innovative strategies for electrocatalytic water splitting. *Acc. Chem. Res.* 51 (7): 1571–1580.

64 Vincent, K.A., Parkin, A., and Armstrong, F.A. (2007). Investigating and exploiting the electrocatalytic properties of hydrogenases. *Chem. Rev.* 107 (10): 4366–4413.

65 Noori, M.T., Rossi, R., Logan, B.E., and Min, B. (2024). Hydrogen production in microbial electrolysis cells with biocathodes. *Trends Biotechnol.* [Internet, cited 2024 Apr 15]; Available from: https://www.sciencedirect.com/science/article/pii/S0167779923003669.

66 Reuillard, B., Costentin, C., and Artero, V. (2023). Deciphering reversible homogeneous catalysis of the electrochemical H_2 evolution and oxidation: role of proton relays and local concentration effects**. *Angew. Chem. Int. Ed.* 62 (36): e202302779.

67 Ginovska, B., Gutiérrez, O.Y., Karkamkar, A. et al. (2023). Bioinspired catalyst design principles: progress in emulating properties of enzymes in synthetic catalysts. *ACS Catal.* 13 (18): 11883–11901.

68 De Lacey, A.L., Fernández, V.M., Rousset, M., and Cammack, R. (2007). Activation and inactivation of hydrogenase function and the catalytic cycle: spectroelectrochemical studies. *Chem. Rev.* 107 (10): 4304–4330.

69 Fontecilla-Camps, J.C., Volbeda, A., Cavazza, C., and Nicolet, Y. (2007). Structure/function relationships of [NiFe]- and [FeFe]-hydrogenases. *Chem. Rev.* 107 (10): 4273–4303.

70 Murphy, B.J., Sargent, F., and Armstrong, F.A. (2014). Transforming an oxygen-tolerant [NiFe] uptake hydrogenase into a proficient, reversible hydrogen producer. *Energy Environ. Sci.* 7 (4): 1426–1433.

71 Hexter, S.V., Grey, F., Happe, T. et al. (2012). Electrocatalytic mechanism of reversible hydrogen cycling by enzymes and distinctions between the major classes of hydrogenases. *Proc. Natl. Acad. Sci. USA* 109 (29): 11516–11521.

72 Fourmond, V., Baffert, C., Sybirna, K. et al. (2013). The mechanism of inhibition by H_2 of H_2-evolution by hydrogenases. *Chem. Commun.* 49 (61): 6840–6842.

73 Roseboom, W., De Lacey, A.L., Fernandez, V.M. et al. (2006). The active site of the [FeFe]-hydrogenase from Desulfovibrio desulfuricans. II. Redox properties, light sensitivity and CO-ligand exchange as observed by infrared spectroscopy. *JBIC* 11 (1): 102–118.

74 Lu, Y. and Koo, J. (2019). O_2 sensitivity and H_2 production activity of hydrogenases—a review. *Biotechnol. Bioeng.* 116 (11): 3124–3135.

75 Yaropolov, A.I., Karyakin, A.A., Varfolomeev, S.D., and Berezin, I.V. (1984). Mechanism of H_2-electrooxidation with immobilized hydrogenase. *Bioelectrochem. Bioenerg.* 12 (3): 267–277.

76 Garcin, E., Montet, Y., Volbeda, A. et al. (1998). Structural bases for the catalytic mechanism of [NiFe] hydrogenases. *Biochem. Soc. Trans.* 26 (3): 396–401.

77 Rüdiger, O., Abad, J.M., Hatchikian, E.C. et al. (2005). Oriented immobilization of desulfovibrio gigas hydrogenase onto carbon electrodes by covalent bonds for nonmediated oxidation of H_2. *JACS* 127 (46): 16008–16009.

78 Cosnier, S., Holzinger, M., and Le Goff, A. (2014). Recent advances in carbon nanotube-based enzymatic fuel cells. *Front. Bioeng. Biotechnol.* 2. [Internet, cited 2024 Apr 15] Available from: https://www.frontiersin.org/articles/10.3389/fbioe.2014.00045.

79 Alonso-Lomillo, M.A., Rüdiger, O., Maroto-Valiente, A. et al. (2007). Hydrogenase-coated carbon nanotubes for efficient H_2 oxidation. *Nano Lett.* 7 (6): 1603–1608.

80 Kihara, T., Liu, X.Y., Nakamura, C. et al. (2011). Direct electron transfer to hydrogenase for catalytic hydrogen production using a single-walled carbon nanotube forest. *Int. J. Hydrogen Energy* 36 (13): 7523–7529.

81 Morra, S., Valetti, F., Sarasso, V. et al. (2015). Hydrogen production at high faradaic efficiency by a bio-electrode based on TiO_2 adsorption of a new [FeFe]-hydrogenase from *Clostridium perfringens*. *Bioelectrochemistry* 106: 258–262.

82 Liu, Y., Webb, S., Moreno-García, P. et al. (2023). Facile functionalization of carbon electrodes for efficient electroenzymatic hydrogen production. *JACS Au* 3 (1): 124–130.

83 Ruth, J.C., Milton, R.D., Gu, W., and Spormann, A.M. (2020). Enhanced electrosynthetic hydrogen evolution by hydrogenases embedded in a redox-active hydrogel. *Chem. A Eur. J.* 26 (32): 7323–7329.

84 Hardt, S., Stapf, S., Filmon, D.T. et al. (2021). Reversible H_2 oxidation and evolution by hydrogenase embedded in a redox polymer film. *Nat. Catal.* 4 (3): 251–258.

85 Fourmond, V. and Léger, C. (2021). Theoretical understanding of the penetration of O_2 in enzymatic redox polymer films: the case of unidirectional catalysis and irreversible inactivation in a film of arbitrary thickness. *ChemElectroChem* 8 (13): 2607–2615.

86 Rutz, A., Das, C.K., Fasano, A. et al. (2023). Increasing the O_2 resistance of the [FeFe]-hydrogenase CbA5H through enhanced protein flexibility. *ACS Catal.* 13 (2): 856–865.

87 Fasano, A., Guendon, C., Jacq-Bailly, A. et al. (2023). A chimeric NiFe hydrogenase heterodimer to assess the role of the electron transfer chain in tuning the enzyme's catalytic bias and oxygen tolerance. *JACS* 145 (36): 20021–20030.

88 Greening, C., Kropp, A., Vincent, K., and Grinter, R. (2023). Developing high-affinity, oxygen-insensitive [NiFe]-hydrogenases as biocatalysts for energy conversion. *Biochem. Soc. Trans.* 51 (5): 1921–1933.

89 Tapia, C., Milton, R.D., Pankratova, G. et al. (2017). Wiring of photosystem I and hydrogenase on an electrode for photoelectrochemical H_2 production by using redox polymers for relatively positive onset potential. *ChemElectroChem* 4 (1): 90–95.

90 Zhao, F., Wang, P., Ruff, A. et al. (2019). A photosystem I monolayer with anisotropic electron flow enables Z-scheme like photosynthetic water splitting. *Energy Environ. Sci.* 12 (10): 3133–3143.

91 Wang, P., Frank, A., Zhao, F. et al. (2021). Closing the gap for electronic short-circuiting: photosystem I mixed monolayers enable improved anisotropic electron flow in biophotovoltaic devices. *Angew. Chem. Int. Ed.* 60 (4): 2000–2006.

92 Tian, L., Németh, B., Berggren, G., and Tian, H. (2018). Hydrogen evolution by a photoelectrochemical cell based on a Cu_2O-ZnO-[FeFe] hydrogenase electrode. *J. Photochem. Photobiol., A* 366: 27–33.

93 Tapia, C., Zacarias, S., Pereira, I.A.C. et al. (2016). In situ determination of photobioproduction of H_2 by in $_2S_3$ -[NiFeSe] hydrogenase from *Desulfovibrio vulgaris* Hildenborough using only visible light. *ACS Catal.* 6 (9): 5691–5698.

94 Luna-López, G., del Barrio, M., Fize, J. et al. (2023). Photobio-electrocatalytic production of H_2 using fluorine-doped tin oxide (FTO) electrodes covered with a NiO-In_2S_3 p-n junction and NiFeSe hydrogenase. *Bioelectrochemistry* 150.

95 Luna-López, G., Sainz, R., Coito, A.M. et al. (2023). Hybrid biological/inorganic photocathode for H_2 production based on a NiFeSe hydrogenase immobilized on electrodeposited $CuGaS_2$. *Catal. Today* 423: 114281.

96 Zhao, Y., Anderson, N.C., Ratzloff, M.W. et al. (2016). Proton reduction using a hydrogenase-modified nanoporous black silicon photoelectrode. *ACS Appl. Mater. Interfaces* 8 (23): 14481–14487.

97 Lee, C.Y., Park, H.S., Fontecilla-Camps, J.C., and Reisner, E. (2016). Photoelectrochemical H_2 evolution with a hydrogenase immobilized on a TiO_2-protected silicon electrode. *Angew. Chem. Int. Ed.* 55 (20): 5971–5974.

98 Edwardes Moore, E., Andrei, V., Zacarias, S. et al. (2020). Integration of a hydrogenase in a Lead halide perovskite photoelectrode for tandem solar water splitting. *ACS Energy Lett.* 5 (1): 232–237.

99 Witkowska, M., Jedrzejczak, R.P., Joachimiak, A. et al. (2023). Promising approaches for the assembly of the catalytically active, recombinant Desulfomicrobium baculatum hydrogenase with substitutions at the active site. *Microb. Cell Fact.* 22 (1): 134.

100 Davis, V., Heidary, N., Guiet, A. et al. (2023). Immobilization of O_2-tolerant [NiFe] hydrogenase from Cupriavidus necator on tin-rich indium oxide alters the catalytic bias from H_2 oxidation to proton reduction. *ACS Catal.* 13 (9): 6312–6327.

101 Kim, H.J., Park, H.S., Hyun, M.S. et al. (2002). A mediator-less microbial fuel cell using a metal reducing bacterium, *Shewanella putrefaciens*. *Enzym. Microb. Technol.* 30 (2): 145–152.

102 Bond, D.R. and Lovley, D.R. (2003). Electricity production by Geobacter sulfurreducens attached to electrodes. *Appl. Environ. Microbiol.* 69 (3): 1548–1555.

103 Schröder, U. (2012). Editorial: microbial fuel cells and microbial electrochemistry: into the next century! *ChemSusChem* 5 (6): 959–959.

104 Lu, M. and Li, S.F.Y. (2012). Cathode reactions and applications in microbial fuel cells: a review. *Crit. Rev. Environ. Sci. Technol.* 42 (23): 2504–2525.

105 Sydow, A., Krieg, T., Mayer, F. et al. (2014). Electroactive bacteria--molecular mechanisms and genetic tools. *Appl. Microbiol. Biotechnol.* 98 (20): 8481–8495.

106 Perona-Vico, E., Feliu-Paradeda, L., Puig, S., and Bañeras, L. (2020). Bacteria coated cathodes as an in-situ hydrogen evolving platform for microbial electrosynthesis. *Sci. Rep.* 10 (1): 19852.

107 Geelhoed, J.S. and Stams, A.J.M. (2011). Electricity-assisted biological hydrogen production from acetate by Geobacter sulfurreducens. *Environ. Sci. Technol.* 45 (2): 815–820.

108 Fu, Y., Zhang, Y., Li, B. et al. (2019). Extracellular electron transfer of *Shewanella oneidensis* MR-1 for cathodic hydrogen evolution reaction. *Electrochim. Acta* 305: 528–533.

109 Villano, M., De Bonis, L., Rossetti, S. et al. (2011). Bioelectrochemical hydrogen production with hydrogenophilic dechlorinating bacteria as electrocatalytic agents. *Bioresour. Technol.* 102 (3): 3193–3199.

110 Wang, Y., Guo, W.Q., Xing, D.F. et al. (2014). Hydrogen production using biocathode single-chamber microbial electrolysis cells fed by molasses wastewater at low temperature. *Int. J. Hydrogen Energy* 39 (33): 19369–19375.

111 Dai, J., Huang, Z., Zhang, H. et al. (2023). Promoted Sb removal with hydrogen production in microbial electrolysis cell by ZIF-67-derived modified sulfate-reducing bacteria bio-cathode. *Sci. Total Environ.* 856: 158839.

112 Logan, B.E., Rossi, R., Ragab, A., and Saikaly, P.E. (2019). Electroactive microorganisms in bio-electrochemical systems. *Nat. Rev. Microbiol.* 17 (5): 307–319.

113 Jafary, T., Yeneneh, A.M., Daud, W.R.W. et al. (2022). Taxonomic classification of sulphate-reducing bacteria communities attached to biocathode in hydrogen-producing microbial electrolysis cell. *Int. J. Environ. Sci. Technol.* 19 (8): 7297–7306.

114 Foster, S.L., Bakovic, S.I.P., Duda, R.D. et al. (2018). Catalysts for nitrogen reduction to ammonia. *Nat. Catal.* 1 (7): 490–500.

115 Montoya, J.H., Tsai, C., Vojvodic, A., and Nørskov, J.K. (2015). The challenge of electrochemical ammonia synthesis: a new perspective on the role of nitrogen scaling relations. *ChemSusChem* 8 (13): 2180–2186.

116 Wang, B., Zhang, Y., and Minteer, S.D. (2023). Renewable electron-driven bioinorganic nitrogen fixation: a superior route toward green ammonia? *Energy Environ. Sci.* 16 (2): 404–420.

117 Chen, J.G., Crooks, R.M., Seefeldt, L.C. et al. (2018). Beyond fossil fuel–driven nitrogen transformations. *Science* 360 (6391): eaar6611.

118 Burén, S. and Rubio, L.M. (2018). State of the art in eukaryotic nitrogenase engineering. *FEMS Microbiol. Lett.* 365 (2): fnx274.

119 Lancaster, K.M., Roemelt, M., Ettenhuber, P. et al. (2011). X-ray emission spectroscopy evidences a central carbon in the nitrogenase iron-molybdenum cofactor. *Science* 334 (6058): 974–977.

120 Stein, L.Y. and Klotz, M.G. (2016). The nitrogen cycle. *Curr. Biol.* 26 (3): R94–R98.

121 Gu, W. and Milton, R.D. (2020). Natural and engineered electron transfer of Nitrogenase. *Chemistry* 2 (2): 322–346.

122 Yang, Z.Y., Ledbetter, R., Shaw, S. et al. (2016). Evidence that the pi release event is the rate-limiting step in the nitrogenase catalytic cycle. *Biochemistry* 55 (26): 3625–3635.

123 Harris, D.F., Lukoyanov, D.A., Kallas, H. et al. (2019). Mo-, V-, and Fe-nitrogenases use a universal eight-electron reductive-elimination mechanism to achieve N_2 reduction. *Biochemistry* 58 (30): 3293–3301.

124 Harris, D.F., Lukoyanov, D.A., Shaw, S. et al. (2018). Mechanism of N_2 reduction catalyzed by Fe-nitrogenase involves reductive elimination of H_2. *Biochemistry* 57 (5): 701–710.

125 Harwood, C.S. (2020). Iron-only and vanadium nitrogenases: fail-safe enzymes or something more? *Ann. Rev. Microbiol.* 74: 247–266.

126 Zheng, Y., Harris, D.F., Yu, Z. et al. (2018). A pathway for biological methane production using bacterial iron-only nitrogenase. *Nat. Microbiol.* 3 (3): 281–286.

127 Jasniewski, A.J., Sickerman, N.S., Hu, Y., and Ribbe, M.W. (2018). The Fe protein: an unsung hero of nitrogenase. *Inorganics* 6 (1): 25.

128 Rutledge, H.L. and Tezcan, F.A. (2020). Electron transfer in nitrogenase. *Chem. Rev.* 120 (12): 5158–5193.

129 Danyal, K., Dean, D.R., Hoffman, B.M., and Seefeldt, L.C. (2011). Electron transfer within nitrogenase: evidence for a deficit-spending mechanism. *Biochemistry* 50 (43): 9255–9263.

130 Milton, R.D. and Minteer, S.D. (2019). Nitrogenase bioelectrochemistry for synthesis applications. *Acc. Chem. Res.* 52 (12): 3351–3360.

131 Hoffman, B.M., Lukoyanov, D., Dean, D.R., and Seefeldt, L.C. (2013). Nitrogenase: a draft mechanism. *Acc. Chem. Res.* 46 (2): 587–595.

132 Lukoyanov, D., Khadka, N., Yang, Z.Y. et al. (2016). Reductive elimination of H_2 activates nitrogenase to reduce the N≡N triple bond: characterization of the E_4(4H) Janus intermediate in wild-type enzyme. *JACS* 138 (33): 10674–10683.

133 Raugei, S., Seefeldt, L.C., and Hoffman, B.M. (2018). Critical computational analysis illuminates the reductive-elimination mechanism that activates nitrogenase for N_2 reduction. *Proc. Natl. Acad. Sci.* 115 (45): E10521–E10530.

134 Moshiri, F., Kim, J.W., Fu, C., and Maier, R.J. (1994). The FeSII protein of Azotobacter vinelandii is not essential for aerobic nitrogen fixation, but confers significant protection to oxygen-mediated inactivation of nitrogenase in vitro and in vivo. *Mol. Microbiol.* 14 (1): 101–114.

135 Gallon, J.R. (1992). Tansley Review No. 44. Reconciling the incompatible: N_2 fixation and O_2. *New Phytol.* 122 (4): 571–609.

136 Ambrosio, R., Ortiz-Marquez, J.C.F., and Curatti, L. (2017). Metabolic engineering of a diazotrophic bacterium improves ammonium release and biofertilization of plants and microalgae. *Metab. Eng.* 40: 59–68.

137 Liu, C., Sakimoto, K.K., Colón, B.C. et al. (2017). Ambient nitrogen reduction cycle using a hybrid inorganic–biological system. *Proc. Natl. Acad. Sci.* 114 (25): 6450–6455.

138 Lu, S., Guan, X., and Liu, C. (2020). Electricity-powered artificial root nodule. *Nat. Commun.* 11 (1): 1505.

139 Chen, S., Jing, X., Yan, Y. et al. (2021). Bioelectrochemical fixation of nitrogen to extracellular ammonium by pseudomonas stutzeri. *Appl. Environ. Microbiol.* 87 (5): e01998–e01920.

140 Wang, B., Xiao, K., Jiang, Z. et al. (2019). Biohybrid photoheterotrophic metabolism for significant enhancement of biological nitrogen fixation in pure microbial cultures. *Energy Environ. Sci.* 12 (7): 2185–2191.

141 Zeng, Y., Bai, H., Yu, W. et al. (2023). Increased nitrogenase activity in solar-driven biohybrids containing non-photosynthetic bacteria and conducting polymers. *Angew. Chem. Int. Ed.* 62 (30): e202303877.

142 Moser, C.C., Keske, J.M., Warncke, K. et al. (1992). Nature of biological electron transfer. *Nature* 355 (6363): 796–802.

143 Kano, K. and Ikeda, T. (2000). Fundamentals and practices of mediated bioelectrocatalysis. *Anal. Sci.* 16 (10): 1013–1021.

144 Chen, H., Cai, R., Patel, J. et al. (2019). Upgraded bioelectrocatalytic N_2 fixation: from N_2 to chiral amine intermediates. *JACS* 141 (12): 4963–4971.

145 Milton, R.D., Cai, R., Sahin, S. et al. (2017). The in vivo potential-regulated protective protein of nitrogenase in Azotobacter vinelandii supports aerobic bioelectrochemical dinitrogen reduction in vitro. *JACS* 139 (26): 9044–9052.

146 Plumeré, N., Rüdiger, O., Oughli, A.A. et al. (2014). A redox hydrogel protects hydrogenase from high-potential deactivation and oxygen damage. *Nature Chem.* 6 (9): 822–827.

147 Fourmond, V., Stapf, S., Li, H. et al. (2015). Mechanism of protection of catalysts supported in redox hydrogel films. *JACS* 137 (16): 5494–5505.

148 Lanzilotta, W.N., Ryle, M.J., and Seefeldt, L.C. (1995). Nucleotide hydrolysis and protein conformational changes in Azotobacter vinelandii Nitrogenase iron protein: defining the function of aspartate 129. *Biochemistry* 34 (34): 10713–10723.

149 Danyal, K., Inglet, B.S., Vincent, K.A. et al. (2010). Uncoupling nitrogenase: catalytic reduction of hydrazine to ammonia by a MoFe protein in the absence of Fe protein-ATP. *JACS* 132 (38): 13197–13199.

150 Brown, K.A., Harris, D.F., Wilker, M.B. et al. (2016). Light-driven dinitrogen reduction catalyzed by a CdS:nitrogenase MoFe protein biohybrid. *Science* 352 (6284): 448–450.

151 Hickey, D.P., Lim, K., Cai, R. et al. (2018). Pyrene hydrogel for promoting direct bioelectrochemistry: ATP-independent electroenzymatic reduction of N_2. *Chem. Sci.* 9 (23): 5172–5177.

152 Lee, Y.S., Yuan, M., Cai, R. et al. (2020). Nitrogenase bioelectrocatalysis: ATP-independent ammonia production using a redox polymer/MoFe protein system. *ACS Catal.* 10 (12): 6854–6861.

153 Li, D., Xu, X., Li, Z. et al. (2020). Detection methods of ammonia nitrogen in water: a review. *TrAC, Trends Anal. Chem.* 127: 115890.

154 Ohki, Y., Munakata, K., Matsuoka, Y. et al. (2022). Nitrogen reduction by the Fe sites of synthetic [Mo_3S_4Fe] cubes. *Nature* 607 (7917): 86–90.

155 Huang, J., Gao, Y., Chang, Y. et al. (2024). Machine learning in bioelectrocatalysis. *Adv. Sci.* 11 (2): 2306583.

156 Chen, H., Simoska, O., Lim, K. et al. (2020). Fundamentals, applications, and future directions of bioelectrocatalysis. *Chem. Rev.* 120 (23): 12903–12993.

157 Wu, R., Ma, C., and Zhu, Z. (2020). Enzymatic electrosynthesis as an emerging electrochemical synthesis platform. *Curr. Opin. Electrochem.* 19: 1–7.

158 Lewis, J.C., Coelho, P.S., and Arnold, F.H. (2011). Enzymatic functionalization of carbon–hydrogen bonds. *Chem. Soc. Rev.* 40 (4): 2003–2021.

159 Aranda, C., Carro, J., González-Benjumea, A. et al. (2021). Advances in enzymatic oxyfunctionalization of aliphatic compounds. *Biotechnol. Adv.* 51: 107703.

160 Zhang, W., Fernández-Fueyo, E., Ni, Y. et al. (2018). Selective aerobic oxidation reactions using a combination of photocatalytic water oxidation and enzymatic oxyfunctionalizations. *Nat. Catal.* 1 (1): 55–62.

161 Ortiz de Montellano, P.R. (2010). Hydrocarbon hydroxylation by cytochrome P450 enzymes. *Chem. Rev.* 110 (2): 932–948.

162 Yuan, M., Abdellaoui, S., Chen, H. et al. (2020). Selective electroenzymatic oxyfunctionalization by alkane monooxygenase in a biofuel cell. *Angew. Chem. Int. Ed.* 59 (23): 8969–8973.

163 Horst, A.E.W., Bormann, S., Meyer, J. et al. (2016). Electro-enzymatic hydroxylation of ethylbenzene by the evolved unspecific peroxygenase of *Agrocybe aegerita*. *J. Mol. Catal. B: Enzym.* 133: S137–S142.

164 Sayoga, G.V., Bueschler, V.S., Beisch, H. et al. (2023). Electrochemical H_2O_2 - stat mode as reaction concept to improve the process performance of an unspecific peroxygenase. *New Biotechnol.* 78: 95–104.

165 Amao, Y. (2018). Formate dehydrogenase for CO_2 utilization and its application. *J. CO_2 Util.* 26: 623–641.

166 Lee, S.Y., Lim, S.Y., Seo, D. et al. (2016). Light-driven highly selective conversion of CO_2 to formate by electrosynthesized enzyme/cofactor thin film electrode. *Adv. Energy Mater.* 6 (11): 1502207.

167 Zhang, L., Ong, J., Liu, J., and Li, S.F.Y. (2017). Enzymatic electrosynthesis of formate from CO_2 reduction in a hybrid biofuel cell system. *Renewable Energy* 108: 581–588.

168 Bassegoda, A., Madden, C., Wakerley, D.W. et al. (2014). Reversible interconversion of CO_2 and formate by a molybdenum-containing formate dehydrogenase. *JACS* 136 (44): 15473–15476.

169 Singh, R.K., Singh, R., Sivakumar, D. et al. (2018). Insights into cell-free conversion of CO_2 to chemicals by a multienzyme cascade reaction. *ACS Catal.* 8 (12): 11085–11093.

170 Kuwabata, S., Tsuda, R., and Yoneyama, H. (1994). Electrochemical conversion of carbon dioxide to methanol with the assistance of formate dehydrogenase and methanol dehydrogenase as biocatalysts. *JACS* 116 (12): 5437–5443.

171 Abdellaoui, S., Macazo, F.C., Cai, R. et al. (2018). Enzymatic Electrosynthesis of alkanes by Bioelectrocatalytic Decarbonylation of fatty aldehydes. *Angew. Chem. Int. Ed.* 57 (9): 2404–2408.

172 Otto, K., Hofstetter, K., Röthlisberger, M. et al. (2004). Biochemical characterization of StyAB from pseudomonas sp. strain VLB120 as a two-component Flavin-diffusible monooxygenase. *J. Bacteriol.* 186 (16): 5292–5302.

173 Hildebrand, F. and Lütz, S. (2007). Electroenzymatic synthesis of chiral alcohols in an aqueous–organic two-phase system. *Tetrahedron: Asymmetry* 18 (10): 1187–1193.

174 Chen, H., Prater, M.B., Cai, R. et al. (2020). Bioelectrocatalytic conversion from N_2 to chiral amino acids in a H_2/α-keto acid enzymatic fuel cell. *JACS* 142 (8): 4028–4036.

175 Rosenbaum, M.A. and Henrich, A.W. (2014). Engineering microbial electrocatalysis for chemical and fuel production. *Curr. Opin. Biotechnol.* 29: 93–98.

176 Xiu, S., Yao, J., Wu, G. et al. (2019). Hydrogen-mediated electron transfer in hybrid microbial–inorganic systems and application in energy and the environment. *Energy Technol.* 7 (8): 1800987.

177 May, H.D., Evans, P.J., and LaBelle, E.V. (2016). The bioelectrosynthesis of acetate. *Curr. Opin. Biotechnol.* 42: 225–233.

178 Deutzmann, J.S., Kracke, F., Gu, W., and Spormann, A.M. (2022). Microbial electrosynthesis of acetate powered by intermittent electricity. *Environ. Sci. Technol.* 56 (22): 16073–16081.

179 Dessì, P., Buenaño-Vargas, C., Martínez-Sosa, S. et al. (2023). Microbial electrosynthesis of acetate from CO_2 in three-chamber cells with gas diffusion biocathode under moderate saline conditions. *Environ. Sci. Ecotechnol.* 16: 100261.

180 Krige, A., Rova, U., and Christakopoulos, P. (2021). 3D bioprinting on cathodes in microbial electrosynthesis for increased acetate production rate using *Sporomusa ovata*. *J. Environ. Chem. Eng.* 9 (5): 106189.

181 Wang, G., Huang, Q., Song, T., and shun, Xie J. (2020). Enhancing microbial Electrosynthesis of acetate and butyrate from CO_2 reduction involving engineered clostridium ljungdahlii with a nickel-phosphide-modified electrode. *Energy Fuels* 34 (7): 8666–8675.

182 Aryal, N., Tremblay, P.L., Lizak, D.M., and Zhang, T. (2017). Performance of different *Sporomusa* species for the microbial electrosynthesis of acetate from carbon dioxide. *Bioresour. Technol.* 233: 184–190.

183 Liu, H., Song, T., Fei, K. et al. (2018). Microbial electrosynthesis of organic chemicals from CO_2 by clostridium scatologenes ATCC 25775T. *Bioresour. Bioprocess.* 5 (1): 7.

184 Van Eerten-Jansen, M.C.A.A., Heijne, A.T., Buisman, C.J.N., and Hamelers, H.V.M. (2012). Microbial electrolysis cells for production of methane from CO_2: long-term performance and perspectives. *Int. J. Energy Res.* 36 (6): 809–819.

185 Blasco-Gómez, R., Batlle-Vilanova, P., Villano, M. et al. (2017). On the edge of research and technological application: a critical review of electromethanogenesis. *Int. J. Mol. Sci.* 18 (4): 874.

186 Kambara, H., Dinh, H.T.T., Matsushita, S. et al. (2023). New microbial electrosynthesis system for methane production from carbon dioxide coupled with oxidation of sulfide to sulfate. *J. Environ. Sci.* 125: 786–797.

187 Villano, M., Aulenta, F., Ciucci, C. et al. (2010). Bioelectrochemical reduction of CO_2 to CH_4 via direct and indirect extracellular electron transfer by a hydrogenophilic methanogenic culture. *Bioresour. Technol.* 101 (9): 3085–3090.

188 De la Puente, C., Carrillo-Peña, D., Pelaz, G. et al. (2023). Microbial ELECTROSYNTHESIS for CO_2 conversion and methane production: influence of electrode geometry on biofilm development. *Greenhouse Gases Sci. Technol.* 13 (2): 173–185.

189 Mayer, F., Sabel-Becker, B., and Holtmann, D. (2022). Enhanced electron uptake and methane production by corrosive methanogens during Electromethanogenesis. *Microorganisms* 10 (11): 2237.

190 Cheng, S., Xing, D., Call, D.F., and Logan, B.E. (2009). Direct biological conversion of electrical current into methane by electromethanogenesis. *Environ. Sci. Technol.* 43 (10): 3953–3958.

191 Zhang, X., Li, X., Zhao, X., and Li, Y. (2019). Factors affecting the efficiency of a bioelectrochemical system: a review. *RSC Adv.* 9 (34): 19748–19761.

9

Wearable and Implantable Devices

Edmond Magner

Department of Chemical Sciences and Bernal Institute, University of Limerick, Limerick, Ireland

9.1 Introduction

The fundamental principles of bioelectrocatalysis and electron transfer in enzymatic bioelectrocatalysis have been described in Chapters 1 and 2, while the principles and details of biofuel cells are described in Chapter 6. A wide range of studies have described wearable and implantable devices, with a significant increase in the number of publications in recent years. The development of sensors for clinically important analytes, such as glucose, to monitor diabetic conditions and of analytes, such as lactate, for exercise physiology has been a major focus of research. The development of sensors for the detection of glucose is driven by the recognition that management of the concentration of glucose in patients with diabetes can provide significant improvement in their quality of life, with significant decreases in conditions such as retinopathy and neuropathy [1]. Prior to the publication of the results of this clinical trial, the need to manage and control the concentration of blood glucose was well established. The clinical trial results demonstrated that "Intensive therapy effectively delays the onset and slows the progression of diabetic retinopathy, nephropathy, and neuropathy in patients with insulin dependent diabetes mellitus" [1]. These findings underpinned the need for glucose sensors and spurred the development of a wide range of biosensors for the detection of glucose, with the addition of a range of new sensor systems to the systems that were already available in the market. Commercial interest in such devices arises from the awareness that approximately 14% of all health care costs in the United States arise from complications due to diabetes [2]. The successful development of glucose biosensors, discussed in Chapter 4, also spurred the development of sensors for a wide range of other analytes, though to date no system has been developed on the scale of that of glucose sensors. In addition to the development of sensors, the (continual) introduction of new mobile communication devices (smartphones, etc.) has significantly enabled the development of monitoring systems that can be used to collect, analyze, and display the output from sensors. In terms of the development of sensors, a number of reviews [3–5] and special issues [6, 7] have been published. A recent review has described the range of approaches that have been used to monitor the concentration of glucose using wearable devices [4].

The focus of this chapter is to describe recent developments in wearable and implantable devices and, in particular, on the materials used in such applications. The use of devices can be divided into two categories: skin and ocular systems.

Bioelectrocatalysis: From Electron Transfer Processes to Emerging Technological Applications, First Edition.
Edited by Shelley D. Minteer and Matteo Grattieri.

9.2 Skin-Based Systems

One of the best-known examples of a wearable glucose sensor is the GlucoWatch Biographer developed by Cygnus, Inc. [8] that received approval from the U.S. Food and Drug Administration. The device measured the concentration of glucose using an enzyme-based detection system that relied on the process of reverse iontophoresis (Box 9.1) to extract fluid from the skin. While the device was commercialized, it was subsequently withdrawn from use after being launched due to issues such as irritation of the skin, the requirement for quite a long break-in period, and the need to calibrate the device using an invasive glucose test. These issues illustrate the challenges that arise in the successful development of wearable devices and, in particular, the development of devices that require regulatory approval. The regulatory requirements are very stringent, with the requirement that devices provide equivalent performance (accuracy and precision) to an established standard method of analysis. The laboratory-based system developed by Yellow Springs Instruments, Ltd. [9] is such a standard method for the determination of the concentration of glucose. The method of detection relies on the use of membrane bound glucose oxidase with quantification of the concentration of hydrogen peroxide at a platinum electrode.

Box 9.1 Schematic diagram of the process of iontophoresis

On application of a potential, the electric field that is generated between the anode and cathode results in migration of charged species, with accumulation of negatively and positively charged species at the anode and cathode compartments, respectively. Neutral species migrate to both compartments as part of the movement of the solvation spheres of the charged species.

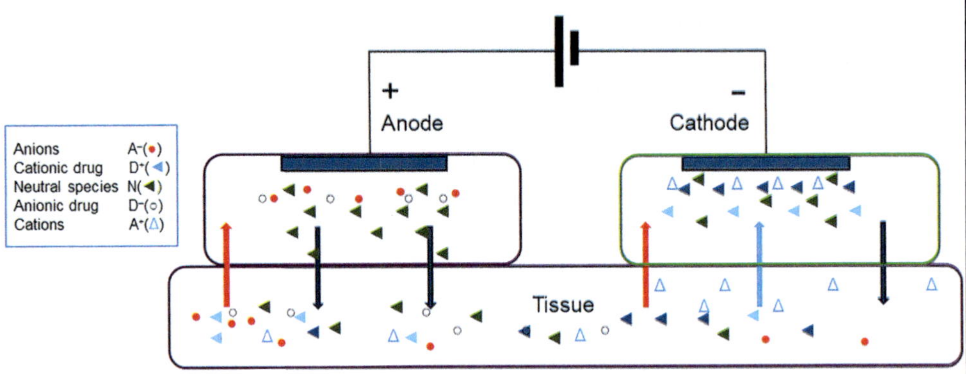

Implanted glucose monitoring systems have been available for over 20 years [10]. A device manufactured by MiniMed utilized a glucose monitor together with a disposable subcutaneous glucose sensor. Data from the monitor was downloaded onto a computer using a customized docking system. The sensor was successfully used to monitor glucose levels, but in its initial use, data was acquired, reviewed, and assessed by a health-care practitioner. Patients were required to take at least four measurements of their blood glucose levels, data that was used to calibrate the sensor. Significant improvements have been made since. The most recent version of the MiniMed system enables continuous monitoring of glucose levels for a period of up to seven days and can be combined with an insulin pump. The sensor is inserted under the skin, and the data can be transmitted directly to the insulin pump.

Currently, two continuous glucose monitoring (CGM) sensors have a significant market share: the Freestyle Libre and G7™, manufactured by Abbott Laboratories and Dexcom, respectively. Both systems rely on the use of a sensor that measures the concentration of glucose in interstitial fluid and relays the data to a device such as a smartphone or a dedicated handheld reader. The Freestyle Libre™ system relies on the use of an osmium-modified polymer that is used to mediate electron transfer between the reduced form of glucose oxidase and the electrode, while the Dexcom G7™ system relies on the detection of hydrogen peroxide produced by the glucose oxidase-catalyzed oxidation of glucose by oxygen. Each device has particular advantages in terms of the length of time that the device can be implanted in the skin, the level of correlation with blood glucose concentration over the full therapeutic concentration range, frequency of data collection, and ease of use together with cost. While a range of studies have been performed, a detailed comparison of the properties of the sensors would require a large-scale clinical trial (>1000 patients), and the comparisons provided to date tend to be on small patient cohorts. Both systems can be used to continuously monitor the concentration of blood glucose in diabetic patients. However, both systems require daily calibration using a separate measuring device. Such a requirement is indicative of the complexity of the problems associated with the development of clinically approved monitoring devices. The complexity of the devices is evident in the large suite of patents that have been granted to the companies involved, running well into hundreds of patents, indicative of the challenges inherent in developing devices for clinical applications. Such caveats illustrate the challenges in developing devices that possess the accuracy, precision, and reliability required for approval in a clinical environment. Examples of recent reports in developing devices for potential clinical applications are described below.

An autonomous wearable device (Figure 9.1, [11]) for the *in situ* analysis of glucose and chloride in sweat samples has been described as a method of testing the concentrations of glucose and electrolytes in patients with diabetes and cystic fibrosis, respectively.

A gold electrode array was prepared on polyethylene terephthalate, with fabrication of an Ag/AgCl electrode for the detection of Cl^- and a Na^+ ionophore-modified electrode for the detection of Cl^-. The concentration of glucose was detected using glucose oxidase immobilized with chitosan and carbon nanotubes together with a Prussian blue mediator. The use of a device for the measurement of electrolytes is of use as a screening tool for cystic fibrosis, with the possibility of conducting such testing more rapidly. The device could be used to extract sweat using iontophoresis with sweat extraction occurring on demand or at defined times. The device utilized hydrogels to minimize discomfort to the patient during the iontophoretic process. Fluid extraction at a rate of $100\,nL\,min^{-1}\,cm^2$ was achieved, providing a sufficient volume of fluid for analysis. Testing of the device was performed on patients with cystic fibrosis, and the device was able to detect elevated levels of Na^+ and Cl^-. Preliminary evaluation of the device was performed on testing the concentration of glucose (seven patients) and while changes in the concentration of glucose in sweat samples were observed that were concomitant with changes in the concentration of glucose in blood, a direct correlation was not obtained, again illustrating the complexity of the challenges [2, 12] in accurately measuring the concentration of glucose to the level of accuracy required for an approved medical device.

More recent reports have focused on using alternative device configurations to help address these challenges. An iontophoretic approach was successfully used in an epidermal biosensor to monitor levels of vitamin C (Figure 9.2, [13]). The sensor was composed of a printed tattoo patch using screen-printed electrodes. Two electrodes were used for iontophoretic stimulation of sweat, with carbon paste, Ag/AgCl, and carbon paste working, reference, and counter electrodes, respectively. The device monitored the concentration of ascorbic acid using ascorbic acid oxidase. A range of

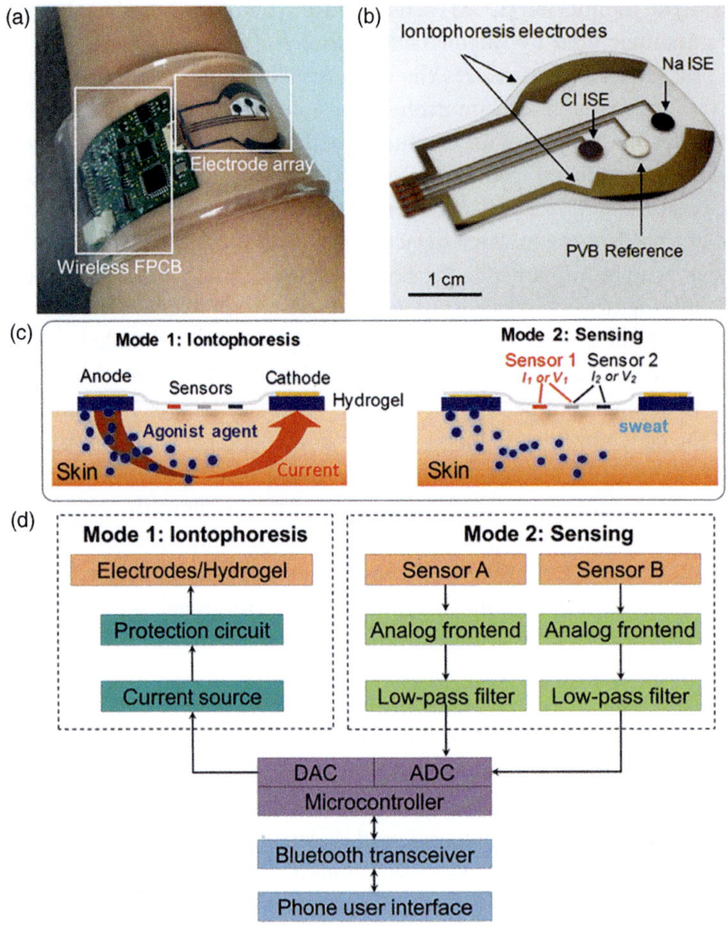

Figure 9.1 Images of (a) sweat extraction and sensing system and (b) iontophoretic and sweat sensor electrodes for the detection of Na^+ and Cl^-. (c) Illustrations of iontophoretic and sensing modes. (d) Block diagram of the iontophoretic and sensing circuits. *Source:* Emaminejad et al. [11]. Copyright 2017 National Academy of Sciences.

materials were examined for use as the working electrode: a commercially available carbon ink, unmodified screen-printed carbon, carbon nanotubes, and carbon inks prepared with 1% platinum, palladium, or rhodium. The optimum response was obtained using a rhodium-modified carbon paste electrode, with the sensor response based on the use of the enzyme ascorbate oxidase and determining the concentration of ascorbate via the reduction of oxygen concentration in sweat samples. The sensor could monitor the level of vitamin C on ingestion of fruit juices and vitamin tablets. The concentration of vitamin C was also monitored in tear and saliva samples to ascertain the effect of different fluids on the response. Similar correlations with ingestion of vitamin C were observed for both types of fluid. This type of sensor format demonstrates the feasibility of using the approach for wearable disposal sensors; however, its use in the detection of, e.g. glucose would need to be demonstrated and, in particular, the ability to surmount the challenge of correlating the signal response with the concentration of glucose in blood.

An approach that avoids the use of iontophoresis was described in a sensor that used sweat from a fingertip [14]. The sweat sample was absorbed onto a polyvinyl alcohol hydrogel layer and the

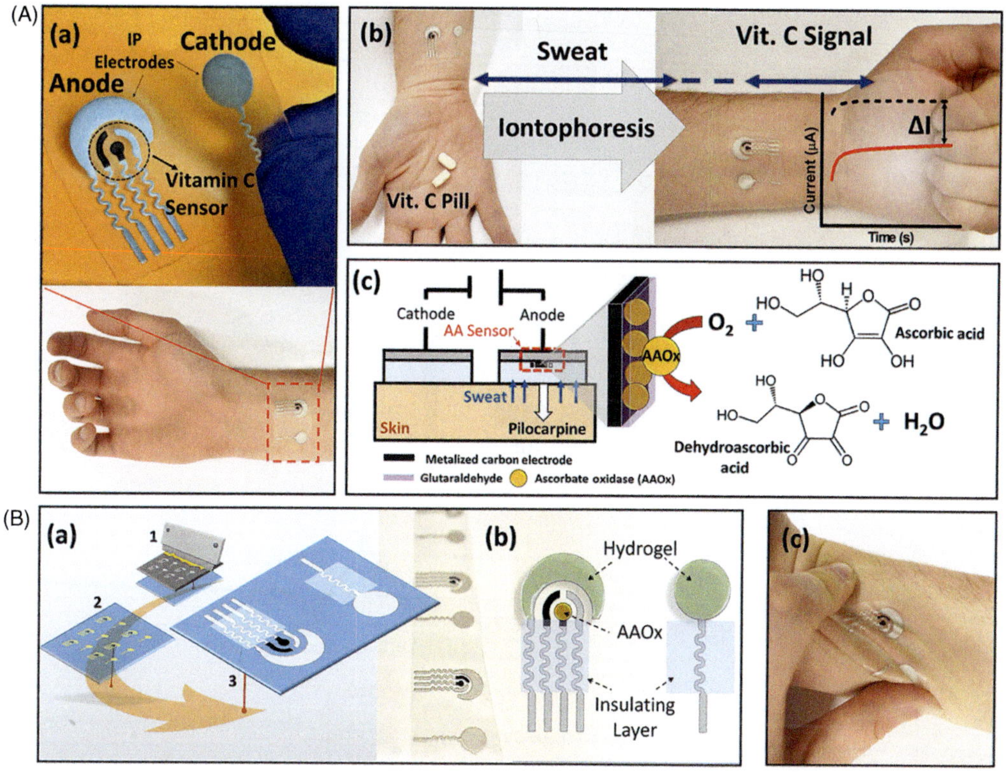

Figure 9.2 Diagram of a sensor for the determination of ascorbic acid in sweat. (A) (a) Electrode design, (b) protocol used for the detection of ascorbic acid in sweat stimulated before (black dotted line) and after (red solid line) taking vitamin C tablets, and (c) illustration of localized sweat stimulation and of the enzymatic reaction. (B) Schematic diagrams of (a) the biosensor and (b) the hydrogel and enzyme layers. (c) Image of epidermal sensor. *Source:* Sempionatto et al. [13]. Copyright 2021 American Chemical Society.

concentration of glucose was then determined using a Prussian blue-mediated glucose oxidase sensor, with the detection of H_2O_2 produced by the action of the enzyme. The response obtained (Figure 9.3f) indicates that a good correlation between the sensor response and the concentration of glucose was not observed, and incorporation of a measurement algorithm was required to produce an acceptable level of correlation. Such a requirement presents significant challenges to the development of a clinically reliable sensor that would meet the requirements of regulatory bodies as it is likely that calibration of the sensor response would be required for each individual patient.

In addition to monitoring the concentration of glucose, the detection of ketone compounds such as β-hydroxybutyrate is of interest as an additional measure of the physiological status of diabetic patients. While the concentration of glucose is the major focus in the development of sensors for diabetic patients, a sensor for the detection of ketones is of value as the measurement of glucose provides a measure at a particular time, the measurement of the concentration of ketones provides a long-term perspective to clinical staff of the effectiveness of management of diabetes in patients [15]. A commercial, finger stick system for the detection of ketones has been developed [16]. To circumvent the need to take a blood sample, a wearable device would be of interest. A microneedle system was developed to monitor the concentration of ketones that was based on using a gold-coated platinum working electrode that had been modified with the enzyme β-hydroxybutyrate dehydrogenase, toluidine blue O as a redox mediator, nicotinamide adenine

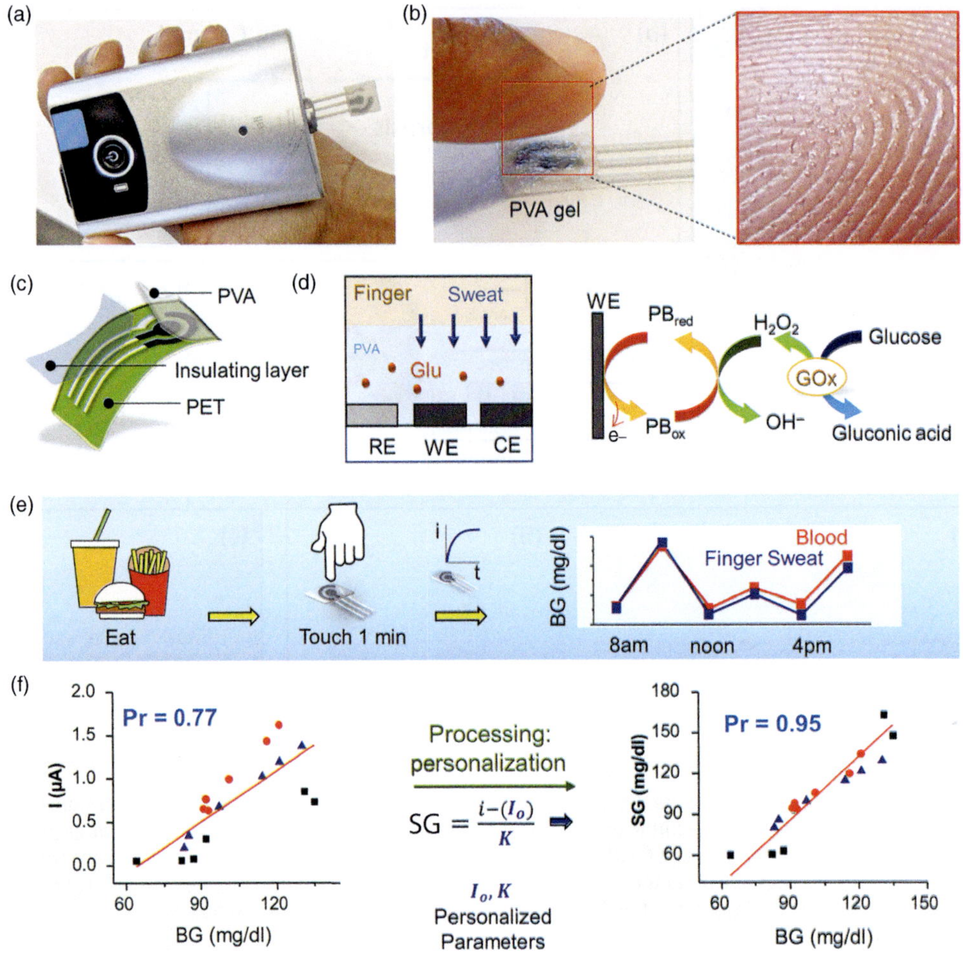

Figure 9.3 Schematic illustration of a glucose sensor on a fingertip. (a) Image of a portable hand-held potentiostat and sensor. (b) Image of the touch sensor device and fingertip. Schematic illustrations of (c) the touch sensor, (d) the method of collecting sweat, (e) the process used to detect glucose in sweat, and (f) illustration of the conversion of the sweat signal into a reading of the concentration of glucose in blood glucose level. *Source:* Sempionatto et al. [14]. Copyright 2021 American Chemical Society.

dinucleotide, carbon nanotubes, chitosan, and poly(vinylchloride). The skin-worn microneedle sensing device operated with a miniaturized electrochemical analyzer connected wirelessly to a mobile electronic device for data analysis. Cytotoxicity and skin penetration studies indicated no potential harmful effects. On comparison of the data from a small pilot study vs. a standard method of analysis, a good correlation was obtained, indicating that the approach held promise in terms of tracking ketoacidosis. However, large-scale trials would be needed to verify the method in a manner that would be suitable for compliance with regulatory requirements.

9.2.1 Electrode Material

The use of carbon nanotubes has been a major focus of research. A carbon nanotube-based system was used for the preparation of a biofuel cell. The cell was comprised of an anode prepared using glucose oxidase and multiwalled carbon nanotubes and a cathode prepared from laccase and multiwalled carbon nanotubes. The electrodes were prepared as disks under hydraulic pressure and then connected in a cell that delivered a power density of 1.3 mW cm^{-2} and could operate in a stable manner for a period of 1 month. This approach demonstrated that self-powered biofuel cells with a reasonably high power density could be prepared. However, when such cells were implanted into rats, a significant decrease in the power density was observed, demonstrating that the development of implantable self-powered systems holds significant challenges, as evident in the challenges that face implantable sensors.

A wide range of reports have described the use of paper-like materials prepared from carbon nanotubes. Buckypapers (BPs) are thin self-supporting sheets of entangled nanotubes (CNTs) that are held together by a combination of π–π stacking and interweaving interactions. Typically, sheets have a thickness ranging from 200 nm to 200 mm and are prepared by filtration of dispersions of CNTs or by rolling methods (Figure 9.4). A range of studies have been performed using this material [3]. The material is very versatile in terms of the potential shapes that can be prepared and is amenable to large-scale production, producing materials with high surface areas and conductivity that can act as effective supports for enzymes. As an example, a wearable biofuel cell prepared from BP displayed a maximum power output of 450 μW [17]. The cell relied on the use of lactate from sweat as a fuel; as such, the system is limited by the generation of sweat on the skin. More recently,

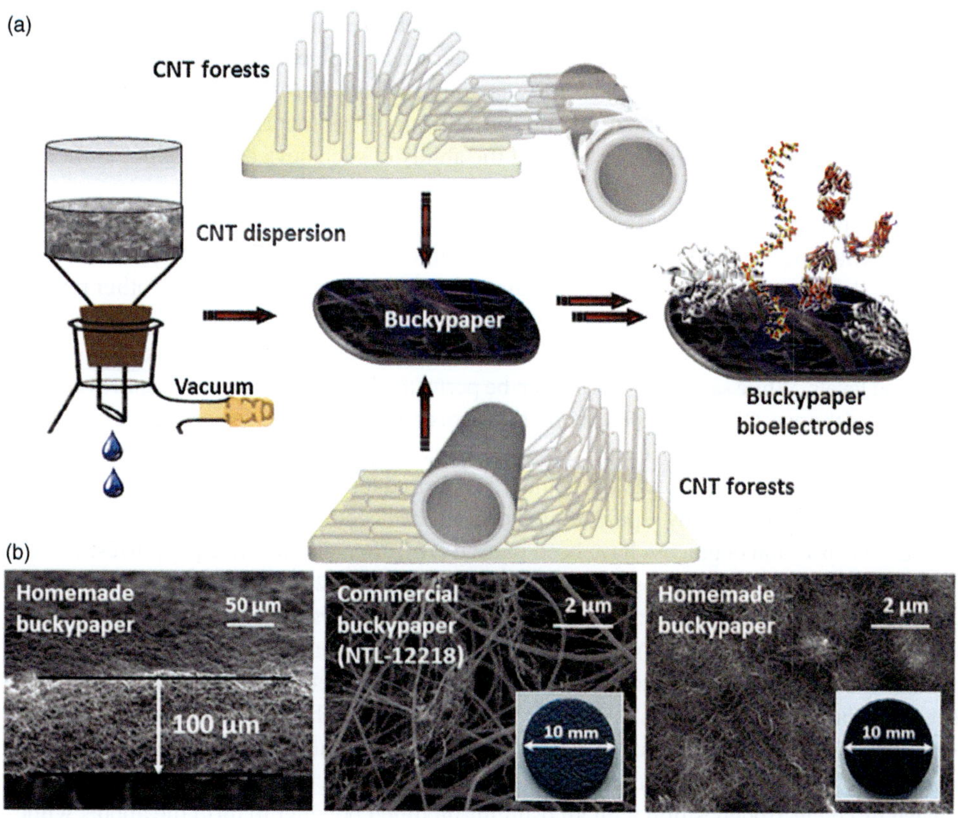

Figure 9.4 (a) Schematic illustration of the preparation of buckypaper using vacuum filtration and rolling methods, with subsequent functionalization with biological species and (b) scanning electron microscopy (SEM) images and photographs of laboratory and commercially prepared samples of buckypaper. *Source:* Gross et al. [3]. Copyright 2008 Royal Society of Chemistry.

a system that could be used to store charge has been described that utilized carbon nanotubes combined with polypyrrole to act as a capacitive material with a power output of 1.7 mW cm^{-2} that was reduced to 340 μW cm^{-2} when the electrode was placed on skin.

Another example of the use of an implanted electrode is the use of microporous gold electrodes to support enzymes in a biofuel cell [18]. The electrodes were prepared using a template approach with bilirubin oxidase used at the cathode and glucose oxidase coupled with an osmium polymer at the anode. The cell was able to operate for over a day when implanted in a rat, with wireless communication between the implanted cell and a monitoring station that could take place in a wireless manner using an implanted radio frequency identification system. The system was lightweight and allowed the animal to move freely in its cage, while also allowing long-term measurements to be made. Good agreement between the *in vitro* and *in vivo* measurements was obtained.

9.3 Ocular Systems

The use of ocular systems was examined with the aim of ascertaining if a biofuel cell in tears (or simulated tears) could provide the level of power output required in a sensor or monitoring device. A prototype biofuel cell (Figure 9.5) was fabricated using gold wire electrodes modified with gold nanoparticles. The enzymes cellobiose dehydrogenase (*Corynascus thermophilus*) and bilirubin oxidase (*Myrothecium verrucaria*) were then immobilized at a thiol-modified gold anode and an unmodified gold cathode. Direct electron transfer was possible using these configurations with the cell producing a cell voltage of 0.5 V, a maximum power density of 1 μW cm^{-2} (at a cell voltage of 0.5 V), and an operational half-life of over 20 hours. This report demonstrated proof of principle of a contact lens-type device, in terms of the ability of such devices to produce sufficient power from tears.

While proof of principle was clearly established, translating this proof into an operational device has significant challenges, requiring robust testing in a large clinical trial. In some reports, a correlation between the concentration of glucose in tears and in blood has been described; other reports indicate that such correlations can be difficult to achieve. In assessing such systems, it is essential that irritation or stimulation of the eye does not occur to avoid dilution of the concentration via the increased fluid levels. While sampling of tears can be performed, such an approach requires care to avoid irritation of the eye. Given this constraint, the use of contact lens-type devices where irritation of the eye post-insertion of the device can be minimized is attractive. It should be noted, however, that the development of such devices to the standards required for approval as a medical device is a significant challenge, as evident by reports that efforts on the development of smart lens devices to monitor the concentration of glucose have been paused [20], after significant capital investment.

A range of reports on contact lens systems have been described [20]. A nanoporous gold-based biofuel cell was prepared in a contact lens format. Silver-gold alloys were sputtered onto a flexible polyethylene terephthalate (PET) substrate. Electrochemical dealloying enabled the removal of Ag with the formation of a gold nanoporous layer that could then be used as an enzyme support. The advantage of the approach is that the porous nature of the gold layer provides a high surface area support that can act as an antifouling material by excluding large proteins, while the PET substrate possesses the flexibility required for use with contact lenses. Lactate oxidase was immobilized by co-electrodeposition with an osmium-modified polymer to form the anode, while bilirubin oxidase was immobilized on a diazonium-modified nanoporous gold electrode. A biofuel cell was assembled (Figure 9.6) using commercially available contact lenses yielding a power output of 2.4 μW cm^{-2}. On testing in artificial tear solution, the power output decreased to 1.7 μW cm^{-2}, a decrease that was ascribed to the effect of ascorbate as an interfering species at the cathode. After

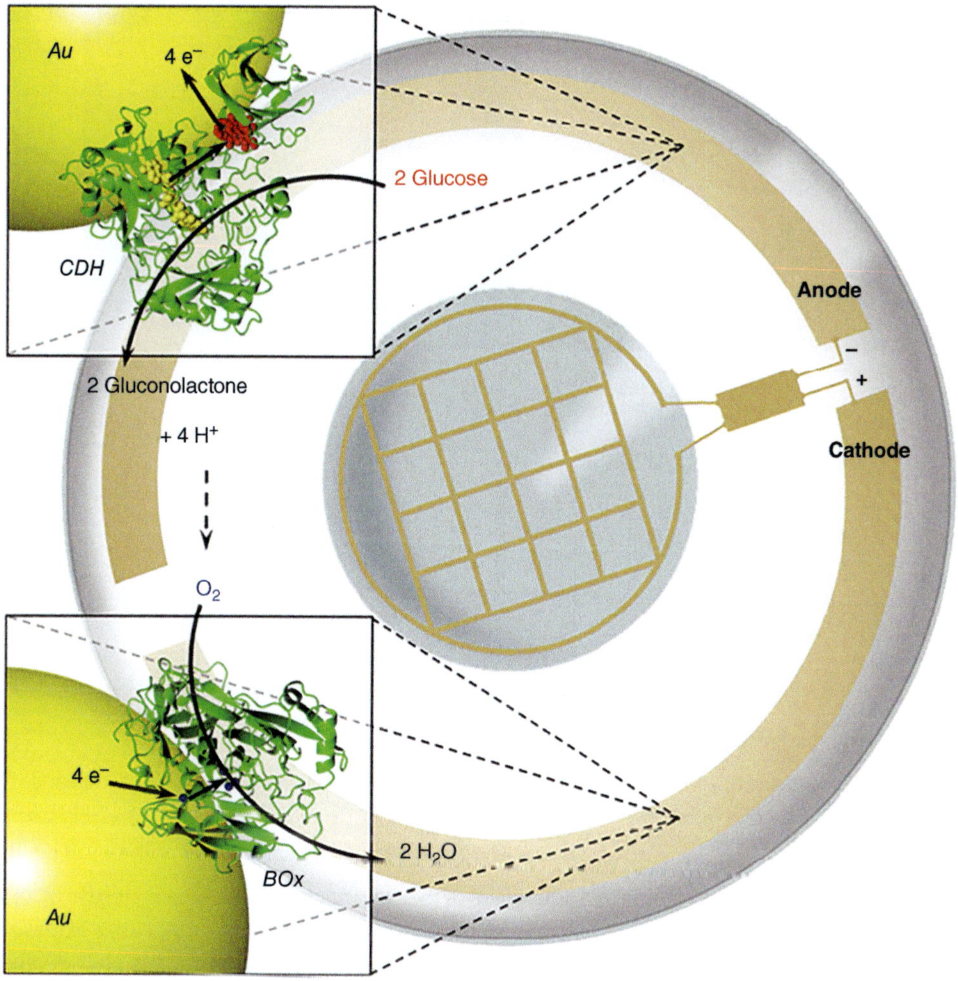

Figure 9.5 Schematic diagram of a biofuel cell embedded in a contact lens, illustrating the oxidation of glucose to gluconolactone (catalyzed by cellobiose dehydrogenase) at the anode and the reduction of oxygen to water (catalyzed by bilirubin oxidase) at the cathode. *Source:* Falk et al. [19]. Copyright 2012 Elsevier.

five hours of continuous operation, the power output decreased by 80%, due to a loss of signal that is likely due to the stability of the cathode. The response of the anode remained stable, indicating that the material, nanoporous gold, is a suitable material for use as an electrode material and as a support for enzymes. Again, the development of such devices to a level of performance that would be compliant with the standards required by regulatory agencies represents a significant challenge.

Recently, Keum et al. [20] utilized a PET support to prepare a three-electrode cell for the detection of glucose. The device was prepared on a silicone hydrogel with glucose oxidase immobilized on a platinum working electrode together with a platinum counter electrode and a Ag/AgCl reference electrode. The electrochemical response to glucose was determined via the oxidation of hydrogen peroxide that was produced by glucose oxidase. The system was self-contained with a wireless power supply and was used to successfully monitor the concentration of glucose *in vitro* in tear samples over a 24-hour period. The system was integrated with a controlled drug delivery system

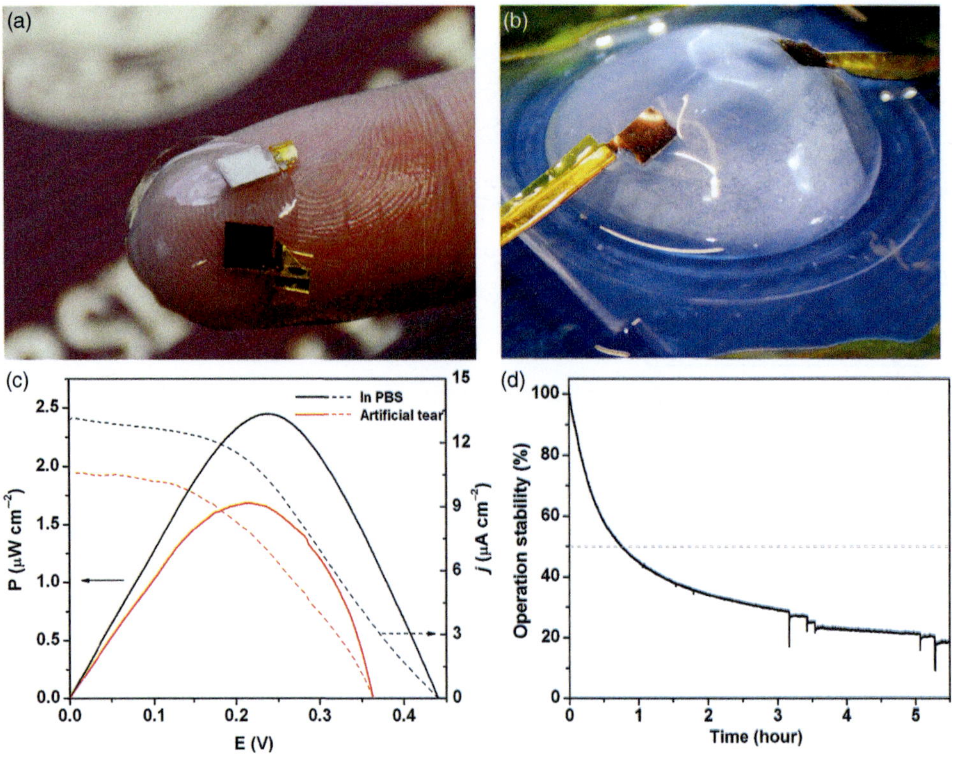

Figure 9.6 Images of (a and b) a nanoporous gold modified contact lens, plots of the (c) polarization and power curves, and (d) the operational stability of the cell in artificial tears. *Source:* Xiao et al. [21]. Copyright 2018 American Chemical Society.

that was then tested on diabetic rabbits. An advantage of this approach is that it avoids the use of an enzyme-based cathode, obviating issues with the stability of an enzyme such as bilirubin oxidase by the use of abiotic reference and counter electrodes. While the sensor responded to changes in the concentration of glucose in blood, the degree of correlation between glucose levels in blood and in tears was not clearly defined, indicating the challenges that pertain to this approach. In preparing such sensors, stable reference electrodes are crucial to performance. A screen-printed Ag/AgCl reference electrode that incorporated a polydimethylsiloxane layer to prevent electrolyte leakage has been described. The electrode can provide a stable potential for up to 27 days, in comparison with a stability of less than a day for uncoated electrodes [22]. A good response was obtained in a range of samples (water, beverage, etc.) but testing was not conducted on clinical samples.

9.4 Textile Material-Based Systems

The use of textile materials as a support for sensors is an attractive approach as the sensors can be incorporated directly into clothing and used to monitor analytes directly [23, 24]. Textiles are of interest due to their ease of manufacture, comfort, and wearability. Using a styrene-based material as support, gold-modified fibers were manually threaded into a t-shirt fabric and used to prepare

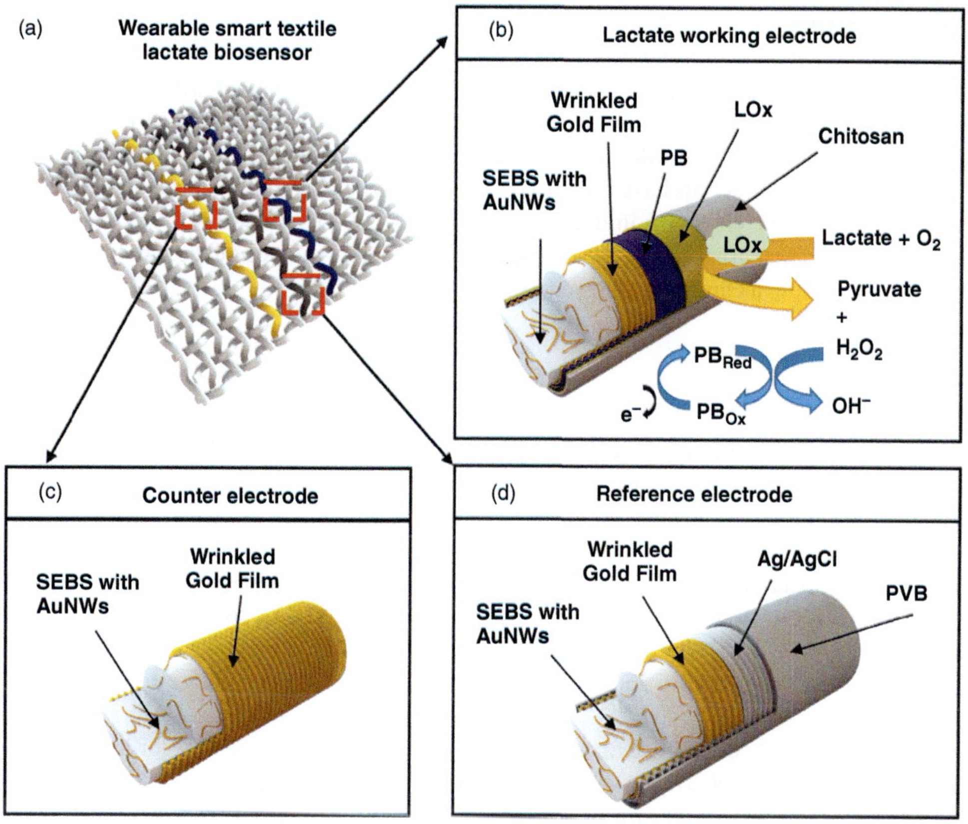

Figure 9.7 Schematic illustrations (a) of a wearable smart textile electrochemical biosensor for the detection of lactate showing the configuration of the (b) working, (c) counter, and (d) reference electrodes. *Source:* Wang et al. [25]. Copyright 2021 Elsevier.

working, reference, and counter electrodes, as shown in Figure 9.7 [25]. Lactate sensors were prepared using lactate oxidase together with a Prussian blue-based mediator layer. The sensors could operate with a small volume (0.1 mL) of sweat and displayed a response time of 1.5 seconds. A sensitivity of 19.13 μA mM^{-1} cm^2 in buffer was obtained in buffer samples, decreasing to 14.6 μA mM^{-1} cm^2 in artificial sweat, a decrease that was attributed to the adsorption of interfering species in the latter. The response of the sensor was maintained on tensile strain of 100% without any visible structural design.

The use of nonenzymatic systems for the detection of glucose has been described. The attraction of such systems lies in the stability of the enzyme-free method of detection, with the challenge of losing the specificity of the enzymatic system. An example of the approach taken in this area is the use of copper oxides (Cu_2O) with the {100} and {111} facets of Cu_2O providing the best electrocatalytic activity [26]. Using this system as a portable wireless fuel cell, the concentration of glucose was determined in a proof-of-concept system. However, the system was not tested on simulated body fluids such as artificial sweat or interstitial fluid nor on skin, indicating that the feasibility of utilizing this approach in a clinical system has yet to be determined.

9.5 Conclusions

The development of materials and the use of rapid printing systems together with the ability to wirelessly connect sensors with communication devices enable the development of new sensor systems in a rapid time frame. These developments enable rapid turnaround and production of new devices. A wide range of materials have been used to produce sensors for analytes such as glucose, lactate, Na^+, and Cl^-. The challenge in developing sensors for these analytes is to overcome the significant challenges that arise in clinical samples, challenges that will need to be met in order to translate the sensors into a clinical setting.

References

1 The Diabetes Control and Complications Trial Research Group (2021). The effect of intensive treatment of diabetes on the development and progression of long-term complication in insulin-dependent diabetes mellitus. *N. Engl. J. Med.* 329: 977–986.

2 Magner, E. (1998). Trends in electrochemical biosensors. *Analyst* 123: 1967–1970.

3 Gross, A.J., Holzinger, M., and Cosnier, S. (2018). Buckypaper bioelectrodes: emerging materials for implantable and wearable biofuel cells. *Energy Environ. Sci.* 11: 1670–1678.

4 Huang, X., Yao, C., Huang, S. et al. (2024). Technological advances of wearable device for continuous monitoring of *in vivo* glucose. *ACS Sens.* 9: 1065–1088. https://doi.org/10.1021/acssensors.3c01947.

5 Kim, J., Campbell, A.S., de Avila, B.E., and Wang, J. (2019). Wearable biosensors for healthcare monitoring. *Nat. Biotechnol.* 37: 389–406.

6 Chen, X., Kim, D.H., and Lu, N. (2024). Introduction: wearable devices. *Chem. Rev.* 124: 6145–6147. https://doi.org/10.1021/acs.chemrev.4c00271.

7 DeMello, A. and Bakker, E. (2023). Wearable sensors. *ACS Sens.* 8: 1368–1370. https://doi.org/10.1021/acssensors.3c00437.

8 Tierney, M.J., Tamada, J.A., Potts, R.O. et al. (2001). Clinical evaluation of the GlucoWatch biographer: a continual, non-invasive glucose monitor for patients with diabetes. *Biosens. Bioelectron.* 16: 621–629.

9 Han, J., Nichols, J.H., Rice, M., and Klonoff, D.C. (2020). The end of the road for the YSI 2300 analyzer: where do we go now? *J. Diabetes Sci. Technol.* 14: 595–600. https://doi.org/10.1177/1932296819886603.

10 Mastrototaro, J.J. (2000). The MiniMed continuous glucose monitoring system. *Diabetes Technol. Ther.* 2: S13–S18.

11 Emaminejad, S., Gao, W., Wub, E. et al. (2017). Autonomous sweat extraction and analysis applied to cystic fibrosis and glucose monitoring using a fully integrated wearable platform. *Proc. Natl. Acad. Sci.* 114: 4625–4630.

12 Magner, E. (2001). Detection of ferricyanice as a probe for the effect of hematocrit in whole blood sensors. *Analyst* 126: 861–865.

13 Sempionatto, J.R., Khorshed, A.A., Ahmed, A. et al. (2020). Epidermal enzymatic biosensors for sweat vitamin C: toward personalized nutrition. *ACS Sens.* 5: 1804–1813. https://doi.org/10.1021/acssensors.0c00604.

14 Sempionatto, J.R., Moon, J.M., and Wang, J. (2021). Touch-based fingertip blood-free reliable glucose monitoring: personalized data processing for predicting blood glucose concentrations. *ACS Sens.* 6: 1875–1883. https://doi.org/10.1021/acssensors.1c00139.

15 Moonla, C., Reynoso, M., Casanova, A. et al. (2024). Continuous ketone monitoring via wearable microneedle patch platform. *ACS Sens.* 9: 1004–1013. https://doi.org/10.1021/acssensors.3c02677.

16 Forrow, N.J., Sanghera, G.S., Walters, S.J., and Watkin, J. (2005). Development of a commercial amperometric biosensor electrode for the ketone d-3-hydroxybutyrate. *Biosens. Bioelectron.* 20: 1617–1625.

17 Chen, X., Yin, L., Lv, J. et al. (2019). Stretchable and flexible buckypaper-based lactate biofuel cell for wearable electronics. *Adv. Funct. Mater.* 29: 1905785.

18 Di Trocchio, L., Carucci, C., Sindhu, K.R. et al. (2021). Wireless *in vivo* biofuel cell monitoring. *IEEE J. Electromagn. RF Microwaves Med. Biol.* 5: https://doi.org/10.1109/JERM.2020.2998325.

19 Falk, M., Andoralova, V., Blum, Z. et al. (2012). *Biosens. Bioelectron.* 37: 38–45.

20 Keum, D.H., Kim, S.K., Koo, J. et al. (2020). Wireless smart contact lens for diabetic diagnosis and therapy. *Sci. Adv.* 6: eaba3252.

21 Xiao, X., Siepenkoetter, T., Conghaile, P.Ó. et al. (2018). Nanoporous gold-based biofuel cells on contact lenses. *ACS Appl. Mater. Interfaces* 10: 7107–7116. https://doi.org/10.1021/acsami.7b18708.

22 Dawkins, R.C., Wen, D., Hart, J.N., and Vepsäläinen, M. (2021). A screen-printed Ag/AgCl reference electrode with long-term stability for electroanalytical applications. *Electrochim. Acta* 393: 139043.

23 Lv, J., Yin, L., Chen, X. et al. (2021). Wearable biosupercapacitor: harvesting and storing energy from sweat. *Adv. Funct. Mater.* 31: 2102915.

24 Tian, H., Wang, H., Yang, W. et al. (2023). Hierarchical Fermat helix-structured electrochemical sensing fibers enable sweat capture and multi-biomarker monitoring. *Mater. Horiz.* 10: 5192.

25 Wang, R., Zhai, Q., An, T. et al. (2021). Stretchable gold fiber-based wearable textile electrochemical biosensor for lactate monitoring in sweat. *Talanta* 222: 121484.

26 Jiang, Y., Xia, T., Shen, L. et al. (2021). Facet-dependent Cu_2O electrocatalysis for Wearable enzyme-free smart sensing. *ACS Catal.* 11: 2949–2955. https://doi.org/10.1021/acscatal.0c04797.

Index

Bioelectrocatalysis: From Electron Transfer Processes to Emerging Technological Applications, First Edition.
Edited by Shelley D. Minteer and Matteo Grattieri.

d

e

f

The Electrochemical Society Series

The Electron Microprobe
Edited by T. D. McKinley, K. F. J. Heinrich, and D. B. Wittry

Chemical Physics of Ionic Solutions
Edited by B. E. Conway and R. G. Barradas

High-Temperature Materials and Technology
Edited by Ivor E. Campbell and Edwin M. Sherwood

Alkaline Storage Batteries
S. Uno Falk and Alvin J. Salkind

The Primary Battery (in Two Volumes)
Volume I
Edited by George W. Heise and N. Corey Cahoon

Volume II
Edited by N. Corey Cahoon and George W. Heise

Zinc-Silver Oxide Batteries
Edited by Arthur Fleischer and J. J. Lander

Lead-Acid Batteries
Hans Bode
Translated by R. J. Brodd and Karl V. Kordesch

Thin Films-Interdiffusion and Reactions
Edited by J. M. Poate, M. N. Tu, and J. W. Mayer

Lithium Battery Technology
Edited by H. V. Venkatasetty

Quality and Reliability Methods for Primary Batteries
P. Bro and S. C. Levy

Techniques for Characterization of Electrodes and Electrochemical Processes
Edited by Ravi Varma and J. R. Selman

Electrochemical Oxygen Technology
Kim Kinoshita

Synthetic Diamond: Emerging CVD Science and Technology
Edited by Karl E. Spear and John P. Dismukes

Corrosion of Stainless Steels, Second Edition
A. John Sedriks

Semiconductor Wafer Bonding: Science and Technology
Q.-Y. Tong and U. Göscle

Fundamentals of Electrochemistry, Second Edition
V. S. Bagotsky

Fundamentals of Electrochemical Deposition, Second Edition
Milan Paunovic and Mordechay Schlesinger

Uhlig's Corrosion Handbook, Third Edition
Edited by R. Winston Revie

Fuel Cells: Problems and Solutions
Vladimir S. Bagotsky

Lithium Batteries: Advanced Technologies and Applications
Edited by B. Scrosati, K. M. Abraham, W. A. van Schalkwijk, and J. Hassoun

Modern Electroplating, Fifth Edition
Edited by Mordechay Schlesinger and Milan Paunovic

Electrochemical Power Sources: Batteries, Fuel Cells, and Supercapacitors
By V. S. Bagotsky, A. M. Skundin, and Y. M. Volfkovic

Molecular Modeling of Corrosion Processes: Scientific Development and Engineering Applications
Edited by C. D. Taylor and P. Marcus

Atmospheric Corrosion, Second Edition
Christofer Leygraf, Inger Odnevall Wallinder, Johan Tidblad, and Thomas Graedel

Electrochemical Impedance Spectroscopy, Second Edition
Mark E. Orazem and Bernard Tribollet

Electrochemical Systems, Fourth Edition
John Newman and Nitash P. Balsara

LaQue's Handbook of Marine Corrosion
Edited by David A. Shifler

Metallic, Magnetic and Carbon-Based Nanomaterials: Synthesis and Biomedical Applications
Ajit Khosla, Irshad A. Wani, and Mohammad N. Lone

Bioelectrocatalysis: From Electron Transfer Processes to Emerging Technological Applications
Edited by Shelley D. Minteer and Matteo Grattieri

Printed and bound by CPI Group (UK) Ltd, Croydon, CR0 4YY

17/09/2026

14964314-0002